ADAM, WHERE ARE YOU?

Dutch scholar Willem J. Ouweneel's *Adam, Where Are You?* is a timely, thorough, well-researched book that comes down solidly on the side of affirming the antithesis between orthodox Christianity and human evolution. It also does a good job of being thetical as well, affirming the clarity and value of biblical truth and viewing it positively without drastically modifying it with extra-biblical ideas. The book is eminently readable, well-edited, biblically-grounded, mind-enriching, and God-honouring.

DR. JOEL R. BEEKE, president, Puritan Reformed Theological Seminary, Grand Rapids, Michigan

---

Dr. Ouweneel understands what many atheists—and too few Christians—do, that the conflict between the clear reading of Scripture and modern evolutionary theory permits no reconciliation. *Adam, Where Are You?* highlights that if we cannot trust the Bible in Genesis 1–3, we cannot trust it on any point. Anyone who is interested in honestly engaging with the question of origins and the implications for Christian theology will be helped by this book.

DR. JIM MASON, PhD Experimental Nuclear Physics, speaker with Creation Ministries International

---

Adam, Where Are You? by Willem J. Ouweneel is a balanced theological exploration of debates about "the historical Adam," particularly modern theology's dismissal of him to the realms of myth or literary figure. Dr. Ouweneel explores the many and profound implications for Christianity, leading to the inescapable conclusion that every theological twist and turn boils down to a stark choice of traditional biblical theology or liberal apostasy. Compromise positions, so popular today, are inherently untenable. This book will be a great help to Christians searching for a ray of biblical clarity on a foggy path.

DR. JOHN K. REED, geology editor, *Creation Research Society Quarterly*

All Christians who in their scientific work are ashamed of the Name of Christ Jesus, because they desire honour among people, will be totally useless in the mighty struggle to recapture science, one of the great powers of Western culture, for the kingdom of God. This struggle is not hopeless, however, so long as it is waged in the full armour of faith in him who has said, "All authority in heaven and on earth has been given to Me," and again, "Take heart! I have overcome the world."
—Herman Dooyeweerd (1894–1977), Christian philosopher

ADAM, WHERE ARE YOU? —AND WHY THIS MATTERS

A THEOLOGICAL EVALUATION OF THE EVOLUTIONIST HERMENEUTIC

WILLEM J. OUWENEEL

PAIDEIA PRESS

www.reformationalpublishingproject.com

Published by Paideia Press
P.O. Box 1000, Jordan Station, Ontario, Canada L0R 1S0

Book design by Janice Van Eck

Printed in China

ISBN 978-1-989169-04-9

Of Man's First Disobedience, and the Fruit
Of that Forbidden Tree, whose mortal taste
Brought Death into the World, and all our woe,
With loss of Eden, till one greater Man
Restore us, and regain the blissful Seat,
Sing Heav'nly Muse…
—John Milton, *Paradise Lost*

Archangel Uriel (to Adam and Eve):
O! happy pair, and happy still might be,
If not misled by false conceit
Ye strive at more than granted is,
And more desire to know than know ye should.
—Joseph Haydn, *The Creation*
(words from Milton, *Paradise Lost*)

CONTENTS

ABBREVIATIONS

BIBLE VERSIONS

AMP	Amplified Bible
AMPC	Amplified Bible, Classic Edition
CEB	Common English Bible
CEV	Contemporary English Version
CJB	Complete Jewish Bible
DARBY	Darby Translation
DRA	Douay-Rheims 1899 American Edition
ERV	Easy-to-Read Version
ESV	English Standard Version
GNT	Good News Translation
GNV	1599 Geneva Bible
GW	God's Word Translation
HCSB	Holman Christian Standard Bible
JUB	Jubilee Bible 2000
KJV	King James Version
MSG	The Message
NASB	New American Standard Bible
NKJV	New King James Version
NOG	Names of God Bible
NRSV	New Revised Standard Version
RSV	Revised Standard Version
TLB	Living Bible
VOICE	The Voice
WYC	Wycliffe Bible

OTHER SOURCES

EDR Ouweneel, W. J. 2007–2013. *Evangelisch Dogmatische Reeks*. Vaassen/Heerenveen: Medema.

RD Bavinck, H. 2002–2008. *Reformed Dogmatics*. Edited by John Bolt. Translated by John Vriend. 4 vols. Grand Rapids, MI: Baker Academic.

PREFACE

> And they heard the sound of the LORD God walking
> in the Paradise by evening,
> and Adam and his wife hid themselves from the face of the LORD God
> in the midst of the wood of the Paradise.
> And the LORD God called to Adam and said to him,
> "Adam, where are you?"
> Genesis 3:8–9 (translated from the Septuagint)

The title of this book reiterates God's question to Adam, recorded in Genesis 3:9, "Where are you?" These three words translate merely one Hebrew word: *ayyekah*. The Septuagint and the Peshitta include the proper name introducing the question: "Adam, where are you?" The question can be taken literally: "Where have you hid yourself?" Or figuratively: "In what condition are you?" Of course, God already knew the answers to both questions. But he wanted the first man to voice the answers himself.

In this book, we are examining God's question in terms of both a figurative and an extended meaning: Adam, where can we find you today? Did you disappear among the "bushes" of myth, saga, and archetype? Have you disappeared behind a "bush" heavily influenced by evolutionary thinking, where you are perhaps still historical but scarcely resemble your biblical portrait? Or can we still find you where you have been for thousands of years: in the genuine world of real history? Can we still trust what Moses, Jesus, and Paul have told us about you, Adam, or must we listen today to the voices of modern science in order to understand what those ancient men really meant? Were you directly created by God, some 6,000–10,000 years ago, or are you the product of millions years of evolutionary development?

The issue of creation versus evolution has captivated me personally throughout my life, and I suppose I will never escape it in this life. I was raised to believe a literal interpretation of Genesis (in my Plymouth Brethren family, which included British authors Frederick F. Bruce, David W. Gooding, and John Lennox[1]). However, in high school my Greek language teacher recommended the book *Creatie en Evolutie* (*Creation and Evolution*), written by biology professor Jan Lever (Kuyperian Reformed) from the Free University of Amsterdam, who defended the doctrine of general (particles-to-people) evolution.[2] For quite some time, I was enamored with this (less than) Reformed viewpoint. During a university course on animal physiology, one of my professors said, "There are two ways to imagine how the world originated: through creation or through evolution. Creation is religion and as such cannot have a place in our course. So what remains is evolution." A stunning *non sequitur*, indeed!

In 1968, an older friend gave me my first creationist book, written by Christian zoology professor Hannington Enoch from India, entitled *Evolution or Creation*.[3] In my youthful enthusiasm, I bought almost all the books (mostly American) that were mentioned in his bibliography. Among them was the creationist standard work of the time, written by hydrologist Henry M. Morris and theologian John C. Whitcomb, entitled *The Genesis Flood*.[4] This book was probably the most powerful tool of propaganda for creationism in the 1960s, at least in North America.

In the meantime I obtained my first PhD at the University of Utrecht (the Netherlands, 1970), in biology, more specifically, developmental biology. For ten years (1966–1976), I conducted biological research at the Hubrecht Institute of the Royal Netherlands Academy of Art and Sciences. For my accomplishments there, I was nominated for a national honour by the Dutch queen at the time, Queen Juliana. I worked with so-called homeotic mutants of the fruit fly, mutants that turned out to be somewhat significant for evolutionary genetics.[5]

During this time I myself began writing about creationism, first a booklet on Genesis 1,[6] then a book entitled *Operatie Supermens* (*Operation Superman*).[7] The retired theology professor Willem H. Gispen (Kuyperian

---

1 See Bruce (1990); Gooding and Lennox (2014a; 2014b); Lennox (2009; 2011).
2 Lever (1956).
3 Enoch (1976).
4 Morris and Whitcomb (2011).
5 See Ouweneel (1975b).
6 Ouweneel (1974).
7 Ouweneel (1975a).

Reformed) reportedly said that this book overcame the dominance of evolutionary theory in the Netherlands. If this were true, its success was only temporary; today, I know only a handful of Dutch people with doctorates in the natural sciences who remain critical of the theory of general evolution.[8]

0.1 SPREADING MY WINGS

After producing these initial writings, I authored five more books relating to creationism; I will not bother the reader with their Dutch titles. During the 1970s and 1980s, I began giving lectures on creationism, mostly in the Netherlands, but also in other European countries, and occasionally in North America. The time seemed to be ripe for these lectures, which were being given to audiences numbering up to a thousand. Some people became Christians after hearing such lectures, people for whom the theory of evolution had been a real stumbling block to faith in God.

My books and lectures attracted the attention of three Reformed men (one Dutch Reformed and two Kuyperian Reformed) in the Netherlands: theologian W. J. J. (Willem) Glashouwer, mathematician J. A. (Koos) van Delden, and businessman F. J. (Frits) Kerkhof. In 1974 they had founded the Stichting ter Bevordering van Bijbelgetrouwe Wetenschap (Foundation for the Advancement of Science Faithful to the Bible). I became the first editor of their new magazine, *Bijbel en wetenschap* (*Bible and Science*). In 1977 the four of us founded the Evangelische Hogeschool (Evangelical College), specifically for the purpose of offering a one-year course between secondary school and college/university, a major component of which was the teaching of creationism. I continue to teach there on occasion.

My views were kindly received among Plymouth Brethren friends, although some preferred the "gap theory" advocated by Irish Brethren Bible teacher William Kelly, and by geography professor André Gibert, member of the French Brethren. According to them, millions of years of the geological periods constitute the "gap" between the events of Genesis 1:1 and Genesis 1:2, a period that was followed by the six literal days of creation, which occurred some six thousand years ago.

I began to visit Brethren Bible Conferences in the United States, and took the opportunity to attend creationist conferences as well. In 1975 I lectured at the Third National Creation Science Conference, sponsored by the Bible Science Association and the Twin Cities Creation Science Association. In 1977 I attended the Fifth Annual Creation Convention in

---

[8] Specifically, Drs. Willem Binnenveld, Peter Borger, Herman Bos, Juri van Dam, Hans Degens, and Wim de Jong.

Philadelphia. During those years I also published various English-language articles in the *Creation Research Society Quarterly*. Later I became disaffected by American creationism for various reasons. I began to consider the idea that we Europeans, with our longstanding philosophical and theological tradition, should develop our own position on the creation–evolution controversy. To me, the most beautiful result of this was perhaps the first European Creationist Congress held in 1984.[9]

Meanwhile, my scientific interests had changed direction. During those years I had published several books on (Christian) psychology.[10] Gradually becoming more interested in the philosophical prolegomena of my research, and of biology and psychology in general, I began to study philosophy. In 1986 I earned a doctorate in philosophy, more specifically, in philosophical anthropology, from the Free University of Amsterdam (cf. chapter 6 below). From 1990 to 1998, I was professor of the philosophy of the natural sciences at Potchefstroom University for Christian Higher Education in South Africa. I made no secret of my anti-evolutionist leanings, so that I soon came into conflict with some of the biologists there who, it turned out, were strong advocates of J. Lever (see above).

0.2 THEOLOGY

While at Potchefstroom University, I had the opportunity to satisfy a longstanding desire. I had been studying academic theology from an early age, and in 1993 I completed these studies with a doctorate in theology from the University of the Orange Free State in Bloemfontein (South Africa). My special focus was the philosophical and bibliological prolegomena of systematic theology. In 1995 I became professor of systematic theology at the Evangelical Theological Faculty in Leuven (Belgium), where I worked part-time until my retirement in 2014.

In summary, I have been professionally active in all three fields in which I earned these doctorates: biology, philosophy, and theology. In my view, each of these three disciplines is necessary to evaluate the creation–evolution debate properly. Biology is required for evaluating properly the scientific value of evolutionary theory, while philosophy is required for evaluating properly the underlying fundamental questions and ideologies, whereas theology is required for evaluating properly the meaning of the debate in terms of the Bible.

---

[9] See Andrews, Gitt, and Ouweneel (2000); also cf. Andrews (1981); Gitt (1993; 1999; 2006). For a narrative history of these Dutch discussions, see Blanke (2014).

[10] See Ouweneel (1984; its more popularly written version is the English translation [2014b]).

I never lost my interest in the creation–evolution controversy, although for many years my primary interests lay in different areas of scholarship. For instance, from 2007 to 2013, I produced the twelve volumes of my Evangelical Dogmatic Series, in which I presented my view of Christian truth. The major contents of these volumes are being translated and re-worked for publication in English.

I can honestly say that throughout the years, I have never abandoned the conviction that I developed in the 1970s. With the knowledge that we now have of biological processes and of the fossil record, it is impossible for me to believe in general evolution (i.e., evolution from the first living cells to human beings). Please notice that I say this—insofar as I know myself—on a purely biological and not a theological basis. I do have theological objections—these are the subject of this book—but I would like to keep my biological objections to evolution distinct from my theological objections to evolution, as much as possible (but this is challenging, because I have only one heart or centre of being).

Do not expect this book to provide biological alternatives to general evolution. To a large number of questions I simply do not have definitive answers, and many tentative answers that I do have I would rather keep to myself. How old is the universe? How old is the earth? What are the limits to micro-evolution (which is really nothing more than biological variation)? What is the proper exegesis of many details of Genesis 1–3? In the last decade or so, I have occasionally described myself as an "origins agnostic," but this turned out to be an unfortunate term because it led to various misunderstandings (such as, Ouweneel is an "agnostic" of a special variety). This unfortunate misunderstanding is akin to what you get when you call yourself "Catholic but not Roman," or "Reformed without the Three Forms of Unity." I might not be sure about many answers, but I am neither a biological agnostic (since I do not believe in general evolution), nor a theological agnostic (since I believe that there were a historical Adam and Eve, and a historical Fall, all of them involving real events in time and space).

0.3 EYES OPENED AGAIN

I have been greatly helped in recent years by books like those authored by D. O. Lamoureux, K. W. Giberson, R. Carlson and T. Longman, D. C. Harlow, M. Tinker, D. and L. Haarsma, P. Enns, D. O. Alexander, N. T. Wright, W. T. Cavanaugh and J. K. A. Smith, and G. van den Brink.[11] Their works

[11] Lamoureux (2008; 2009); Giberson (2009); Carlson and Longman (2010); Harlow (2010); Tinker (2010); Haarsma and Haarsma (2011); Enns (2012); Alexander (2014); Wright (2014); Cavanaugh and Smith (2017); van den Brink (2017).

have opened my eyes, so to speak, but in a way that was opposite to what these authors intended. When I began reading about the evolutionary views of these authors (and of some contributors to the volume edited by Barrett and Caneday[12]), I realized: *this is not the way*. Some of these writers are not liberal theologians in the usual sense of that term; they claim to retain their orthodoxy, for instance, by emphasizing their faith in the resurrection of Christ. Some of them do their best to maintain some notion of a historical Adam and of a historical Fall, though firmly believing in human evolution (or provisionally accepting it as the best theory available). But in my view, their reading of Genesis 1–3 through the lens of the theory of evolution is disastrous.

Along this route, these authors often reach conclusions—no matter how tentative—diametrically opposed to what thousands of Jewish and Christian expositors have historically understood these chapters to be teaching. By itself, this fact *proves* nothing, but it *reveals* a lot. In my view, these authors are treating the Bible with the techniques of a ventriloquist: they make the Bible say what they wish it to say; the Bible itself is not really speaking, but it is they who speak. Christians ought to listen to the Bible, and not "speak through" the Bible.

I am not judging the motives of these authors; I am saying merely that I view their conclusions as devastating. I am firmly convinced that their views surrender a biblical view not only of humanity and of sin, but in the end also of Scripture, of salvation, and of Christ himself. When liberal theologians spoke openly of myths in the early chapters of the Bible, we shrugged our shoulders; after all, there was a religious chasm between them and us. Now scientists and theologians who claim to be orthodox Christians are saying these things. This means that the discussion is threatened by far more potential pitfalls and booby traps.

Indeed, the discussion is far from easy. Those who study the important book *Four Views on the Historical Adam* (see note 12) may discover for themselves how often the various contributors talk past each other, how they misunderstand and misrepresent each other, despite their usually courteous attitudes. Yet, in the end the reader comes to the conclusion—at least I did—that one now understands a few things better than before one began reading. All of our books are only minor contributions to a process in which, despite the great differences of opinion, we ultimately hope to have made some progress in understanding Scripture, history, and science.

The essence of this book can be summarized very simply: if we believe human evolution, can we still retain the biblical message of Genesis 1–3?

---

[12] Barrett and Caneday (2013).

The authors mentioned above say: Yes. I (and many others) say: No. The subject of this book is not the theory of evolution itself. Rather, it is whether, *if* we accept human evolution, orthodox Christianity can be salvaged. Some say that you can have both: human evolution and the Bible. Others say that you will have to choose. If human evolution is true, the Bible is false. If the Bible is true, the theory of human evolution is false.

This book is not just a polemical book, combating the views of others. Parts of it have a more thetical than antithetical character. As Jude indicated, sometimes we find "it necessary to write appealing to you to contend for the faith that was once for all delivered to the saints," although it is much more pleasant "to write to you about our common salvation" (Jude 3).

I wish to thank those who have read earlier versions of the manuscript, or parts of it, and have given me their highly appreciated comments. In alphabetical order they are: Canadian theologian and apologist Dr. Joseph Boot, Dutch biologists Dr. Peter Borger and Dr. Wim de Jong, Australian philosopher Kerry J. Hollingsworth, American theologian Dr. Nelson D. Kloosterman (who also functioned as my appreciated editor), German biologist Dr. Siegfried Scherer, and South African philosopher (expert in evolutionary thought) Dr. Danie F. M. Strauss. My warm thanks also go out to those involved in the publication work: in addition to Dr. Boot, these were Ryan Eras and John Hultink, who also gave their valuable comments. It goes without saying that I alone am responsible for the contents of the final product.[13]

WILLEM J. OUWENEEL
Huis ter Heide (Netherlands)
Winter 2017–2018

[13] After this manuscript was completed, an important book appeared on the scene, published by Crossway, entitled *Theistic Evolution: A Scientific, Philosophical, and Theological Critique*, edited by J. P. Moreland, Stephen C. Meyer, Christopher Shaw, Ann K. Gauger, and Wayne Grudem. Contributors to the book include two dozen highly credentialed scientists, philosophers, and theologians from Europe and North America. The book consists of four parts: (1) the failure of Neo-Darwinism; (2) the case against universal common descent and for a unique human origin; (3) the philosophical critique of theistic evolution; and (4) the biblical and theological critique of theistic evolution. Unfortunately, it was impossible to incorporate insights from this book into my own work, but I wholeheartedly recommend *Theistic Evolution* as a very valuable resource for this discussion. We are confederates in the same conflict. May the Lord be pleased to bless our joint efforts!

1

STATING THE PROBLEM

For the time is coming when people will not endure sound teaching, but having itching ears they will accumulate for themselves teachers to suit their own passions and will turn away from listening to the truth and wander off into myths (2 Timothy 4:3–4).

THESIS
Today, some Christians try to find a middle road between orthodox theology and liberal theology: they are orthodox with regard to the central parts of the gospel, but liberal when it comes to Genesis 1–3. This does not work; ultimately, this entire way of thinking ends in progressive phases of liberal theology, as it always has.

1.1 LAST DAY MYTHS

1.1.1 WILSON AND LEWIS

Christians, especially those living in the last days (2 Tim. 3:1), have often wondered about the "myths" that the apostle Paul refers to in 1 Timothy 1:4; 4:7, and 2 Timothy 4:4. Myths are ancient popular stories, especially about the origin of the world. Traditionally, gods and goddesses, heroes and heroines, plants and animals play important roles in these stories. In modern times, the gods can easily be replaced by "principles" thought to have a virtually supernatural power (even if those who proclaim such "myths" usually dislike a term like "supernatural"—and have an even greater dislike of the term "myth" itself, unless, of course, it is applied to the views of others). About what myths might the apostle Paul be thinking? Ancient "creation myths" (pardon the term) in which orthodox Christians, Jews, and Muslims still believe, or some modern "evolution myth"?[1] Or was he referring to specific myths of a very different nature?

Expositors are generally careful in specifying what kind of myths Paul might have had in mind. Perhaps his intention was very general; he may have used the Greek word *mythos* in the sense of the Latin term *fabula*, "fable," that is, any fictitious story, or simply any falsehood.[2] Yet, the Greek term *mythos* intrigues me. Do we not live in a time when modern science is working hard to remove all kinds of myths from our thinking? How could Paul suppose that in a time like ours—a time many Christians believe is close to the "last days"—especially Christians would "wander off" to follow certain "myths"? *Myths*, of all things! We are reminded here of the King James rendering of 1 Timothy 6:20, "O Timothy, keep that which is committed to thy trust, avoiding profane and vain babblings, and oppositions of science [Gk. *gnōsis*] falsely so called: …."

In 1987, Danish-Swedish biologist Søren Løvtrup, who is not a creationist,[3] published his book entitled *Darwinism: The Refutation of a Myth* [ie., Darwinism]. And if we may believe (non-creationist) Andrew N. Wilson (famous for his biography of Queen Victoria), this mythology began with Darwin himself. In a recent book, Wilson depicted Darwin as a "Victorian mythmaker," as the title says.[4] In a preview he is quoted as saying,

> Funnily enough, in the course of my researches, I found both pride and prejudice [an allusion to Jane Austen's novel] in bucketloads among the ardent Darwinians, who would like us to believe that if you do not worship Darwin, you are some kind of nutter. He has become an object of veneration comparable to the old heroes of the Soviet Union, such as Lenin and Stalin, whose statues came tumbling down all over Eastern Europe twenty and more years ago.[5]

I am fond of recalling the writings of C. S. Lewis, arguably the greatest Christian apologist of the twentieth century (d. 1963). Lewis wrote an essay

---

[1] Consider the titles of scientists like Milton (2000): *Shattering the Myths of Darwinism*; Wells (2002): *Icons of Evolution: Science or Myth?*; Wells (2011): *The Myth of Junk DNA*; or Mejsnar (2014): *The Evolution Myth*. Others identify supposed "myths" advocated by the creationist side; see https://www.newscientist.com/article/dn13620-evolution-24-myths-and-misconceptions/; see also https://thelogicofscience.com/2015/08/31/10-common-myths-about-evolution/.

[2] Earle (1978, 351); cf. Collins (2011, 28–41).

[3] For the terms "creationist" and "evolutionist," see Appendix 1.

[4] Wilson (2017).

[5] *Evening Standard* (Aug. 4, 2017).

called "The Funeral of a Great Myth."[6] The myth to which he is referring is what he called popular evolutionism. Lewis considered the theory of evolution to be "a genuine scientific hypothesis," held by practicing biologists. I must observe, however, that most practicing biologists are specialists in their fields. Very few of them have an expert grasp of both morphology and genetics, as well as developmental biology, let alone of geology and paleontology. Very few are in a position to speak authoritatively when they declare evolution to be "a genuine scientific hypothesis." But that is not the point I wish to make right now. Lewis' "great myth" involves not evolution as a strictly scientific hypothesis but evolution as the faith of virtually all modern intellectuals (see chapter 2). This myth is believed especially by those who are not at all in a position to judge the general scientific status of such an encompassing hypothesis. This myth is believed by most biological specialists, most other natural scientists, most philosophers and theologians, and so many other educated people.[7]

Throughout the years I have observed that biologists are generally more willing to listen to and engage opposing arguments than are theologians. This is because biological generalists are somewhat aware of the weaknesses in the theory of evolution, whereas theologians generally are not. This is understandable, since usually theologians have not studied evolution academically. Yet—or must I say, therefore?—many theologians speak with a conviction that is often astonishing. Since the nineteenth century, the majority of theologians have been converted to the theory of evolution—which they are not qualified to assess and they are unlikely to surrender.

1.1.2 KIDNER AND KELLER

Consider, for example, British Old Testament scholar Derek Kidner, whose work was published about fifty years ago.[8] He noted that God is said to have fashioned Job with his hands out of clay (Job 10:8–9), and wondered if this could explain Genesis 2:7 ("God formed the man of dust from the ground"). If in reality Job had been formed in his mother's womb, why could the same not be true of Adam, with motherhood extended further back into history? He continued,

---

[6] Contained in Lewis (1972); see http://fpb.livejournal.com/297710.html.

[7] Note that, for Lewis, "mythical" is not necessarily "ahistorical"; elsewhere he spoke of "myths" in the Bible in the sense of grand divine narratives, which he nonetheless viewed as historically true (1996, 63–64). It is in this sense that we must understand Vanhoozer's title (2012, *Remythologizing Theology*), a book written especially in response to Bultmann (2012).

[8] Kidner (1967, 28n2); regarding his view, see Timothy Keller (n.d.), who seems to agree with him.

> Man in Scripture is much more than *homo faber*, the maker of tools: he is constituted man by God's image and breath, nothing less.... The intelligent beings of a remote past, whose bodily and cultural remains give them the clear status of "modern man" to the anthropologist, may yet have been decisively below the plane of life which was established in the creation of Adam.... Nothing requires that the creature into which God breathed human life should not have been of a species prepared in every way for humanity.

I cannot agree with Kidner, nor with Timothy Keller, who seems to agree with Kidner on this point. The comparison with Job fails simply because the Bible explicitly calls Adam "the first man" (1 Cor. 15:45–49), whereas there are numerous allusions to Job's mother (Job 1:21; 3:10; 19:17; 31:18). Adam was the "one man" from whom all humans descended (Acts 17:26),[9] and his wife became "the mother of all living" (Gen. 3:20). Kidner's argument became still more imaginative: he suggested that, after the special creation of Eve—but how does this event fit into his picture?—

> God may have now conferred his image on Adam's collaterals, to bring them into the same realm of being. Adam's "federal" headship of humanity extended, if that was the case, outwards to his contemporaries as well as onwards to his offspring, and his disobedience disinherited both alike.[10]

Of what benefit is all this imagination—described by Kidner but endorsed by Keller—which fails to line up with the written Word? It is nothing but an attempt to accommodate evolutionary thinking, and at the same time to remain as close to Scripture as possible. Keller gives this advice:

> My conclusion is that Christians who are seeking to correlate Scripture and science must be a "bigger tent" than either the anti-scientific religionists or the anti-religious scientists. Even though in this paper I argue for the importance of belief in a literal Adam and Eve, I have shown here that there are several ways to hold that and still believe in God using EBP [i.e., evolutionary biological processes].[11]

---

9 Admittedly, some expositors (such as John Gill, http://biblehub.com/acts/17-26.htm) have thought this could refer to Noah; cf. J. H. Walton in Barrett and Caneday (2013, 105).

10 Kidner (1967, 30).

11 Keller (n.d., 13).

This view is shared by many theologians today. Throughout this work, I will repeatedly refer to this view as AEH, shorthand for "the view that *A*dam was an *E*volved *H*ominid,"[12] a humanlike being produced by evolutionary biological processes. AEH advocates wish to leave room for human evolution, while simultaneously somehow retaining the notions of a historical Adam and a historical Fall.

1.1.3 EVANGELICAL "HEROES"

It is impossible to survey the entire body of literature about Genesis 1–3 that has appeared between 2007 and today (2017). I will simply take note of several publications that advocate a more liberal[13] view of these chapters.[14] In addition to Timothy Keller, I would select three of the best known and appreciated evangelical "heroes." To all four of these, David's statement seems appropriate: "How the heroes have fallen in the heat of battle" (2 Sam. 1:25 CJB). Remarkably, the following three are British-born Anglican theologians.

(a) British Anglican theologian John Stott (d. 2011). According to *New York Times* writer David Brooks, if evangelicals were to have chosen a pope, it would probably have been Stott. Yet, in his book *Understanding the Bible*,[15] he defended theistic evolutionism, including the notion of pre-Adamic hominids.

(b) British-born Canadian Anglican theologian James I. Packer is another evangelical "hero" who defends theistic evolutionism. In 2008 he penned an endorsement for theistic evolutionist Denis Alexander,[16] and later a positive Foreword to a book by theistic evolutionist Melvin Tinker.[17]

(c) British Anglican theologian N. T. Wright. I will discuss his view more extensively in §1.3.

Let me mention a fourth British Anglican theologian, Alister McGrath. In a 2011 interview,[18] he did not clearly choose a specific position, but he clearly defended the permissibility of theistic evolutionism, and of the Big Bang theory. Despite his caution, we find this equally disappointing.

---

[12] The term "hominid" refers to a family of primates that includes both humans and apes.

[13] I use the term "liberal" to identify theology that does not submit to Scripture, but instead subjects Scripture to the methods and insights of modern science.

[14] See Blocher (1984; 1997); Hyers (1984); Barr (1993); Peacocke (1993); Sailhamer (1996); McGrath (1997); Ward (1998); Waltke (2001); Rana and Ross (2005); and Collins (2006).

[15] Stott (1999, 54–56).

[16] See Alexander (2014); for comments on Alexander, see David Anderson in Nevin (2009, chapter 5).

[17] Tinker (2010).

[18] https://www.bethinking.org/christian-beliefs/the-universe-is-not-an-accident.

More acceptable is the view of Calvinist Baptist theologian John Piper, who allowed for an old earth, but also said,

> I don't believe in evolution as the way that Adam came to be a human. I think God created Adam from the dust of the ground. I think he was unique and that he is the father of all humanity—Adam and Eve—and that he is not the product of a long evolutionary process. I can't make that jive with the way the text reads.
>
> And I think that it's very important that Adam be a historical figure, because that's the way he is treated by the other biblical writers. The heart passage in Romans 5 collapses, and the whole nature of God's making with Adam a covenant and then him failing and then Christ being a second Adam comes to naught, if he's not a historical person.[19]

The words of natural scientist John A. Bloom are fitting at this point:

> In light of the comparative literature and cultural knowledge available to us, we cannot presume that it was impossible for the writer of Genesis to explain to his ancient audience that God empowered a small group of already-living beings that shared some common features with apes to develop into mankind. God could have told the story of a hominid-origin for mankind in a manner the early Hebrews could have grasped. The fact that he did not must have some significance.[20]

At the outset of my examination, I would ask AEH advocates this straightforward question: If God selected some hominids from a much larger population in order to enter into a covenant relationship with them, why did he not simply tell us? He selected Noah, Abram, and Israel from large multitudes of people; if he did this with Adam, too, explaining this to us would not have been difficult. Why did he speak in Genesis as though he had created Adam and Eve directly, without any other humans around? Our question will continue to surface throughout this book.

---

[19] http://www.desiringgod.org/interviews/do-you-accept-old-earth-and-evolution.
[20] http://www.asa3.org/ASA/education/origins/humans-jb.htm.

1.2 RECENT PUBLICATIONS

1.2.1 TWO BOOKS

Let me illustrate recent developments by mentioning two theological books, both of which appeared in 2017, both of which endeavor to recast the origins story of Genesis 1–3. The first one was edited by Catholic theologian William T. Cavanaugh and Protestant philosopher James K. A. Smith, entitled *Evolution and the Fall*.[21] There are clear differences among the various contributors to this book; yet, overall the book entertains the idea that God selected some hominids, and placed them in paradisal (literally or figuratively) circumstances, where some kind of a "fall" occurred. These selected hominids lived much earlier than most Bible readers have traditionally supposed, and when they fell, they were certainly not the only hominids alive on earth. In fact, there never was any historical Adam in the biblical sense of a first human whose creation and fall constituted the beginning of human history. The editors and some of the contributors have fully embraced the theory of evolution.

The second book is a Dutch publication, written by Reformed theologian Gijsbert van den Brink, entitled *En de aarde bracht voort* (*And the Earth Brought Forth*).[22] van den Brink explains the theory of evolution in great detail, and draws conclusions similar to those in the essays mentioned earlier: if evolution is true, Genesis 1–3 must be read in a way totally different from the traditional Christian reading. Once again, the question is: If we accept the theory of evolution, can we retain the historical Adam and the historical Fall?

The appearance of these two books within one year is a very significant phenomenon—together with many similar books that have appeared during the last decade.[23] Within the theological world, anyone holding to (young- and old-earth) creationism seems about to be classified as a fossil. Theologians who were once orthodox and conservative are now publicly defending the theory of evolution,[24] even though they seem to be aware

---

[21] Cavanaugh and Smith (2017).

[22] van den Brink (2017); at the time of this writing, an English translation of this work is rumored to be underway; the title of the Dutch volume alludes to Gen. 1:12. For a preview of the book's argument, see van den Brink (2012; 2015, 111–13).

[23] See the works of Giberson (2009), Northcott and Berry (2009), Carlson and Longman (2010), Tinker (2010), Haarsma and Haarsma (2011), Enns (2012), Young (2012), Wright (2014), and Venema and McKnight (2017); regarding the latter, see the excellent comment by Carter (2017).

[24] Consider the striking example of Canadian biologist and theologian Denis Lamoureux (2009), who declared in Barrett and Caneday (2013, 39), "I'm a born-again Christian." Yet,

of the consequences. These consequences are both exegetical, involving a drastically new interpretive approach to Genesis 1–3, and doctrinal, involving especially the New Testament teaching about the fall of the first humans (see especially Rom. 5:12–21; see chapters 8–10). What biblical truth remains if the historical Fall is effectively dismissed? The consequences of this new thinking are comprehensive, affecting the doctrines of creation (ktiseology), sin (hamartiology), humanity (anthropology), salvation (soteriology), and Christ (Christology).

1.2.2 TRADITION

I have used the word "traditional" four times, but an appeal to tradition cannot be the full answer, of course. Traditions have sometimes been mistaken. Yet, van den Brink insists that he is still Reformed, obligating us to consider what the Three Forms of Unity have to say on the matter. These Confessions are the minimum standard for defining what is Reformed. The Heidelberg Catechism says (Q/A 6) that God created people good, and it confesses humanity's fall into sin (Q/A 7): "Q. Then where does this corrupt human nature come from? A. The fall and disobedience of our first parents, Adam and Eve, in Paradise. This fall has so poisoned our nature that we are all conceived and born in a sinful condition." The Belgic Confession confesses that "Adam and Eve had plunged themselves ... into both physical and spiritual death," (Art. 17), and describes "our first parents, Adam and Eve" (Art. 23). If consistency with these Confessions is required for being identified as Reformed, then van den Brink is no longer Reformed, precisely because he is suggesting that there never was a first man, Adam, created directly by God, who also formed Eve from his side, and that there was no discrete historical event known as the Fall, when these two human beings fell into sin.

Even so, the most cherished confessional tradition cannot be the final authority regarding the existence of a historical Adam. We need a thorough theological assessment. This book deals not so much with the biological aspects of the issue (except in chapter 2), but rather with the theological problems involved. Those who claim that these denials do not affect orthodox Christianity[25] are deceiving themselves and others.

In the first place, I claim that an essentially different hermeneutic is operating here, an essentially novel way of reading the Word of God,

---

van den Brink (2017, 133) sees a close connection between Lamoureux and the very liberal German theologian Rudolf Bultmann.

[25] Cf. van den Brink (2017, 20–21).

different from what Jews and Christians have practiced for many centuries. I plan to substantiate this claim in chapter 3.

Second, there is the problem of historicity (see chapter 4). What do we mean when we say that the early chapters of Genesis are, or are not, historical? These Bible chapters may not be historical in the same sense as journalism or historiography is historical, but then, in what sense are they historical? I will argue that these chapters tell us about events that really happened, and about the way they happened.

Third, what does the New Testament teach about creation and the Fall? What light does it shed on the exegesis of Genesis 1–3? Can the authors mentioned maintain their new exegesis in the light of New Testament teaching? I believe the answer is No (see chapters 5–10). Only when this preliminary analysis has been completed will we be able, fourth, to provide our exegesis of Genesis 1–3 (already touched upon in chapter 5). Our basic premise is simply that Christians must read the Old Testament in the light of the New Testament (see especially chapter 10).

1.2.3 A COUNTER-TESTIMONY

The theological scene is not altogether dominated by theistic evolutionism. In the last decade publications have appeared that defend, or at least allow, a more traditional reading of Genesis 1–3.[26] Most of these publications will be quoted again throughout this book. A serious battle is being waged by powerful combatants on both sides.

1.3 N. T. WRIGHT ON ADAM

1.3.1 THE CHOOSING OF ADAM

In his work *Surprised by Scripture*, N. T. Wright surprised me with the following statement:

> Just as God chose Israel from the rest of humankind for a special, strange, demanding vocation, so perhaps what Genesis is telling us

[26] See Schwertley (2000); Wells (2002); Collins (2003; 2006; 2010; 2011); Poythress (2006; 2012; 2013; 2014); Lennox (2009); M. Reeves in Nevin (2009); Pruitt (2009; 2010a; 2010b); Richards (2010); Strauss (2010); R. Strimple in Muether and Olinger (2011); Versteeg (2012); Gaffin (2012; 2015); Collins in Barrett and Caneday (2013, 126–175); W. D. Barrick in Barrett and Caneday (2013, 188–227); Trueman (2013); Madueme and Reeves (2014), especially Part I; Kelly (2015); Phillips (2015); Rossiter (2015); Mortenson (2016); VanDoodewaard (2016); Immink (2017); Paul (2017). See also many articles on www.alliance.org/mos, www.discovery.org, and logos.nl.

is that *God chose one pair from the rest of early hominids for a special, strange, demanding vocation*. This pair (call them Adam and Eve if you like) were to be the representatives of the whole human race.[27]

The parallelism between Adam and Israel is well-known, and was familiar also to Second Temple Jews, as Wright pointed out. We may even add the parallel with Noah here. Adam lived in the Garden of Eden, and soon fell. After the Flood, Noah lived in another garden, a vineyard, and soon fell (Gen. 9:20–21). Israel lived in the Holy Land, which is explicitly compared to Eden (Gen. 13:10; Isa. 51:3; Ezek. 36:25), and soon fell as well (Exod. 32). One could add that the New Testament church began, so to speak, in a garden (John 19:41; 20:15), and soon fell (see Acts 20:29–30; cf. Appendix 3).

However—and this is the point—no Second Temple Jew would have entertained the idea that Adam and Eve had been chosen from early hominids (or from early humans). They would have rightly supposed that, if Adam and Eve had been chosen from a larger population, God would have told us so (unless people are referring to the large population of *all created beings*). The Second Temple Jew would have referred to Genesis 2:7, to indicate where Adam came from: "…then the Lord God formed the man of dust from the ground and breathed into his nostrils the breath of life, and the man became a living creature." Every Jew in those days would have concluded from this (as I do) that Adam came forth directly from the hand of God. If there was anything like a choosing of Adam—as a parallel with the choosing of Noah, Abraham, and Israel—we may say that, as unique image-bearer of God, Adam was chosen from all the created beings that God had made.

Let me quote some rabbis from a time not long after the Second Temple period: "Our Rabbis taught: Man was created alone [i.e., only one man was created, no other people being around]."[28] Important theological consequences are drawn from this fact that the first man was entirely alone in the world. One consequence is that every human being descended from this one man and his wife. There is an interesting parallel with Abraham, of whom the Lord says,

> Look to Abraham your father,
> And to Sarah *who* bore you;

[27] Wright (2014, 37); italics original. Wright was not the first to make such a suggestion; cf. Gavin B. McGrath (1997), who stated that "God took two hominids to become the first human beings, Adam and Eve."

[28] Bab. Talmud: Sanhedrin 38a, b.

> For I called him alone,
> And blessed him and increased him" (Isa. 51:2 NKJV).

Naturally, the rabbis would not have doubted that Adam was literally alone in the world, whereas Abraham was "alone" merely as the only one called by God. Yet, they would have seen the parallel: from the one Adam all humans descended, and from the one Abraham all Israelites descended.

A bit later in the same Talmud tract we read that Rabbi Meir used to say: "The dust of the first man was gathered from all parts of the earth." Rabbi Jochanan ben Chanina said that the day of Adam's creation

> consisted of twelve hours. In the first hour, his dust was gathered; in the second, it was kneaded into a shapeless mass; in the third, his limbs were shaped; in the fourth, a soul was infused into him; in the fifth, he arose and stood on his feet; in the sixth, he gave [the animals] their names; in the seventh, Eve became his mate; in the eighth, they ascended to bed as two and descended as four [i.e., Cain and his twin sister were fathered]; in the ninth, he was commanded not to eat of the tree [of the knowledge of good and evil], in the tenth, he sinned; in the eleventh, he was tried, and in the twelfth he was expelled [from Eden] and departed.

This is perhaps the most literal reading imaginable of Genesis 2 and 3. But if we find it exaggerated, let us remember that the ancient rabbis are bad allies for someone who ignores the details of Genesis 2 and 3, and understands Adam to be one of many ancient hominids. By way of commentary on the statement of Rabbi Jochanan, I would add:

(a) I would prefer to say that Adam and Eve did not fall into sin before the Sabbath day ended: on the seventh day God "rested and was refreshed" (Exod. 31:17), but soon after they "wearied" God with their sins (cf. Isa. 43:24b; Mal. 2:17).

(b) Cain could not have been fathered before the Fall. We read that afterward Adam "knew" his wife, and thus fathered Cain (Gen. 4:1). If Cain had been fathered before the Fall, he could not have said with David, "In sin did my mother conceive me" (Ps. 51:5b).

In fact, the literal view of the Talmudic rabbis is basically still the view of modern orthodox rabbis. I have asked several of them, from Europe and America, how literally they understood Genesis 1–3. The answer was invariably: Very literally. I asked them whether they were not confused by the views of modern science on origins. Their answers went something like this: Let science do its job, while we do ours, which is to believe what God tells us in Genesis 1–11, and to understand it just as literally as what he tells us in Genesis 12–50.[29] Genesis 1–3 is God's inspired revelation about the origins of the world and of humanity, and about the latter's fall into sin.

The point is not simply to reject the idea that God might have chosen one pair of early hominids from the rest. That notion is bad enough. But that bad starting point generates numerous additional questions: Was this pair created in the image and likeness of God, differently than the other hominids were made (Gen. 1:26–27)? Or were they *all* created in God's image? What happened to all those other hominids after Adam's selection *and* after his fall? What does Wright do with Genesis 2:7? And what does he do with the way the New Testament writers handle Genesis 1–3? How could Wright entertain the idea that Genesis 2:7, or the statements by Paul, *allow for* the notion that God *chose* a few early hominids? These and similar questions will continue to arise throughout our study.

1.3.2 LACK OF CLARITY

Moreover, Wright is very unclear about his subject. According to usual taxonomy, hominids are a primate family that includes orangutans, gorillas, chimpanzees, and gibbons, in addition to humans. Is this what Wright is talking about? Modern science claims that humans separated from apes at least 5 or 10 million years ago. Wright speaks of *early* hominids; in current geological chronology, they existed tens of millions of years ago. Is this where Wright wishes to assign Genesis 2 and 3? Or did Wright make a mistake and did he simply mean to refer to "humans"? According to modern science, humans are hominids, but not *early* hominids. However, the timeline of modern science posits the existence of *early* humans millions of years ago. Early representatives of the genus *Homo* are thought to have existed several millions of years ago. Early representatives of the species *Homo sapiens* are thought to have existed several hundred thousand years ago. Early representatives of the subspecies *Homo sapiens sapiens* are thought to have existed several tens of thousands of years ago.[30] (All

---

[29] Kevin DeYoung (http://www.alliancenet.org/mos/1517/why-it-is-wise-to-believe-in-the-historical-adam#.WaaZAq2iGMJ): "There is a seamless strand of history from Adam in Genesis 2 to Abraham in Genesis 12. You can't set Genesis 1–11 aside as prehistory, not in the sense of being less than historically true as we normally understand those terms. Moses deliberately connects Abram with all the history that comes before him, all the way back to Adam and Eve in the garden." C. J. Collins in Barrett and Caneday (2013, 157; cf. 160): "*Genesis 1–11 Sets the Stage for Genesis 12–50*." Yet, the same author notes that there is a literary distinction between Gen. 1–11 and 12–50: "The narrator slows down in the Abraham story" (248).

[30] For a creationist assessment of human fossils, see Lubenow (2004).

of this reflects the timeline of modern science, which for Wright appears unproblematic.)

We are puzzled that Wright does not bother to specify what exactly he is talking about, or to explain the implications of what he is writing. How can someone write about God "perhaps" selecting "one pair from the rest of early hominids" without attempting to explain some of the implications of such a drastic claim?

If Adam and Eve were early hominids (or early humans), Wright should have proposed some kind of timeline for their life. If he or anyone else wishes to maintain credibility by keeping the time span between Adam and Abraham as brief as possible, what happened to the other representatives of *Homo*, or of *Homo sapiens*, or of *Homo sapiens sapiens*? If he does not set us on some kind of historical footing here, he may be simply perpetuating some kind of new "creation myth." To express it in language parallel with Acts 14:16, must we assume that God simply "allowed all the other hominids, or humans, or *Homo* representatives, to walk in their own ways"? But the apostle Paul says explicitly that Adam was the "first man" (1 Cor. 15:45, 47).

1.3.3 THREE OPTIONS

Here, then, are our three options. Either Paul did not know any better—but why then believe the many other things he said and wrote (see chapter 10)? In this case, "every nation of mankind" perhaps did indeed descend from Wright's early hominids (or early humans), so that God was obliged—to mention just one option—to let all other *Homo* (*sapiens*) representatives die out (perhaps in the Noahic Flood?). Or Paul deliberately deceived us. Or Paul spoke the truth. To be honest, I prefer the latter option, and I intend to show why.

I also prefer Luke, who provides us with a long genealogy in chapter 3:23–38, which ends: "...the son of Enos, the son of Seth, the son of Adam, the son of God." Why the apposition "the son of God"? Because Adam had no earthly father,[31] and this was because he was the first human on earth, as Paul said. However, to be consistent with AEH, Luke should have written: "Adam, the son of some other unknown human (or hominid)," because Adam was supposedly not the first man at all; at most he was a descendant of hominids who had existed for millions of years. In 1 Corinthians 15:45 Paul refers to Genesis 2:7, apparently because he believed that Adam was not born but "formed" (he says so in 1 Tim. 2:13) from the dust of the ground:

---

[31] Interestingly, we are told that King Melchizedek, too, was "without father or mother or genealogy" (Heb. 7:3), but few expositors think that the author means this literally; he is presenting Melchizedek merely as a type of the Son of God (see v. 3).

> Thus it is written, "The first man Adam became a living being"; the last Adam became a life-giving spirit.... The first man was from the earth [Gk. *ek gēs*, genitive of source], a man of dust; the second man is from heaven [Gk. *ex ouranou*, genitive of source]. As was the man of dust, so also are those who are of the dust, and as is the man of heaven, so also are those who are of heaven. Just as we have borne the image of the man of dust, we shall also bear the image of the man of heaven (1 Cor. 15:45–49).

I do not think that Paul's words are capable of a figurative interpretation. If we are being told today not to take the phrase "man of dust" literally, then why should we believe in a literal "man of heaven"? Or if Adam were not at all "the first," why should we believe in him who is called "the last" (v. 45)? Or if Adam were not a real, historical figure—which is not what Wright is claiming—then according to 1 Corinthians 15 what ground remains under the feet of the Christian faith? The same concern could be raised with regard to Romans 5; but all these matters will be discussed more extensively later, especially in the last chapter.

1.4 JAMES K. A. SMITH ON ADAM

1.4.1 THE ADAMIC PRIMEVAL POPULATION

Let us briefly examine any new insights from the contributors to the collection edited by Cavanaugh and Smith. Most of these contributors appear to be AEH advocates as well (though I am not sure about each of them). For now we will look at the essay by James K. A. Smith, entitled "What Stands on the Fall?"[32] At the end of his exposé we are told this:

> From out of this [evolutionary] process there emerges a population of hominids who have evolved as cultural animals with emerging social systems, and it is this early population (of, say, 10,000) that constitutes our early ancestors. When such a population has evolved to the point of exhibiting features of emergent consciousness, relational aptitude, and mechanisms of will—in short, when these hominids have evolved to the point of exhibiting moral capabilities—our creating God "elects" this population as his covenant people. The "creation"

[32] Smith in Cavanaugh and Smith (2017, 48–64).

> of humanity, on this picture, is the first election—the first of many (Noah, Abraham, Jacob, et al.)….
>
> This original humanity is *not* perfect (the catholic theological tradition has never claimed that[33]). They are able to carry out this mission—God's law would not be established where obedience is not possible—but they are also characterized by moral immaturity, since moral virtue requires habituation and formation, requires time. So while they are able to carry out this mission, there are no guarantees, and also no surprises when they fall. Since we're dealing with a larger population in this "garden," so to speak, there is not one discrete event at time T1 where "the transgression" occurs. However, there is still a temporal, episodic nature of a Fall. We might imagine a Fall-in-process, a sort of probationary period in which God is watching.[34]

The similarities to the view of Wright are obvious. Again we hear about a long pre-history of human evolution and about the origin of hominids. Again there is development—I suppose under God's providential guidance, according to Smith—in the sense of an evolutionary emerging of a very special (human-like) population of hominids. One phrase strikes us here immediately: "features of emergent consciousness, relational aptitude, and mechanisms of will—in short, when these hominids have evolved to the point of exhibiting moral capabilities." I would rather say that animals possess "emergent consciousness, relational aptitude, and [to a certain extent] mechanisms of will"—but they do *not* exhibit moral capabilities (see §§6.1.1, 6.2.2, and 6.3). However, for Smith, the first three features seem to imply the fourth. This is a remarkable example of anthropomorphizing animals.

In Smith's thinking, when this human-like population of hominids has developed, God elects *this* population. The difference with Wright is that the latter continued speaking of a *pair* of hominids ("call them Adam and Eve if you like"). With Smith, the entire early human population is identified with Adam and Eve. In other words, in Smith's presentation there is no longer a place for the individual traditionally understood as the historical Adam; there is no longer one pair of humans at the beginning of human history as depicted in the Bible (one coming from dust, the other from

---

33 This is an interesting remark (cf. Thomas Aquinas, *Summa Theologiae*, http://www.newadvent.org/ summa/1095.htm). Did the Reformed tradition claim perfection? Notice the Heidelberg Catechism (Q/A 6): "God created [Adam and Eve] good and in his own image, that is, in true righteousness and holiness"; see Ouweneel (2016, ad loc.).

34 Smith in Cavanaugh and Smith (2017, 61); he italicized the entire quotation.

Adam's side). Consequently, there is no historical Fall (the first humans, Adam and Eve, who ate of the forbidden tree).

It is important to notice that not only do Smith's ideas run counter to the biblical presentation, but also he cannot claim to have modern science on his side. Modern science knows nothing about some well-defined population (perhaps 10,000 people) of early humans ("cultural animals with emerging social systems," "evolved to the point of exhibiting moral capabilities") who differ considerably from earlier hominids, and from all other existing hominids in existence (apparently not so "cultural," not so "social," not so "moral"). In no evolutionary presentation has such a distinct and discrete population ever existed, one that was ostensibly ripe for God's covenant dealings. If human history went the way that modern science and AEH advocates believe it did, God would have had a hard time identifying a population adequately prepared to be accepted as his covenant people, in contrast with all the other hominids, only then to call this population "Adam and Eve."

1.4.2 A "LAPSARIAN MYTH"

To be sure, only with severe exegetical gyrations can Smith maintain the notion of a historical Fall, even if it were a Fall that occurred over an extended period of time, and among a large population of people. Moreover, for Smith, the Fall was not from a state of innocence to a state of evil (which eating from the tree of the knowledge of good and evil implies). Rather, the Fall was (part of) a process of maturation (notice the term "moral immaturity" in the quotation above). The Fall was a process, so to speak, of climbing the ladder of ethical development by trial and error. It is the opposite of the biblical picture, in which people who were created "very good" fell to a level that is very bad. In the Bible, the Fall was a descent (as the word "fall" suggests, of course) from good to bad. For AEH advocates, or at least for James Smith, the Fall—the term is carefully retained!—moved in the opposite direction: it was an ascent. In the Bible the Fall was degeneration, while for AEH advocates it was a development. In the Bible the Fall constituted deprivation, while for AEH advocates it consists of maturation.[35]

Stating the matter in literary terms, here we have not only a "creation myth," but a "lapsarian myth" as well. The apostle Paul had a very different view of the opening chapters of the Bible. Paul tells us that Adam was not born of a woman, as was every human being *after* Adam (1 Cor. 11:8–12). In AEH thinking, Paul is wrong on this point; Wright's Adamic pair and Smith's

---

[35] Not all AEH advocates agree on this point; van den Brink (2017, 230; cf. 233) believes that the idea of "maturation" does not do justice to Genesis 3.

Adamic population were all born of females. Paul also maintains the order of formation as described in Genesis 2, namely, Adam first, then Eve (1 Tim. 2:13). Consider especially 1 Timothy 2:14: "Adam was not deceived, but the woman was deceived and became a transgressor." I admit that this verse has caused much controversy (v. 15 even more), but one thing is clear: Paul distinguishes the way Adam fell from the way Eve fell. This agrees entirely with Genesis 3: Eve was deceived by the serpent (cf. 2 Cor. 11:3), and Adam followed her in her deception; but this sequence definitely does not agree with the portrait given by Smith and other AEH advocates, because the sequence of an entire population falling into sin during an extended period of time does not allow for the accuracy of such narrative details.

How can Smith in this way ignore almost everything that the New Testament has to say about Adam and Eve, their creation and their fall? He gives us a clue on the very first page of his article: "the traditional or orthodox doctrine of the Fall has proven more difficult [than the doctrine of human origins itself] to reconcile with the picture of human origins that emerges with evolutionary accounts of human origins."[36] The clue here is the word "reconcile."[37] This will be a key point in our chapter 3 on hermeneutics. Never before in world history has the theological community faced such enormous pressure to reconcile the Scriptures (as understood in the traditional and orthodox way) with the conclusions of the natural sciences. Even the Galileo controversy would not be an example of this kind of reconciliation. Some believe in a heliocentric model of the universe, others (also today[38]) believed in a geocentric model (moreover, it can be argued that neither is actually correct—but that, if viewed from different perspectives, both are [almost] correct); however, one's views on this point do not at all conflict with any historical, poetic, or prophetic portion of the Bible, nor any subject within systematic theology.[39]

1.4.3 A HERMENEUTICAL BREAK

The project of AEH advocates—Wright, Smith, van den Brink, Carlson and Longman, Haarsma and Haarsma, and so many others—entails a definite

---

[36] Smith in Cavanaugh and Smith (2017, 48–49).

[37] Enns (2012, 147) uses a comparable word, "synthesis": "*A true rapprochement between evolution and Christianity requires a synthesis, not simply adding evolution to existing theological formulations.*"

[38] See Nussbaum (2002).

[39] The Galileo story is often adduced by AEH advocates as an example of how on earlier occasions theology had to adapt to science; see Falk (2004, 26–31); van den Brink (2017, 113–14, 138–39). For a response to this claim, see Steve Fuller in Nevin (2009, 119–121); Paul (2017, chapter 5).

break with the church's hermeneutical past. Please note, this does not mean that we cannot use extra-biblical sources to better understand certain passages in the Bible. Theologians have always used such sources. Think of historians using ancient secular historiographies to shed light on certain historical aspects of the Bible. Therefore, opponents of AEH inappropriately appeal to the Westminster Confession at this point, which says, "The infallible rule of interpretation of Scripture is the Scripture itself."[40] I suppose that Wright, Smith, and van den Brink (to limit myself to these three for now) all subscribe to this article. This is fully in line with the ancient adage: Holy Scripture is its own interpreter (Lat. *Sacra Scriptura sui ipsius interpres*), proclaimed in particular by Augustine,[41] and later by Martin Luther.[42] Such a valid maxim does not at all exclude the use of extra-biblical sources in connection with interpreting the Bible.

You may use as many aids as you wish in order to understand the Bible text. However, you cannot, on the basis of extra-biblical sources, interpret the Bible to say the opposite of what it says.[43] In the new "creation myth" and "lapsarian myth" being propagated by AEH advocates, this is exactly what is happening. These myths do not help us understand Romans 5, 1 Corinthians 11 and 15, or 1 Timothy 2 any better. On the contrary, these Bible passages become seriously distorted, as I intend to show. We can no longer speak of a single human pair formed at the beginning of human history, directly created by God, who fell into sin shortly after their creation. Smith is fair enough not to claim that extra-biblical sources—read: modern science—have helped us understand the Bible text much better and more profoundly. Rather, he wonders how he can "reconcile" Genesis 1–3 with modern evolutionary theory. This supposed need to reconcile the two implies that they conflict, and this is precisely our claim. This conflict permits no "reconciliation"; this conflict is a life-and-death battle (see chapter 3).

Whether the universe is geo- or heliocentric does not conflict with any subject within systematic theology. The vicissitudes experienced by King Nebuchadnezzar or Emperor Augustus are important for understanding the background of Daniel 1–4 or Luke 2, respectively; but again, this history does not at all conflict with any subject within systematic theology.[44] However, whether there was a historical Adam—in the biblical sense of the term,

---

[40] Chapter I.9 (http://www.reformed.org/documents/wcf_with_proofs/).

[41] See his *De doctrina Christiana*.

[42] Luther (1897, 7:97.23).

[43] One example is that according to many liberal expositors, although Dan. 6 speaks of "Darius the Mede," we can find no evidence in extra-biblical sources that this person existed.

[44] Therefore, the appeal to this argument by van den Brink (2017, 237) is beside the point.

not in the sense of a hominid chosen from a population of hominids—is of vital importance. It touches upon the doctrine of human creation (e.g., did God personally form Adam and Eve, or not?), the doctrine of original sin, and concomitantly the doctrine of salvation (see chapters 8–10). Bryan D. Estelle wrote: "My thesis is simple: by questioning the historicity of Adam [I would add, as this expression is traditionally understood], one must revise the doctrine of original sin with serious modifications."[45] He added that theistic evolutionists[46] also had to admit this, and referred to Daniel C. Harlow as an example. Harlow had offered the somewhat facile comment: "Once the doctrine of original sin is reformulated, the doctrine of the atonement may likewise be deepened."[47] There you see the direction this new hermeneutic is taking us. Heliocentrism is a debatable theory, but it does not conflict with any Christian doctrine. By contrast, theistic evolutionism affects *everything*, including the doctrine of salvation.[48]

Why is it that many atheists and agnostics understand this and so many Christians don't? Staunch evolutionist and atheist G. Richard Bozarth wrote,

> It becomes clear now that the whole justification of Jesus' life and death is predicated on the existence of Adam and the fruit he and Eve ate. Without the original sin, who needs to be redeemed? Without Adam's fall into a life of constant sin terminated by death, what purpose is there to Christianity?... None. What all this means is that Christianity cannot lose the Genesis account of creation... the battle must be waged for Christianity is fighting for its very life.[49]

In a broadcast on Channel 4, January 16, 2006, Richard Dawkins said,

> Oh but of course the story of Adam and Eve was only ever symbolic, wasn't it? Symbolic?! So Jesus had himself tortured and executed for

---

[45] Estelle (2012, 9).

[46] Some "theistic evolutionists" do not like this term because they do not wish to be associated with evolution*ism* (cf. chapter 2); some prefer the title "evolutionary creationists" (§1.5.2). This phrase is strange because "creation" has to do with beginnings, and "evolution" (insofar as it exists) belongs more to the domain of divine providence; cf. van den Brink (2017, 61).

[47] Harlow (2010, 192).

[48] In passing, I mention here that David Anderson in Nevin (2009, chapter 5) views theistic evolutionism as a modern variety of ancient Gnosticism: "The Gnostic empire is not dead, and one of its seats of strength is amongst evangelicals who seek to weld Darwinism on to the biblical story" (92).

[49] Bozarth (1979).

a symbolic sin by a non-existent individual? Nobody not brought up in the faith could reach any verdict other than barking mad!"

Bozarth wrote in the same magazine:

Evolution destroys utterly and finally the very reason Jesus' earthly life was supposedly made necessary. Destroy Adam and Eve and the original sin, and in the rubble you will find the sorry remains of the Son of God. If Jesus was not the redeemer who died for our sins, and this is what evolution means, then Christianity is nothing.[50]

1.5 GIJSBERT VAN DEN BRINK ON ADAM

1.5.1 FIVE INTERPRETATIONS

Both the book *Evolution and the Fall* and van den Brink's book *En de aarde bracht voort* (*And the earth brought forth*) immediately identify their leading theme. For James Smith, evolution is the key to understanding the Fall (in a fundamentally new way, that is). If I understand him correctly, for Gijsbert van den Brink, "the earth bringing forth" (Gen. 1:12 and 24) is a phrase that seems to be his way of describing evolution. The earth has been bringing forth for hundreds of millions of years by now, and is still bringing forth every day. Genesis 1 is no longer understood as a description of early *human* history—the world being prepared in six days for the appearance of the first humans—but at most it is understood, as I would put it, as the Bible writer's literary-artistic impression of the evolution of life, of which human evolution constitutes less than one percent.

A large part of van den Brink's book is devoted to an exposition of current evolutionary doctrine in a manner designed to lead orthodox Christians to discover its plausibility. I am thankful, though, that a number of Dutch natural scientists[51] responded immediately to van den Brink's book by identifying nine major flaws in current evolutionary theory.[52] They conclude:

---

[50] Bozarth (1979, 30).

[51] Rafael Benjamin, Willem Binnenveld, Peter Borger, Herman Bos, Hans Degens, Wim de Jong, Juri van Dam, Jan van Meerten, Ben Vogelaar, and Tom Zoutewelle.

[52] *Universele gemeenschappelijke afstamming: waarschijnlijk juist? Niet voorbijgaan aan de bezwaren* (https://logos.nl/universele-gemeenschappelijke-afstamming-waarschijnlijk-juist/). The flaws concern the origin of life, the subject of information, orphan genes, the supposed molecular clock, the lack of change in many species, the matter of "functionless" DNA, pseudo-genes, mutation repair systems, and the exceptionality of humanity. They might have mentioned many more such flaws.

> Finally, in his book, van den Brink is seeking agreement between evolution and orthodoxy. He does that in a thorough way. But the reasons why he began to view the theory of evolution as a probable history of origins are extremely flimsy; in his description of creationist thought, he shows little understanding of this movement and is prejudiced, whereas a critical note with regard to the scientific contents of the theory of evolution is lacking.[53]

Next, our attention is drawn to the theological consequences of this presentation of evolutionary thinking, in particular for the exegesis of Genesis 3. Generally following the (non-exhaustive) list of British biologist Denis Alexander,[54] van den Brink distinguishes five views concerning the historicity of Adam; they vary from very unhistorical to very historical (cf. Appendix 1).[55]

1. *The ahistorical interpretation* (Appendix 1: views 1 and 8). Genesis 2 and 3 are mythical; their message is purely theological. The Garden of Eden suggests how life with God might have been, the Fall symbolizes the ruin that all humans have caused. Adam is Everyman, his fall is Everyman's fall. The story is not about degeneration but about maturation.

Van den Brink's objection: This approach is hardly reconcilable with the essential doctrine of the Fall.

My response: Does not the very same criticism hold for van den Brink's attempt to reconcile the two positions?

2. *The pre-historical interpretation* (Appendix 1: views 10 and 12). This approach is a little more historical because it dates the beginning of humanity and its early experiences in the Middle East during the Neolithicum (some 13,000 years ago). Several AEH advocates assume that Adam and Eve really lived, not as the first pair of humans but presumably as Neolithic farmers, perhaps tribal heads of a Neolithic community. Their fall into sin affected not only later generations but also their contemporaries.

---

53 Benjamin et al. (2017). The course of the Dutch discussion on van den Brink's book can be followed on logos.nl. As an indirect answer to van den Brink, the recent book by another Reformed theologian in the Netherlands, Mart-Jan Paul (2017), is highly relevant, fighting against the theory of evolution and for the literal interpretation of Genesis 1–3. These two books, appearing shortly after each other, written by two men from the same conservative Reformed group, has led to new controversies in the Netherlands concerning the creation–evolution issue.

54 Alexander (2014); for a somewhat different summary of options, see Collins (2011, 121–31); see also the four different views about Adam (the evolutionary creation view, the archetypal creation view, the old-earth creation view, and the young-earth creation view) presented in Barrett and Caneday (2013).

55 For references documenting these views, see van den Brink (2017, 214–22).

Van den Brink's objection: This approach is not in accordance with common views of human evolution, which claim that early hominids had appeared hundreds of thousands of years earlier.

My response: This objection is not really helpful. Every AEH advocate must fix a date somewhere on the line of supposed human evolution, as does van den Brink himself, and this date will necessarily be arbitrary.

3. *The primeval-historical interpretation* (Appendix 1: views 9 and 11). This approach maintains that Genesis 2 and 3 speak literally of the first members of *Homo sapiens*, as presented to us through divine revelation. Their story is not normal historiography, though, but primeval history (*Urgeschichte*), which preceded ordinary history. This takes us back to the origins of modern Man, perhaps some 200,000 years ago, somewhere in Africa. The location of the story has been moved, however, to the agricultural setting of the Middle East to make it more understandable for its first readers.

Van den Brink's objection: similar to the previous one.

My response: similar to the previous one.

4 and 5. *The old-earth and the young-earth historical interpretation* (Appendix 1: views 2 and 3–6, respectively). This is the conservative approach to Genesis 2 and 3. Young-earth creationism dates the events some 6,000–10,000 years ago.[56] Old-earth creationism is divided on the matter; because the earth is supposedly billions of years old, the first humans may have lived 6,000–10,000 years ago, but possibly much longer ago.

Van den Brink's objection: These approaches are clearly in conflict with what modern evolutionary biology teaches us about human evolution, and with current methods of dating the world.

My response: Do we not have here the crux of the matter? The choice is between the traditional understanding of Genesis 1–3—please note, this is also the New Testament view of these chapters!—and the understanding provided by certain so-called scientists. To put it a bit bluntly: we read Genesis 1–3 either through Jesus' and Paul's glasses or through Darwin's glasses.

1.5.2 REFORMED TESTIMONY

Van den Brink locates his own approach somewhere between interpretations 2 and 3, but wishes to retain what he sees as the positive elements in all five approaches. For instance, from approach 1 he adopts the idea that

---

[56] For a rather recent defense of young-earth creationism, see Mortenson and Ury (2008); W. D. Barrick in Barrett and Caneday (2013, 197–227); and Snelling (2014).

Genesis 2–3 "indeed has a paradigmatic or archetypal[57] function." If this implies belittling the historical character of these chapters, then I think this appears to beg the question.[58] C. John Collins rightly remarked, "The paradigmatic gets its power from the historical: this event [i.e., the historical Fall] has set a pattern by which we can understand temptation and sin."[59] In other words, what is the value and significance of an archetypal event that never occurred (or never occurred the way it is described)?

Along with approaches 4 and 5, van den Brink wants to maintain the notion of a historical Adam—but to do so in a way very different from approaches 4 and 5. On the contrary, advocates of approaches 4 and 5 would probably argue that defenders of views 1, 2, and 3 do not at all advocate a historical Adam in any biblical sense. I know of no creationist who would appreciate van den Brink's seeming sympathy for approaches 4 and 5.

Indeed, throughout his book van den Brink is transparent in seeking to retain the sympathy of conservative Christian readers, and I respect that. He identifies as Reformed, and he wishes to remain so. Unfortunately, I believe his self-identification is misleading. For instance, he appeals to the Reformed theological characterization of Adam and Eve as *covenant heads* (Rom. 5). Although wishing to uphold this characterization, the author adds that this does not require us to believe that all humans have descended from Adam and Eve.[60] This, however, is a case of wanting to have one's cake and eat it too. van den Brink argues against Paul, for whom Christ is the covenant head of all his spiritual offspring, just as Adam is the covenant head of all his physical offspring.

By acknowledging the possibility that evolution may be true, van den Brink is definitely not consistent with Reformed theological tradition. If Louis Berkhof may be seen as the virtual embodiment of Reformed theology, we should listen to him:

> Scripture teaches that the whole human race descended from a single pair. This is the obvious sense of the opening chapters of Genesis. God

---

[57] Cf. Walton (2015, 199); J. H. Walton in Barrett and Caneday (2013, 89–118) saw many details in Gen. 2–3 as more archetypal than historical: Adam as an archetype is "'Everyman,' representing all" (90); but he also says, "I do not use 'archetypal' as an alternative to 'historical' Adam and Eve" (237; cf. 238); he called Christ archetypal as well (107), yet acknowledged his historicity. Cf. W. D. Barrick in Barrett and Caneday (2013, 135, 212), with reference to Wenham (1987, 91).

[58] van den Brink (2017, 225).

[59] C. J. Collins in Barrett and Caneday (2013, 132).

[60] van den Brink (2017, 226).

> created Adam and Eve as the beginning of the human species.... The same truth is basic to the organic unity of the human race in the first transgression [Gen. 3], and of the provision for the salvation of the race in Christ, Rom. 5:12, 19; I Cor. 15:21–22.[61]

Elsewhere, Berkhof emphasized that the consequences of Adam's sin affected "all his descendants. This is evident from the fact that, as the Bible teaches, death as the punishment of sin passes on from Adam to all his descendants. Rom. 5:12–19; Eph. 2:3; 1 Cor. 15:22."[62]

Notice what van den Brink is doing. He is suggesting that older Reformed theologians would have had no difficulty assuming that Adam's sin had consequences for both his descendants and his contemporaries. However, the very idea that Adam had contemporaries would have shocked Berkhof.

Much more must be said about van den Brink's theological suggestions, especially concerning the Fall, but this must wait until later. For the time being, the following general statement will suffice: *It is impossible to maintain the notion of human evolution, and at the same time to do justice to what both the Old Testament and the New Testament say about human creation and the Fall.* One cannot play for both teams.

Regarding this conflict of positions, van den Brink interestingly rejects what he calls concordism, the attempt to "bend" (supposed) scientific and exegetical data toward each other (cf. §3.5.1).[63] I fully agree. My difficulty is my inability to see the essential difference between such concordism and what is called "perspectivism," which seems to appeal more to van den Brink.[64] Perspectivism seeks to reconcile the notion of human evolution with a historical understanding of Genesis 1–3. The attempt of perspectivists may well appear (to some) more intelligent than that of traditional concordists, but the governing principle of both is the same: an amalgamation of two things that do not and cannot belong together (see extensively, §3.5). To express the matter with a variation on Matthew 19:6, what God has not joined together, let man keep carefully separate.

---

[61] Berkhof (1949, 204); for a modern Reformed testimony see Phillips (2015); VanDoodewaard (2016); and Paul (2017).

[62] Berkhof (1949, 270).

[63] van den Brink (2017, 118).

[64] Ibid., 120–28.

1.6 OTHER MODELS

1.6.1 HAARSMA AND HAARSMA

The models from Denis Alexander that are quoted by van den Brink (§1.5.1) can be compared to the five models suggested by Deborah and Loren Haarsma.[65]

(a) *Adam and Eve as recent ancestors* (ca. 10,000 years ago; Appendix 1: views 2–6). This is the traditional approach. The Haarsmas see several theological problems in this model (like "Who was Cain's wife? Who was Cain afraid of?"[66]), but these supposed problems will be addressed in the remainder of this book. The main objection of the Haarsmas is that this model cannot be reconciled "with the evidence God gives us in nature"[!]. In my view, this is a mistake; you cannot identify God's general revelation in nature with the findings of the natural sciences in such a naïve, simplistic way (see §3.1.2).[67] What the Haarsmas should have said is that the model cannot be reconciled with the theory of evolution, which is something quite different.

In each of the following models, the theory of evolution is presupposed.

(b) *Adam and Eve as recent (selected) representatives* (ca. 10,000 years ago; Appendix 1: view 12). Humanity is about 150,000 years old; about 10,000 years ago, God selected a pair of humans, Adam and Eve, to represent all humanity. "When Adam and Eve, humanity's representatives, chose to sin, their sinful status was applied to all human beings."[68] This radically conflicts with Scripture, which teaches that all humans are sinners because they are descendants of Adam, not descendants of his supposed contemporaries.

(c) *Adam and Eve as a pair or a group of ancient ancestors* (perhaps 150,000 years ago; Appendix 1: views 9 and 11). This is a typical AEH model: about 150,000 years ago, God selected a pair (or a group) of evolved hominids "and miraculously modified them into the first humans, Adam and Eve. This was certainly a spiritual transformation, and it included enough mental and physical changes that they became a separate species. All modern humans are descended from this pair and have inherited their sinful status."[69] We have three criticisms of this claim. (1) If one begins with the theory of evolution, why then does one suddenly appeal to some divine, miraculous intervention? Evolutionists vigorously reject a "God of the gaps" (Lat. *deus*

[65] Haarsma and Haarsma (2011, 254–68); see also Appendix 1.

[66] Haarsma and Haarsma (2011, 255); cf. Collins (2011, 112–13).

[67] For an extensive explanation of this claim, see Ouweneel (2012b, chapters 1–4).

[68] Haarsma and Haarsma (2011, 256).

[69] Ibid., 260.

ex machina) solution.[70] (2) This description of Adam and Eve ("enough mental and physical changes") betrays a profound ignorance about what humans are, and about the essential sense in which they differ from animals (see extensively, chapter 6). (3) How could anyone seriously believe that an Adam or Eve of about 150,000 years ago fits within biblical chronology?

(d) *Adam and Eve as a group of ancient representatives* (perhaps 150,000 years ago; Appendix 1: a variety of views 9 and 11). This view is similar to (b), but this time the beginning of the world is estimated to be about 150,000 years ago. Again, in my view, both the idea of selection and the long time span do not fit the biblical data.

(e) *Adam and Eve as merely symbolic* (Appendix 1: view 8). This view is like (d) but no longer senses a need or desire to accord Adam and Eve any historical status. If viewpoint (a) represents the traditional approach, then (b), (c), and (d) are variations of the AEH approach, while (e) is what I call the liberal view,[71] which has been around since the Enlightenment.

Interestingly, the Haarsmas tell us that they are not satisfied with any of these five viewpoints.[72] However, because they wish to create room for the theory of human evolution, option (a) is irrelevant to them. In that case, it is inconsequential whether they choose (b), (c), (d), or (e). It's like choosing arsenic instead cyanide. The problem lies at the very starting point: desiring to combine Genesis 1–3 with the theory of human evolution. As we hope to show, this is simply impossible.

Generally, there are only three approaches to the problem that AEH has raised.

(a) The *liberal* position (Appendix 1: views 1, 7, and 8): the theory of evolution is fully accepted, and advocates care little about the historicity of Genesis 1–3; the contents of these chapters are understood symbolically or viewed as mythical; or, if one wishes to avoid this loaded term, these chapters are viewed as ahistorical. Those advocating the liberal position include pure naturalists (no supernatural elements allowed) and deists (the world may well have been created by God, but afterward he was not involved with it any further), as well as Christians who are orthodox with respect to the core of the gospel but liberal with respect to the historicity of Adam and the Fall.

---

[70] This phrase refers to appealing for help to this god when one is unable to solve a particular scientific problem; this god conveniently supplies what is lacking (the "gaps") in one's knowledge.

[71] A rather recent representative of this view is offered by Enns (2012); see §1.6.2.

[72] Haarsma and Haarsma (2011, 270).

(b) The actual AEH position (Appendix 1: views 9–12):[73] the theory of evolution is fully accepted, but advocates defend the historicity of Genesis 1–3. They seek to maintain, in some form or another, the historicity of Adam and the Fall. The attempts to fulfill this (in my view impossible) task are numerous (see [b], [c], [d], and [e] in the previous section). I consider this a bad sign: if one begins by *attempting to reconcile Genesis 1–3* and certain so-called scientists, one faces an impossible task. Of course, AEH advocates do not wish to be called "liberal";[74] yet, surrendering the historicity of Adam and the Fall (in the biblical sense) is certainly a first step in this direction.

(c) The *traditional* position (Appendix 1: views 2–6): the theory of evolution, whatever details of it may be true, is not allowed to compromise accepting the historicity of Genesis 1–3. These chapters may contain all kinds of exegetical problems, but one thing is not doubted: at the beginning of human history we have one pair of humans, directly created by God (the man from dust, the woman from the man's side). They are the ancestors of all humans, and their fall into sin had terrible consequences for themselves and for all their descendants.

1.6.2 BIOLOGOS

The North American organization known as BioLogos, founded in 2007 by physician-geneticist Francis S. Collins (leader of the prestigious Human Genome Project), seems to be today's chief representative of AEH. Its mission is clear: "BioLogos invites the church and the world to see the harmony between science and biblical faith as we present an evolutionary understanding of God's creation."[75] The president of BioLogos is Deborah Haarsma, whom we met in the previous section. Another team member of BioLogos is Darrel R. Falk, whose book *Coming to Peace with Science* is a typical example of AEH thinking.[76] Among members of the Advisory Council we find Denis Alexander, Tremper Longman III, and John Walton,[77] whose works are also referred to on the website. Columnist Kathleen Parker has poignantly observed that the goal of BioLogos is "helping fundamentalists evolve."[78]

---

[73] This is thought to be a "third way"; cf. M. Reeves in Nevin (2009, 47–53).
[74] Cf. Falk (2004, 33).
[75] biologos.org.
[76] Falk (2004).
[77] Alexander (2014); Carlson and Longman (2010); Walton (2009).
[78] "An Evolution for Evangelicals." *The Washington Post*, May 10, 2009.

BioLogos members, who like to call themselves "evolutionary creationists," present themselves as thoroughly orthodox: "While Christians differ on their views of the age of the earth and evolution, we all agree on the essentials of the faith: that all people have sinned and that salvation comes only through the death and resurrection of Jesus Christ."[79] Theologian Joel Beeke rightly remarks,

> However, they do not mean what evangelical and Reformed Christians have meant by this statement. They ... do not affirm the Bible's supreme authority in resolving religious controversies. Instead, the Bible must bow to the changing theories of human science. Ironically, they reject some teachings of the Bible as simply the notions of ancient culture [such as the worldview supposedly underlying Gen. 1–3], while they impose other ideas upon the Bible from modern culture [such as the dogma of evolution]. Instead of absolute divine authority governing our faith, we have only the relative authority of human culture and opinion.[80]

Indeed, in my view, BioLogos members and other AEH advocates are trying to discover the square circle, a metaphor for the impossibility of reconciling the Bible with the theory of evolution. An interesting illustration of this impossibility is the wide-ranging differences among BioLogos members regarding the historical Adam. They tell us on their website: "There have always been multiple views on this in the BioLogos community, including views not presented in the Zondervan book,"[81] and refer to pastor Timothy Keller,[82] pastor Dan Harrell,[83] biologist Denis Alexander,[84] theologian Alister McGrath,[85] and biblical scholar N. T. Wright[86] as examples. Usually, when many different solutions to a problem are suggested,

---

[79] Carlson and Longman (2010) are purely AEH, yet they write (10): "We profess our deep commitment to Christian faith and the biblical teaching about creation," and (15): "We endorse the high view of the Bible as articulated by the 1978 Chicago Statement on Biblical Inerrancy." I think that most writers of this Statement would be stunned to hear that in 2010 even theistic evolutionists openly endorse it!

[80] J. Beeke in Phillips (2015, 37).

[81] The book being referred to is Barrett and Caneday (2013).

[82] Cf. http://biologos.org/uploads/projects/Keller_white_paper.pdf.

[83] http://biologos.org/resources/audio-visual/daniel-harrell-a-pastor-deals-with-adam-and-eve.

[84] Cf. Alexander (2014); see the discussion of his view by Paul (2017, chapter 9).

[85] http://biologos.org/blogs/archive/what-are-we-to-make-of-adam-and-eve.

[86] Cf. Wright (2014).

we may safely conclude either that nobody really understands the problem, or that the problem is a false dilemma.

Among the most progressive collaborators at BioLogos we find people who no longer advocate AEH, but have moved toward liberalism, at least with respect to Genesis 1–3. Peter Enns is identified as a senior fellow of BioLogos, but he rejects the AEH position. He claims that one way in which evolution and Christianity can be reconciled is "*by positing a first human pair (or group) at some point in the evolutionary process.*" This is what I am calling AEH. But Enns says about this position: "I do not think this is the best way to proceed,"[87] and uses the rest of his book to defend the common liberal position on Genesis. However, at the same time—and this is contrary to the past and present practice of ordinary liberals—in his closing chapters Enns confesses his faith in the resurrection of Christ.[88] In this respect, he represents BioLogos very well: orthodox with regard to Christ, liberal with regard to Genesis.[89] At the same time, he opposes the AEH position of many BioLogos people. Remarkably, I agree with the gist of Enns' arguments against AEH, although he and I arrive at opposite conclusions: both of us oppose AEH, he for evolutionary reasons, I for theological reasons.

Enns gave a completely accurate picture of AEH as it is presented by people like Wright, Smith, and van den Brink: "Some understandably seek to merge evolution with Adam in an attempt to preserve what they perceive as the heart of Paul's teaching on Adam, yet without dismissing natural science. In other words, evolution is fine so long as an 'Adam' can be identified somehow, somewhere."[90] He then gives three reasons why this does not work. Only his first counter-argument fits our discussion at the moment.

> One cannot pose such a scenario [i.e., AEH] and say, "Here is your Adam and Eve; the Bible and science are thus reconciled." Whatever those creatures were, they were not what the biblical authors presumed to be true. They may have been the first beings somehow conscious of God, but we overstep our bounds if we claim that these creatures satisfy the requirements of being "Adam and Eve."[91]

---

[87] Enns (2012, xvii; cf. 120, 123).

[88] He does not explain, however, how Christ's death and resurrection function as the core of the gospel in relation to sin and death; see R. B. Gaffin in Versteeg (2012, xvi–xix).

[89] Cf. Falk (2004, 43): "The story of creation [viewed by him in a liberal way] is incomplete without the resurrection message [viewed by him in the orthodox way]."

[90] Enns (2012, 138).

[91] Ibid., 138–39.

I could not agree more. However, the conclusions of our study will be the opposite of Enns' conclusion. Enns argues that, given the facts of evolution, we must forget about the historicity of Genesis 1–3. I argue that we must give up all attempts to import evolutionism to tell us what in Genesis 1–3 we can and cannot accept as historical.

One final comment. Within theistic evolutionism (or evolutionary creationism, as it is also termed), Matthew Barrett and Ardel Caneday distinguish between (a) *non-teleological evolution* (formerly called "deism": God created the world, but afterward does not intervene in the creation) (see my Appendix 1: view 7, but also view 1), (b) *planned* (or *teleological*) *evolution* (evolution has a divine goal; Adam and Eve are a group of hominids or purely symbolic) (Appendix 1: views 8–10), and (c) *directed evolution* (Adam and Eve are evolved hominids, and historical persons) (Appendix 1: views 11–12).[92] Strangely, the authors assign BioLogos to position (b), but Deborah and Loren Haarsma, leading figures in BioLogos, to position (c). The more correct assignment is probably simply that BioLogos members hold to positions (b) or (c) (Appendix 1: views 9–12 are AEH).

1.6.3 A CLEAR CHOICE

Let me conclude this introductory chapter with a few quotations, the first from American dogmatician Wayne Grudem. In his Foreword to a book edited by Norman C. Nevin, he supplied an excellent survey of AEH.

> Adopting theistic evolution leads to many positions contrary to the teaching of the Bible, such as these: (1) Adam and Eve were not the first human beings, but they were just two Neolithic farmers among about ten million other human beings on earth at that time,[93] and God simply chose to reveal himself to them in a personal way. (2) Those other human beings had already been seeking to worship and serve God or gods in their own ways. (3) Adam was not specially formed by God of "dust from the ground" (Gen. 2:7) but had two human parents. (4) Eve was not directly made by God out of a "rib that the Lord God had taken from the man" (Gen. 2:22), but she also had two human parents. (5) Many human beings both then and now are not descended from Adam and Eve. (6) Adam and Eve's sin was not the first sin. (7) Human physical death had occurred for thousands of years before Adam and Eve's sin – it was part of the way living things had always

[92] Barrett and Caneday (2013, 20–22).

[93] Various AEH advocates may phrase the last sentence rather differently.

> existed.[94] (8) God did not impose any alteration in the natural world when he cursed the ground because of Adam's sin....
>
> What is at stake? A lot: The truthfulness of the three foundational chapters for the entire Bible (Genesis 1 – 3), belief in the unity of the human race, belief in the ontological uniqueness of human beings among all God's creatures, belief in the special creation of Adam and Eve in the image of God, belief in the parallel between condemnation through representation by Adam and salvation through representation by Christ [Rom. 5], belief in the goodness of God's original creation, belief that suffering and death today are the result of sin and not part of God's original creation, and belief that natural disasters today are the result of the Fall and not part of God's original creation. Belief in evolution erodes the foundations.

My second quotation comes from Dutch theologian Johannes P. Versteeg.

> As the first historical man and head of humanity, Adam is not mentioned merely in passing in the New Testament. The redemptive-historical correlation between Adam and Christ determines the framework in which—particularly for Paul—the redemptive work of Christ has its place. That work of redemption can no longer be confessed according to the meaning of Scripture, if it is divorced from the framework in which it stands there. Whoever divorces the work of redemption from the framework in which it stands in Scripture no longer allows the Word to function as the norm that determines *everything*. There has been no temptation through the centuries to which theology has been more exposed than this temptation. There is no danger that theology has more to fear than this danger.[95]

Do not miss the blunt evaluation: the advocates of AEH present their view as orthodox, but in reality the view of AEH is one of the greatest dangers threatening orthodox Christianity and biblical theology today.

---

[94] Cf. G. Haslam in Nevin (2009, 71): "If death, disease, decay and destruction were present from the beginning and declared to be 'good' by God [Gen. 1:31], then why undo them? But the Bible asserts that death is the 'last enemy' that will be finally destroyed at the return of Christ [1 Cor. 15:26]."

[95] Versteeg (2012, 67). This is the English translation of Versteeg's *Is Adam in het Nieuwe Testament een 'leermodel'?* (1969). The first English edition appeared in 1978. In his endorsement, Gregory K. Beale wrote: "This book is the best that I know of in demonstrating exegetically that the parallels drawn by Paul between Adam and Christ (as the Last Adam) necessitate viewing not only Christ as a historical figure but also the first Adam as an actual historical figure."

My third quotation is from American theologian Vern S. Poythress.

> The world around us tells us to accept the latest scientific pronouncements as the product of experts who know much better than we do. As Christians, we must not overestimate our knowledge or our expertise. But we have in the Bible a divine message that we can trust. We ought to use its guidance. The Bible criticizes modern science for its idolatry. Assumptions about the nature of law and assumptions about what counts as an explanation or what counts as relevant evidence play a major role in science.[96]

In the remainder of this book I hope to elucidate the truth and the meaning of these highly relevant words cited from various authors without ignoring the underlying dilemma. The choice is this: either we accept the the assertions of certain so-called scientists and allow them to govern our interpretation of Genesis 1–3. In my view, the result will be that we will lose the essence of these chapters, and of the gospel itself. Or we proceed in the opposite direction: we begin with divine revelation, and our thoroughly tested understanding of it (in concurrence with many centuries of Jewish and Christian exegesis)—and proceed in the light of our Bible-provided understanding to evaluate what is viewed as "science falsely so called" (1 Tim. 6:20 KJV).[97] There is no middle path: *the choice is evolution or revelation.* It is self-deception to think we can follow both paths.

---

[96] Poythress (2012, 8; cf. also 2006).

[97] These choices have been well described by Aaron Riches in Cavanaugh and Smith (2017, 120–21; cf. 135), although I find his solution inadequate: it is only Christ who matters (my wording). This may be true but it does not help us very much in the matter that is at issue.

2

THE THEORY OF EVOLUTION AND EVOLUTIONISM

See, this alone I found, that God made man upright, but they have sought out many schemes (Ecclesiastes 7:29).

THESIS
Many people wish to distinguish carefully between the theory of evolution, which they accept, and the ideology of evolutionism, which they reject. This may be formally correct, but it does not work practically: the theory of evolution is inseparable from the framework of evolutionism.

2.1 WHAT IS EVOLUTIONISM?

2.1.1 EVOLUTION HAS CONQUERED THE WORLD

There is such a thing as "word pollution." This occurs when we hear a term being used constantly, whether or not it is fitting. The word "evolution" is used on Internet sites with reference to films and books with titles about the evolution of beauty, of compassion, of a criminal, of desire, of the female action hero, of flight, of the geek, of machine learning, of medicine, of music, of philanthropy, of the population-based cancer registry, of strategy, of trust, of verse, of video game graphics, of the web, of YouTube, and the like. The term "evolution" is pervasive, and its meaning has been debased so badly that it means little more than growth or development, or more simply, change. The term is in the air we breathe.

Whether intentional or not, this linguistic phenomenon leads to the osmosis of evolutionism by the general public. If everything around us is subject to evolutionary processes, then why not the world itself? Why not life itself? Why not Man himself? Since nineteenth-century historicism (see §2.4.3), people have come to believe that all things around us, including human beings, are products of historical development, just like thinking, language, culture, society, the economy, the arts, the sciences, and last but not least, religion. Nothing is static, everything is fluid, all things are constantly developing. Evolution characterizes everything everywhere.

This is why creationists are looked upon strangely nowadays. They appear to believe—which they don't—that all things are static. All things are more or less still the way they were when they were created in the beginning. Of course, sensible creationists do not assert such things. They believe in biological variation, and in natural selection. But some people have the impression that for creationists the cosmos is static, whereas for evolutionists it is dynamic. And since our entire society is dynamic, changing at a pace like never before in history, all can see for themselves that creationists must be wrong. We live in a world that is changing all the time—and modern scientists now tell us that this has been going on for more than 13 billion years. Why do we need physical, biological, or geological *proof* for this evolutionary process? All sensible people nowadays supposedly know that the world is evolving. This ruling ideology is called *evolutionism*. This claim needs no evidence, since we can see all around us that it must be true.

Let me illustrate this state of affairs with two quotations from the religious humanist Sir Julian Huxley (grandson of Thomas Huxley; see §2.4.3), one of the leading evolutionists of the twentieth century. In 1955 he wrote:

> The concept of evolution was soon extended into other than biological fields. Inorganic subjects such as the life-histories of stars and the formation of chemical elements on the one hand, and on the other hand subjects like linguistics, social anthropology, and comparative law and religion, began to be studied from an evolutionary angle, until today we are enabled to see evolution as a universal, *all-pervading process*.[1]

To this he added,

> With the adoption of the evolutionary approach in non-biological fields, from cosmology to human affairs, we are beginning to realize that biological evolution is only one aspect of evolution in general.

[1] Huxley (1955, 272); italics added.

> Evolution in the extended sense can be defined as a directional [!?] and essentially irreversible process occurring in time, which in its course gives rise to an increase of variety and an increasingly high level of organization, in its products. Our present knowledge indeed forces us to the view that *the whole of reality is evolution*—a single process of self-transformation.[2]

In 1959 Huxley gave the Darwin Centennial Convocation address, in which he said:

> Evolutionary man can no longer take refuge from his loneliness in the arms of a divinized father-figure whom he himself has created, nor escape from the responsibility of making decisions by sheltering under the umbrella of Divine Authority, nor absolve himself from the hard task of meeting his present problems and planning his future by relying on the will of an omniscient, but unfortunately inscrutable, Providence.[3]

The leading figures in evolutionary thinking have often spoken this way. *Evolution implies that we no longer need God*. For these leaders, it was inconceivable to separate a strictly scientific theory of evolution from the ideology of *evolutionism*. This is the matter that we must now consider.

2.1.2 DEFINITIONS

Let me first define the term *creationism*.

(a) In the widest sense, every theist (Christian, Jewish, Islamic) is a *creationist* in the sense that he or she believes in the divine creation of the world.

(b) *Theistic evolutionism* holds to the theory of general (amoeba-to-human) evolution, but views this kind of evolution as part of God's general providence; this is sometimes confusingly called *evolutionary creationism* (e.g., by people writing for BioLogos). Some advocates believe in God's providence, but not in supernatural divine interventions in the process of evolution; in practice this view is hard to distinguish from atheistic evolutionism.[4]

---

[2] Ibid., 278; italics added.

[3] Huxley (1960, 249); cf. Ross (2004).

[4] Carter (2017, 43). To the claim of Venema and McKnight (2017, 90–91) that science has revealed to us how God brought his creation into being, Carter replies: "But where is God in any of this? From the big bang, to the origin of life, to the advent of modern man, God is absolutely and unequivocally irrelevant in the naturalistic mind" (44).

(c) Some advocates of *evolutionary creationism* accept the possibility that, at important moments (such as with abiogenesis, the origin of humanity), God supernaturally intervened ("progressive creationism").[5]

(d) *Old-earth creationism* rejects general evolution; this view holds that the earth is billions of years old; humanity may also be (much) older than 6,000–10,000 years old, but not necessarily so. This view occurs in several variations (see Appendix 1).

(e) *Young-earth creationism* rejects general evolution; this view holds that the earth, and thus also humanity, is young (perhaps 6,000–10,000 years old at most); the creation days were ordinary twenty-four hour days.

In this book, the terms "creationist" and "creationism" are usually meant in the sense(s) explained in (d) and (e).

Many Christians believe in the *General Theory of Evolution* but they do not like to be called "evolutionists." They say they accept the theory of evolution but wish to have nothing to do with any "-ism," whether creationism or evolutionism.[6] I think I understand them; let us therefore begin by carefully distinguishing between the following five aspects.

(a) *Evolutionary population theory* (sometimes called the *Special Theory of Evolution*):[7] this is the strictly natural-scientific theory[8]—for instance, Darwinism, neo-Darwinism, neo-neo-Darwinism—concerning descent with modification of the heritable characteristics of organismal populations over successive generations. In principle, this theory is concerned only with fluctuations in allele frequencies, that is, with the proportions in which alleles (the varying forms of genes) occur in a population, and how these proportions change during the course of time. Such shifts are *not* associated with an increase in information, but are entirely explained by the mechanism of gene regulation and recombination of gene variants and selection. Thus, the theory describes genetic variations of the DNA, which are not antagonized by the mutation repair systems in every cell, but do not expand the quantity of genetic information either.[9] The mechanism often (not always) leads to a phenotypic variability that enables organisms to better adapt to their environment. Competition leads to natural selection of those variants that have the highest fitness ("survival of the fittest," a slogan from English philosopher Herbert Spencer). *Of all five varieties of*

---

5 See, e.g., Ross (2004).

6 Cf. van den Brink (2017, 29–30).

7 The terms *General* and *Special Theory of Evolution* were introduced by Kerkut (1960).

8 The phrase "natural sciences" refers to sciences that investigate nature; I prefer this phrase to the shorter term "sciences" because, in my view, the humanities are sciences, too.

9 Cf. DeJong and Degens (2011).

the theory of evolution to be mentioned here, only this one has a solid scientific foundation, and actually has nothing to do with evolution in the meaning that will now follow.

(b) *General Theory of Evolution*: Extrapolation of the strictly natural-scientific evolutionary population theory to the assumption of general evolution, from the first living organisms, supposedly some 4.1 billion years ago,[10] up to present-day humans ("universal common descent"). In order to be truly scientific—which is not the same as being true—the supposed expansion of the DNA with new functionalities ("innovation") must be explained in a fully naturalistic way, that is, by means of natural laws and biotic mechanisms that can be verified by empirical evidence.

Many natural scientists are convinced that a scientific account can be given for such an extrapolation; other scientists are not convinced of that, or are convinced none can be given.[11] The point is that millions of *variations* themselves do not, and cannot, produce *innovation*, the supposed mechanism of accumulation of irreparable, advantageous, inheritable, code-expanding mutations of the DNA (see § 2.1.3).[12] This supposed mechanism is antagonized by the mutation repair systems in every cell,[13] and by the selective disadvantage of irreparable mutations, which are the cause of cancer and hereditary diseases.

(c) *Abiogenesis*: Extrapolation of the General Theory of Evolution (still strictly natural-scientific, though to a large extent speculative) to the assumption that the first organisms, which lived perhaps as long ago as 4.1 billion years, originated from auto-catalytically self-replicating, and compartmentalized arrangements of complex biomolecules. These complex biomolecules are thought to have originated from simple chemical substances by natural processes. Such natural processes, however, induced by putting simple substances into the radiation of the sun, or in the rain, wind, or lightning, are processes of decay (see further in §2.3.1).

(d) *Epistemological analysis* of the unique character of the theory of evolution. Such a reflection is required because the theory of evolution

---

[10] See Bell et al. (2015).

[11] For the latter group, see Sarfati (2004; 2010a; 2010b; 2015); Wells (2002); Collins (2003; 2006; 2010; 2011); Poythress (2006; 2012; 2013; 2014); Lennox (2009); Nevin (2009); Ashton (2012); Gauger et al. (2012); Junker and Scherer (2013); Brown and Stackpole (2014); Carter (2014); Snelling (2014); Denton (2016); Mortenson (2016); Van Bemmel (2017).

[12] Cf. van den Brink (2017, 29–30). See the work of Cornell geneticist John C. Sanford (2014), who has shown that an accumulation of mutations causes the entire genome to deteriorate; most mutations cannot be effectively eliminated from the genome.

[13] See https://www.nobelprize.org/nobel_prizes/chemistry/laureates/2015/advanced-chemistryprize2015.pdf.

is not a normal natural-scientific theory, unlike, for instance, the theories of thermodynamics or quantum mechanics. These latter theories describe processes that can be studied today in a direct way, and can be replicated in the laboratory under experimental conditions. The evolutionary population theory (see [a]) does indeed investigate processes that can be replicated in the laboratory or in field research. But the General Theory of Evolution (see [b]) and the theory of abiogenesis (see [c]) cannot be studied in a laboratory. These are *historical* theories about a supposed *unique* particles-to-people development. This is a process that presumably never happened before, and presumably will never happen again. It is a unique history of life, 99.999999 percent of which occurred before modern natural science originated.

Moreover, on the basis of the evolutionary population theory we may seek to discover at most how this general evolution *might* have occurred—if at all—but, as a matter of principle, we can never demonstrate *that* it occurred in a certain way. We were not around to observe it. At most we might show that, on a very small scale, it sometimes happens in this or that way *today*. The only conceivable source of information on what actually *did* happen is the fossil record in the earth's strata.[14]

2.1.3 BASIC QUESTIONS

An additional fifth aspect remains to be discussed, but this will be done in §2.2. Before we come to this, we must observe that two basic questions are involved here.

1. Are the findings referred to under point (a) in §2.1.2 *sufficient* to explain the supposed overall development referred to under point (b), or to ask it more pointedly: Does the General Theory of Evolution follow automatically from the Special Theory of Evolution? How can we know this for sure? To give an example: how can we know for sure that the genetic processes that led to the origin of the many breeds within the subspecies *Canis lupus familiaris* (the domestic dog) are *sufficient* to explain, and even *demonstrate*, the supposed common ancestry of the family of the *Canidae* (doglike animals) and the family of the *Felidae* (catlike animals), both belonging to the order of the *Carnivora*?

2. Even if one's theory of evolution in the sense of point (a) could in principle plausibly account for the common ancestry of, say, *Canidae* and

---

[14] Regarding the present-day assessment of that record, however, and its supposed relevance for the theory of evolution, see Gauger et al. (2012).

Felidae through modern genome biology,[15] how can we know for sure that this common descent really *happened* this way? In a sense, this is a question that falls outside the domain of strict biology because we cannot reproduce the common descent of *Canidae* and *Felidae* in the laboratory. It supposedly happened only once, in the distant past. There is no fossil evidence to throw any light on this cat-and-dog problem. Again: we were not there. Biologists can only tentatively *reconstruct* (through cladistic methodologies) how it *might* have happened according to the principles of point (a), but we can never know this for sure. The possibility always remains that what is called "evolution" extends no further than what was described under point (a).

Please note that what was called "evolution" in the previous sentence was formerly called "micro-evolution,"[16] but is nothing other than biological *variation*. The word "evolution" is a misnomer here. In living nature, evolution in the sense of *variation* occurs by the mechanism of gene regulation and the recombination of gene variants and selection. This view of evolution is supported by broad empirical evidence, such as the changes in the beaks of Darwin finches. Real evolution is evolution in the sense of *innovation*, which is supposed to occur by a mechanism of accumulation of irreparable, advantageous, inheritable, code-expanding mutations of the DNA.[17] But ordinary biological variation within a population is basically oriented toward (*homeo*)*stasis*, maintenance of the species, not toward *changing* the species.[18] The genetic mechanisms even seem to have been *created* for this purpose, that is, to basically maintain the species, not allowing it to alter all too drastically.

Let us restate our two questions as follows:

1. The *biological* question: *Could* evolution happen, in the sense not only of point (a) but also of point (b)—that is, according to what we know today of population genetics and genome biology?

2. The *historical* question: *Did* evolution happen in the sense of points (a) and (b)? Of this we can never be *scientifically* certain, but quite the

---

[15] In fact, so-called cladistics has already tried this, but Borger (2009) has pointed out the speculative and selective character of this "science."

[16] Evolutionists have sometimes asserted that the distinction between "micro-" and "macro-evolution" is an invention of creationists. This claim is incorrect. In 1980, an important meeting occurred at the University of Chicago that, according to Roger Lewin (1980), focused on the question "whether the mechanisms underlying microevolution can be extrapolated to explain the phenomena of macroevolution." The answer of most participants was a clear "No." Cf. https://evolutionnews.org/ 2007/09/busting_another_darwinist_myth_2/.

[17] Cf. DeJong and Degens (2011).

[18] See the revolutionary article by Gould and Eldredge (1977); cf. Gould (2002).

opposite[19]—although one may be *ideologically* certain of it, to which claim we now turn.

2.2 IDEOLOGY

2.2.1 THEOLOGICAL PREJUDICES

There is a fifth aspect of evolutionary thought that must be mentioned, the one that really makes it an "-ism."

(e) The *origin* of modern evolutionism cannot be severed from *philosophical and ideological developments* in previous centuries, from the eighteenth century until today. In a sense, this applies to other natural-scientific theories as well. However, none of these theories has, from their origin, been so overwhelmingly interwoven with ideological considerations as evolutionism. This is why we speak of it as an "-ism." Terms like "Darwinism" and "neo-Darwinism" also have this suffix "-ism," as do Marxism and Freudianism, which are similar quasi-scientific ideologies. This "-ism" phenomenon is unique within the natural sciences. (An exceptional term like "electromagnetism" also ends with "-ism" but has no ideological connotations whatsoever.)

It is not difficult to explain why and how the theory of evolution became an "-ism." When it came to the origins of things (cosmos, earth, life, animals, humanity), for many centuries *religion* (animist, Jewish, Christian, Islamic, Hindu, Buddhist) had enjoyed a monopoly with regard to explaining these realities. In this context, we must understand that the claims of theologians before the time of Darwin went far beyond simply stating that God was the Creator of the universe, including the world of living organisms. They also made assertions that went far beyond the scope of the Bible. In this sense today we must characterize them as pseudo-scientists, who made their assertions about origins under the guise of theology. Consider these claims:[20]

1. "*There are as many species today as were created in the beginning.*" How did these theologians know this? Their claim arose more directly from their very static (Greek-scholastic) view of the world than from the Bible. Carl Linnaeus (1707–1778) was probably the first to discover and describe the

---

[19] See extensively, Borger (2009); Sanford (2014).

[20] See Clark (1966), Davidheiser (1969), and Macbeth (1972), who have discussed these historical backgrounds. See especially William Paley (1802), whose work deeply impressed Darwin, first positively, later negatively, as well as natural scientist Georges Cuvier; about him see Pietsch (2014).

way new species originate. Since then, such origins have been confirmed many times through direct observation and experimentation.

Please note, first, that we speak of a new species when a group of organisms can no longer interbreed, and produce fertile offspring together with organisms of the species from which it originated. At the same time the concept of species remains a highly problematic idea in modern microbiology (insofar as biologists still work with the concept of species at all, and not with cladograms, and the like). Think, for instance, of horizontal gene transfer; there is so much exchange of DNA between bacterial species that the concept of species has lost much of its usefulness.

Please note, second, that what Darwin in fact discovered was not at all evolution,[21] but biological *variation*. That he, and especially his followers, extrapolated this into a theory of general evolution was nothing but speculation, speculation that was strongly influenced by ideology.

2. "*The species as God created them in the beginning are, by definition, genetically unrelated.*" This is a natural consequence of the preceding claim. If, as pointed out in our response to the preceding claim, new species do indeed arise, then the possibility exists that many species that we know today at a certain time in the past derived from other, related species. For instance, it is generally assumed, by both evolutionists and creationists, that *Canis lupus* (wolf and dog), *Canis latrans* (coyote), *Canis aureus* (golden jackal) and *Canis adustus* (side-striped jackal) are all interrelated. It is even quite reasonable to assume that *Canis* is related to *Lycaon* (African wild dog), and to foxlike animals (*Cerdocyon*, *Lycalopex*, and the true foxes: *Vulpes* and *Urocyon*; note that *-cyon* comes from Greek *kyon*, "dog"). Similarly, one could legitimately assume that all *Equidae* (horse-like animals), such as *Equus quagga* and *Equus zebra* (two zebra species), *Equus hemionus* (onager), *Equus ferus* (wild horse; *E. ferus caballus* is the well-known domesticated subspecies, the common horse), and *Equus africanus* (African wild ass; *E. africanus asinus* is the well-known domesticated donkey) are interrelated, that is, have common ancestors.

Some creationists would say that all these forms belong to one and the same "Genesis kind" (*Canis* and *Equus*, respectively), such that within such a "kind" endless variation is possible, but there are no evolutionary relationships between these "kinds."[22]

---

[21] The notion of evolution (descent of one organismal type from another organismal type) was present in antiquity (Anaximander, Empedocles). See further in §2.4.1.

[22] See Wood and Garner (2009); "Genesis kinds" are sometimes called *baramins* (from Heb. *bara*, "to create," and *min*, "kind"); see Frair (2000). Some creationists speak of baraminology; cf. http://www.baraminology.net.

3. "*We find the species today in the areas where God put them in the beginning.*" Again, how did these theologians know that? This assertion followed from their very static view of the world. Darwin was the first to discover and describe how species migrate, how they can become geographically isolated (as he observed on the Galápagos Islands), and in this way can become genetically estranged from one another, and how in this way new species can originate that cannot interbreed with their earlier relatives. But again, the only thing that Darwin discovered was biological *variation*, and the ways in which new variations can originate.

2.2.2 THE IDEOLOGICAL CONTEXT

From the outset, the modern evolutionary hypothesis had ideological traits simply because it had to confront these and other claims of Greek-scholastic theologians at the time. An interesting fact, and the humour of history, are that Charles Darwin himself never obtained any other academic degree than a basic degree in theology. This is very relevant; his books indicate that he had little understanding of the biology of his day. Nevertheless he became the most renowned natural scientist of the nineteenth century. In addition, he apparently never became a full-fledged atheist, either, unlike some of his closest colleagues.[23]

Darwin's spiritual development illustrates how from the beginning the evolutionary hypothesis was besieged by the theological creation–evolution controversy. Darwin once wrote, "I have at least, as I hope, done good service in aiding to overthrow the dogma of separate creations."[24] Naturally, to this very day this sentence arouses the indignation of creationists, and rightly so. If the motive of a scientist is not purely scientific, but he or she is ideologically driven—for instance, combating a religious dogma—the resulting project cannot be called pure science.

Incidentally, Darwin did not say that he had overthrown the dogma of creation itself. Perhaps he wished to say nothing more than what I have summarized in three points: (a) there are many *more* species than there were in the beginning (if we ignore for a moment the extinct species); (b) many species turn out to *be related* after all; (c) species do migrate, and in this way form new species, whether through geographic isolation or isolation by distance, or simply through natural selection in which the fitness landscape has two optima (balancing polymorphism). (Today biologists

---

[23] Cf. Miller (2007).

[24] Darwin (1871, 42). Geology pioneer Charles Lyell saw as his mission to "free the science [of geology] from the grasp of Moses" (letter to George J.P. Scrope), June 14, 1830; see K. Lyell (1881, 268–71).

use many other, more sophisticated concepts to explain speciation.[25]) Yet, Darwin's is an enormously broad statement. What *scientist* would dare to hope that through his *scientific* theory he would have administered a heavy blow to a certain *theological* idea? Such statements do not have a scientific quality, but an ideological nature.

If Darwin had wished to show only that, in strictly natural-scientific research, religious ideas (such as "separate creations") were not allowed, his view would have been understandable. But was he not basically doing the same—though in the opposite sense—by combating a religious idea with his theory? And was he doing this by means of a strictly scientific theory? Many biographers and historiographers of science have pointed out that Darwin himself was led by his own ideological prejudices, such as his historicist preoccupation with gradual change, drawing from the work of German philosopher Gottfried Leibniz,[26] and by his Malthusian preoccupation with conflict and struggle.[27]

Actually, the theory of evolution becomes evolutionism precisely at the moment it begins to assume ideological traits. And please consider that, in the widest sense of the term, "ideological" always means "religious," if "religion" may be taken to refer to the Ultimate Ground of a person's thinking and being (about this important matter, see §§3.3.2, 6.4, and 6.5). I emphasize these points in response to those who tell us that they accept general evolution as a scientific fact but strongly reject evolution*ism*. Indeed, many *specialist* studies in evolutionary biology do not exhibit any ideological traits. But when it comes to the theory of evolution in its complex entirety, many people fail to see that historically, hardly any evolutionary theory has existed apart from evolution*ism*.

Let us now examine some of these ideological features.

2.2.3 THE ORIGIN OF MODERN EVOLUTIONISM

In order to better understand the historic conflict between Greek-scholastic theologians and (Enlightenment) biologists, let us devote some attention to the history of modern biology. We begin our story with the great German philosopher Immanuel Kant, who was interested in the problem of the

---

[25] See Gavrilets (2003).

[26] Think of Leibniz's slogan "nature does not make a leap" (Lat. *natura non facit saltus*), quoted by Darwin.

[27] See some of my favorite earlier historical and biographical works: Darlington (1959); Sirks and Zirkle (1964); Clark (1966); Davidheiser (1969); Macbeth (1972); Eiseley (1979). In more modern times, it was especially evolutionist Stephen J. Gould (1977; 2002) who has combated Darwin on these points.

origin of our solar system. In 1755, he published his *Universal Natural History and Theory of the Heavens*,[28] written in opposition to the great British pioneer in physics, Isaac Newton. Newton had postulated that the present order of the solar system could not be explained in a purely mechanical way; he therefore believed in a supernatural cause. Kant did develop a purely naturalist theory, which was later adopted by the French mathematician and astronomer Pierre-Simon Laplace. This theory explained the origin of the solar system in terms of the swirling of a primeval nebula.

For our purposes, the most important aspect of Kant's theory is not its modern tenability, but its underlying *philosophical* novelty: a natural phenomenon was no longer explained from supernatural but only from natural causes (without necessarily denying that in a very general way God "upholds the universe by the word of his power," Heb. 1:3). Interestingly, when Napoleon asked Laplace why he had not mentioned God in his theory about the movements of the planets, the latter answered, "I had no need of that hypothesis" (Fr. *Je n'ai pas eu besoin de cette hypothèse*).[29]

Kant applied this naturalist (materialistic-atheist) principle to the solar system, and British geologists James Hutton and Charles Lyell applied it to the origin of the earth's crust.[30] They endeavoured to explain the formation of various earth strata from phenomena that we can observe on earth today. Thus, they developed the principle of *uniformity*, called *uniformitarianism*. This theory claims that the same natural laws[31] and processes that we see functioning in our world today have always functioned in the universe in the past, and that they apply everywhere in the universe. Its motto was: The present is the key to the past.

Not even creationists would have problems with uniformitarianism if it were formulated this way. They, too, believe that the natural laws that God instituted for the cosmos did not change throughout the ages. On the other hand, it should be a problem to the evolutionists that everything seems to be changing in the evolutionary process *except these natural laws*. Why is this? And where did those natural laws come from in the first place?

---

[28] Kant (2000).

[29] Though some consider this event to be apocryphal (see https://en.wikipedia.org/wiki/Pierre-Simon_Laplace#I_had_no_need_of_that_hypothesis), it was reported by Victor Hugo (1972, 217).

[30] See especially Lyell (1998), which Darwin studied during his well-known world voyage on the HMS *Beagle* (see Darwin [1989]).

[31] I do not understand this term in any humanist sense but in the sense of God's constant structural principles for nature as part of his creational order; see Ouweneel (2012b, chapter 3; for the English version, see the forthcoming volume, *The Eternal Word*).

Because of his background, the Jewish philosopher of science, Karl Popper, should have known where these laws came from; but instead he argued:

> Although we can say that the laws of nature do not change, this is dangerously close to saying that there are in our world some abstract connections which do not change (which is quite trivial if we admit that we do not know, but at best conjecture, what these connections are) and that we call them "laws of nature."[32]

In practice, the counterpart of uniformitarianism is the concept of *gradualism*, implying that the earth's crust—and later the flora and fauna—developed gradually, that is, at a very slow rate. The background of this gradualism is quite important. Lyell maintained gradualism in particular to demolish the catastrophist theory of French natural scientist Georges Cuvier. The latter believed that the earth's strata had generally originated through catastrophic events, including Noah's Flood, and he was prepared to allow for supernatural causes in his theories. Lyell wanted to exclude the assumption of supernatural causes from natural science, and ended up throwing out the baby with the bathwater by excluding from science not only supernaturalism, but also catastrophism. That definitely went too far; there is in the earth's strata too much evidence for past catastrophic events. In recent decades, the explanations of the earth's strata being given by modern geologists are devoting much more attention than Lyell did to natural catastrophic events.[33]

2.3 ANTI-SUPERNATURALISM

2.3.1 THE ORIGIN OF LIFE

The nineteenth century saw an increased tendency to exclude all supernatural elements from the natural sciences. Please note that this is not necessarily a sign of hatred against religion, as has been asserted. Such

[32] Popper (1972, 99).

[33] For a secular approach to catastrophism, see Ager (1995); for an introduction from a creationist point of view, see Snelling (2014). One important and obvious indication of catastrophism is furnished by fossils. Under normal circumstances, any dead organism will decay within days or months due to the activity of micro-organisms, and no fossil will be formed. Only a quick, airtight covering of a dead organism will prevent this. As a consequence, the presence of fossils in the countless layers of sediment covering a large part of the earth's surface contradicts the theory that these layers were formed by a slow, gradual process over hundreds of millions of years.

hatred may have motivated some of these children of the eighteenth-century Enlightenment. But aside from that motivation, this was a healthy development. It is a sign of weakness when scientists resort to supernatural causes for scientific explanations. Whenever scientists searched only for *natural* causes, their science was successful, if they only searched long and tenaciously enough. Nowhere was a point reached within the natural sciences *themselves* where the search for natural causes was so inadequate that scientists had to resort to a supernatural explanation. Today, this is common knowledge; virtually all Christian physicists, astronomers, chemists, and biologists agree with it.

However, at the same time we must remember that this does not necessarily apply to the matter of origins, because *origins* are both historical and unique. Things are different here. For instance, we may conduct experiments investigating how the first living cell *might* have originated but, as a matter of principle, this research can never demonstrate how life *did* originate. If the synthesis of a living cell under laboratory conditions ever succeeds, it will be accomplished only by an intelligent team with enormous expertise.[34] We would be truly impressed only if this team were to create a certain environment as it *might* have existed on earth in the distant past, and if a living cell were to originate spontaneously in such an environment. I suppose that scientists would synthesize this living cell piece by piece according to a highly intricate and precise plan. That is, this living cell would, so to speak, be a product of the necessary building blocks *plus human intellect*. What is required is not just energy, but also information. This cell would be a totality of ingredients *plus information constantly being added to it from the outside* (exogenous information).[35] In other words, the end result—the ultimate living cell—would represent a far higher level of information-content than the added information represented by all the biomolecules.[36]

At this point, a remark by physicist Paul Davies may be quite relevant:

> We now know that the secret of life lies not with the chemical ingredients as such, but with the logical structure and organisational arrangement of the molecules… Biological information is not encoded in the laws of physics and chemistry…[and it] cannot come into existence

[34] This is already the case for synthesizing complex biomolecules; cf. the work of Sybren Otto and his team (http://www.otto-lab.com/).

[35] Cf. the title of Gitt (2006): *In the Beginning Was Information*.

[36] See extensively, Wilder-Smith (1981).

> spontaneously.... There is no known law of physics able to create information from nothing.[37]

If scientists were ever able to demonstrate how life might have begun, the only thing they would have demonstrated is that molecular ingredients will not be sufficient; one needs to introduce exogenous information into the system. That is, it cannot be done without, first, a *plan*, and, second, an *intellect* guiding the entire process. Actually, this is a modern description of what Christ-believers have been claiming throughout history. It reminds me of what American astronomer and agnostic Robert Jastrow wrote:

> At this moment it seems as though science will never be able to raise the curtain on the mystery of creation. For the scientist who has lived by his faith in the power of reason, the story ends like a bad dream. He has scaled the mountains of ignorance; he is about to conquer the highest peak; as he pulls himself over the final rock, he is greeted by a band of theologians who have been sitting there for centuries.[38]

2.3.2 INTEREST IN ORIGINS

It is fascinating to see how some more recent approaches in the natural sciences seem to allow more room for the element of intelligence in the way that living nature functions (cf. §5.6.1, on Intelligent Design). David Ash and Peter Hewitt wrote how they can very well imagine a mechanism through which intelligence would guide evolution, namely, through interaction with the DNA.[39] This position really comes down to some kind of theistic evolution. That is, if general evolution has occurred, it is conceivable only if a Higher Intelligence was involved. I have to add, though, that because of their sympathy with the mystical and the occult, Ash and Hewitt have not been taken very seriously. Their work is similar to those of Austrian-American physicist and mystic Fritjof Capra, and of American spiritual teacher Gary Zukav in the subjects of physics and Eastern mysticism, which have been treated the same way.[40] Science and religion (in whatever particular form) still make for an awkward match.

Interestingly, it was precisely the matter of *origins* that fascinated the early natural scientists so much at the time of Enlightenment: Kant thought

---

37 Davies (1999, 27–30).

38 Jastrow (1978, 113–14).

39 Ash and Hewitt (1991).

40 Capra (1975; 1982); Zukav (2001).

about the origin of the solar system, Hutton and Lyell about the origin of the earth's crust, Darwin about the origin of species. In all these cases, the scientists endeavoured to show that any appeal to supernatural factors (meaning: to the Bible) was irrelevant. Assessing this approach is not easy, because both creationists and evolutionists often respond so emotionally to the issues involved. On the one hand, many creationists view(ed) these attempts to exclude the supernatural from science as an attack on religion, more specifically, on God and the Bible. However, this is not necessarily the case. Both Christians and atheists should agree on this basic starting point: appeals to the supernatural or to the divine have no valid role in conducting natural-scientific research.

On the other hand, however, many evolutionists view(ed) these same attempts (to exclude the supernatural from science) as arguments for atheism: if God is unnecessary for our valid scientific explanations, then perhaps this implies that he simply is unnecessary elsewhere, which amounts to saying that he does not exist at all.[41] It is known that the young Joseph Stalin became an atheist through reading Darwin.[42] Or consider this example: one of the pamphlets of the American Association for the Advancement of Atheism, written by its president, Woolsey Teller, bore the title *Evolution Implies Atheism*.[43] This claim is just as absurd as the claim that "Electromagnetism Implies Atheism."

Of course, it is not at all true that evolution itself implies atheism. In the beginning, God instituted the laws that are valid for the cosmos, for inanimate nature, and for living creatures. Science can be described as the search for these laws.[44] But science itself has nothing to say about whether there is a Creator, Lawgiver, and Sustainer behind these natural laws. It is a philosophical absurdity to claim that natural science itself proves either the existence or the non-existence of God. The people making such claims are not scientists *functioning as scientists*, but scientists who speak as atheists and naturalists. Yet, it is striking to see how often evolution has indeed been linked with anti-theistic claims. The *a priori* cause of this was leaving out supernatural causes, not only for methodological reasons but also for ideological reasons.

---

[41] See the quotation from Grudem at the end of this chapter.

[42] Davidheiser (1969, 353–54).

[43] See especially his book, Teller (1945).

[44] See Ouweneel (2014a).

2.3.3 EVOLUTION AND ANTI-THEISM

American law professor Philip E. Johnson was correct when he wrote:

> The literature of Darwinism is full of anti-theistic conclusions, such as that the universe was not designed and has no purpose, and that we humans are the product of blind natural processes that care nothing about us. What is more, these statements are not presented as personal opinions but as the logical implications of evolutionary science.[45]

A striking example of the truth of this claim was one of the best known evolutionists of the twentieth century, George Gaylord Simpson, who wrote, "The process [of evolution] is wholly natural in its operation. This natural process achieves the aspect of purpose without the intervention of a purposer; and it has produced a vast plan, without the concurrent action of a planner."[46] Elsewhere he wrote, "Man is the result of a purposeless and materialistic process that did not have him in mind. He was not planned. He is a state of matter, a form of life, a sort of animal."[47]

This is the point: in evolutionism, biological ideas are strongly interwoven with personal atheistic and naturalistic opinions, *which are presented as science*. The two cannot be separated: in evolutionary thinking, where does *true* biology begin and naturalism and evolutionism end? Remember this: the scientists who have severely attacked (neo-)Darwinism in the past *never stopped believing in evolution*, even when they had no adequate genetic model to explain how it could have happened. This is because of their ideological, naturalist biases.

It is not easy to find a safe and sensible pathway out of such a heated controversy. Here is my own view:

(a) In principle, Kant, Lyell, and Darwin were right: never resort to the supernatural in empirical-scientific research. The proof that they were right is the enormous success of modern natural science, which through observation and experiments searched, and searches, for explanations in which supernatural factors have turned out to be totally irrelevant.

(b) However, precisely this subject of *origins* is different because life originated only once, a very long time ago. No humans were there to observe it. Thus this natural phenomenon has a unique character, to which the normal principles of science do not apply. This was a process that seems

---

[45] Johnson (1991, 40); about this vital matter of evolution versus religion, see also Collins (2003); Poythress (2006); Lennox (2009); Richards (2010).

[46] Simpson (1964, 212).

[47] Simpson (1967, 365).

to be inconceivable without involving some organizing Plan and Intellect. But this consideration is unacceptable for many advocates of evolutionary thought precisely because they not only believe in general (particles-to-people) evolution but they are also evolution*ists* (read: ideological, naturalist materialists).

(c) The exclusion of supernatural explanations from natural science does not prove that the supernatural itself does not exist. As a matter of logic, the most rigorous version of evolutionism can never be a valid argument for atheism. That so many advocates of the theory of general evolution are still atheists is because in practice, the theory of evolution can hardly be severed from evolution*ism*. Agnostics and atheists view evolutionism as *the* only and most comprehensive explanation for the origins of cosmos, humanity, and society.

2.4 NATURAL SELECTION

2.4.1 WHO DISCOVERED IT?

Let us now look a little more closely at Charles Darwin and his teachings. It is quite interesting to see not only how Darwin's ideas developed during his own life, but also how they developed in the world in which he lived. His most central theory, the principle of natural selection, certainly did not originate with Darwin himself. We encounter this idea already with the French scholars Pierre-Louis Moreau de Maupertuis and Denis Diderot (Darwin's own grandfather, Erasmus Darwin, was a Lamarckist).[48] Usually, the British zoologist Edward Blyth is also mentioned in this connection, who, twenty-four years before Darwin, described the idea of natural selection—interestingly, however, in connection with the notion of divine creation.[49]

In Darwin's day, the notion of natural selection was commonly held among biologists because people did not view this at all as a contradiction to their belief in creation. On the contrary, said Blyth and others, God used natural selection as a means of *maintaining* each created species by eliminating the less fit members; such preservation (not alteration) of a species is called *homeostasis* (cf. §2.1.3). Darwin's novel idea was that natural selection was not only conservative but could also be creative, since small hereditary changes could accumulate, and could thus lead to the origin of all species.

---

[48] For some literature, see note 17.
[49] See Dobzhansky (1959).

Creationist Gary E. Parker, in a book that he wrote together with pioneer creationist Henry M. Morris, tells us how evolutionists sometimes ask, in a rather naïve way, whether creationists believe in natural selection at all.[50] Parker's reply: "Of course—it occurred to us first." This is a beautiful example of how a discussion can become blurred from the outset. The natural selection that earlier creationists advocated at the beginning of the nineteenth century was a kind of "purifying" or "balancing" selection as a means of *conserving* the species. Evolutionists should ask the creationists not whether they accept natural selection—of course they do—but whether they accept natural selection as a process through which new species can originate. Sensible creationists would answer this question in the affirmative as well, because they believe in biological variation in the sense explained in §§2.1.1, 2.1.3, and 2.2.1.

So even this point is not yet decisive. The evolutionist should ask the creationist—or the creationist should ask the evolutionist!—whether the other person accepts natural selection as a process through which not just new species, but eventually even new *orders* and *classes* can originate. Stated another way: Could natural selection be the mechanism that explains the supposed *general* (amoeba-to-human) evolution?

It was not so much the by-then-current notion of natural selection itself that impressed young Darwin, but especially the reading of a book written anonymously, later discovered to have been written by British cleric and scholar Thomas R. Malthus.[51] Malthus argued that, because the population grows faster than the means of production, humanity is continually forced to engage in a struggle for life, in which famines, epidemics, and wars are indispensable as natural brakes. While reading this book, as Darwin himself later explained, he was impressed by the idea that in such a struggle for life the most adapted varieties tend to survive, whereas less adapted varieties tend to disappear, and that—this is new!—in this way new species would possibly originate. It took more than twenty years before he felt secure enough to publish this ostensibly new insight, together with all the evidence that he could adduce, in his well-known book entitled *The Origin of Species*.[52] Perhaps he would have delayed publishing this work had he not heard that a younger natural scientist, A. Russell Wallace, had come to conclusions similar to his own.

---

50 Morris and Parker (1987, 48).
51 Malthus (1992; orig.: 1798).
52 Darwin (1859).

2.4.2 DARWIN'S SUCCESS

Darwin's book was enormously successful; its first edition sold out the first day it was available. Historians of science have always wondered how this could be explained. Darwin was not the first to come up with the notion of evolution—this notion itself existed even in antiquity[53]—or with the notion of natural selection, as we have seen. Moreover, his discoveries did not at all constitute a great scientific breakthrough enabling biology to advance by a huge leap. On the contrary, a century after the publication of Darwin's book, the well-known geneticist and evolutionist Cyril D. Darlington wrote something devastating.[54] He claimed that Darwin's conviction that evolution was the result of natural selection working upon small, accidental variations caused a half-century delay in evolutionary research.

There was some truth in this. In 1865, Silesian-Slovakian monk and biologist Gregor Mendel had formulated what was later called the genetic "laws of Mendel." They really signified the beginning of the modern discipline of genetics. However, not only did Mendel publish his results in an obscure magazine,[55] but when they became known they were totally rejected for a third of a century. The reason was that the intermittent phenotypic changes that Mendel described seemed to conflict with Darwin's idea of steady variation. In 1900 several geneticists, among whom Dutch botanist Hugo de Vries, German botanist Carl E. Correns, and Austrian botanist Erich Tschermak, more or less simultaneously rediscovered the laws of Mendel. Soon the tables were turned. From now on, evolution was understood as a series of intermittent mutations according to the ideas of Mendel. In the 1920s and 1930s, this led to what was called "neo-Darwinism," or the "Modern Synthesis," which combined the ideas of natural selection and genetic mutation into a new, enriched model. Great biologists to be mentioned in this connection are Julian S. Huxley, Sewall Wright, Ronald A. Fisher, John B. S. Haldane, Theodosius Dobzhansky,[56] and George G. Simpson.

Let us return to the matter of Darwin's success; how can it be explained? First, Darwin was modest (and clever) enough to write about the origin of species only, and not explicitly about the origin of orders, classes, and phyla (although the reader could easily surmise Darwin's view about the origin of these). Second, he was wise enough to exclude from the book the subject

---

[53] Cf. Osborn (2007).

[54] Darlington (1959).

[55] *Verhandlungen des naturforschenden Vereines in Brünn.*

[56] As a young biologist, I met Dobzhansky at a professional conference in the United States. Afterward I learned that the Eastern Orthodox Dobzhansky was a theistic evolutionist.

of the origin of humanity. In 1871, after the scientific world had fallen prostrate before him, he ventured to publish his second book, *The Descent of Man*.[57] Third, he was prudent enough to avoid, as much as he could, a controversy raging at the time about whether species traits brought about or influenced by the environment could become heritable, as Lamarckism suggested. In principle, Lamarckism[58] can be combined with the idea of some teleology or design—and thus of some Designer—behind empirical reality, whereas Darwinism cannot.

2.4.3 HISTORICISM

Probably the main reason for Darwin's success had little to do with natural science itself, which is the point of this entire chapter. I am claiming that non-scientific factors have dominated evolutionary thinking from its beginning until today. In the development of Western thought, Darwin's book appeared at the right time. Under the influence of the Enlightenment, the public at large had become much more skeptical toward orthodox Christianity, and people were ripe for the idea of general evolution, that is, a supposedly truly scientific theory about origins which no longer required an appeal to some deity. This was also the period dominated by *historicism*, the philosophical movement that attempted to reduce all cultural and religious elements in society to products of historical development. This movement originated with French essayist Michel de Montaigne and Italian historian Giambattisto Vico, and continued with German philosophers Immanuel Kant and especially Georg W. F. Hegel.[59]

In the middle of the nineteenth century, the more educated people felt a strong need for a scientific explanation that could form a foundation for historicism. Or stated more broadly, they longed for a theory that could unite all the ideas of the modern, post-revolutionary age into one scientific perspective: ideas about national communities, about development, about historic changes, about social life, about the struggle for existence. Because of tremendously high regard for natural science, such a theory had to be a natural-scientific theory. Darwin offered that theory just at the right time, and others picked it up—for instance, Friedrich Engels, and later Karl Marx, who were thrilled with it.[60]

---

[57] Darwin (1871).

[58] The evolutionary view of the French natural scientist Jean-Baptiste de Lamarck; a well-known neo-Lamarckian is Steele (1981).

[59] Hegel (2004; orig.: 1822–31).

[60] See extensively Stack (2003).

At this point, two things have to be clearly distinguished. On the one hand, there is the issue of the purely scientific tenability of Darwin's hypothesis, or the tenability of its successors: neo-Darwinism (the Second Evolutionary Synthesis, based on genetics) and neo-neo-Darwinism (the Third Evolutionary Synthesis, based on genome biology). This question can be formulated as follows: To what extent can the supposed large-scale evolutionary change be explained in terms of natural selection, working on heritable variation? Such a problem can be solved by purely biological arguments alone.

On the other hand, there is the ideological context that really caused Darwin's enormous success. His hypothesis was hailed far more for ideological than for strictly biological reasons. And *this ideological framework of evolutionism has remained to this very day*. The clearest example of this is British biologist Richard Dawkins, who has made a name for himself with, on the one hand, a number of books on biology (*The Selfish Gene*, *The Extended Phenotype*, and *The Greatest Show on Earth*[61]), and, on the other hand, his ideological books, like *The Blind Watchmaker*,[62] *River Out of Eden*, *The Devil's Chaplain*, and *The God Delusion*. Actually, these two categories of books by Dawkins can scarcely be distinguished; they are fully intertwined. Here is an evolutionary biologist who does not for a moment distinguish his study of biology from his ideological beliefs. And why should he? I do the very same with my own scientific convictions. But at the same time it powerfully illustrates how much the controversy concerning origins was *and is* ideologically driven. In my view, this reality renders fatuous the claim made by those who embrace evolutionary thinking that they are not evolutionists.

It reminds me of Louis T. More, who, in his book interestingly entitled *The Dogma of Evolution*, argued, "Our faith in the doctrine of Evolution depends upon our reluctance to accept the antagonistic doctrine of special creation [by God]."[63] Speaking of faith, Harold C. Urey, Nobel Laureate for chemistry in 1934, once said, "We all believe as an article of faith that life evolved from dead matter on this planet."[64] David M. S. Watson wrote that evolution "is accepted by zoologists, not because it has been observed to occur or…is supported by logically coherent evidence to be true, but

[61] Dawkins (2009).

[62] Dawkins (1986). For a recent assessment of Dawkins see Van Bemmel (2017, 17–20, 74–78).

[63] More (1925, 304).

[64] Quoted in *Christian Science Monitor*, January 4, 1962, p. 4.

because the only alternative, special creation, is clearly incredible."[65] Much more recently, biologist Jerry Coyne wrote that Darwinism delivered severe blows to "humanity's theistic worldview."[66] *Darwinism was and is one of the most powerful weapons in the battle against any form of theism* (Christian or otherwise).

It was like this from the beginning. Even more than Darwin himself, men like the English biologist and philosopher Herbert Spencer, English biologist Thomas H. Huxley (nicknamed "Darwin's Bulldog"), and German biologist Ernst Haeckel used Darwinism in their fight against Christianity. Spencer saw in Darwinism an ideal that should be broadened to include ethical and religious views as well; he became the father of "Social Darwinism."[67] Huxley advocated "agnosticism" (the term comes from him) with respect to religious questions.[68] Haeckel became one of the most important nineteenth-century advocates of natural-scientific materialism.[69] None of them accepted Darwinism primarily for its purely natural-scientific benefits—they (and certainly Huxley) were not even fully convinced about these benefits—but rather because of their antipathy toward theism (the supernatural, the divine). For them, Darwinism was primarily the philosophy or ideology they so badly needed as an ally in their struggle against the Christian establishment.

In fact, in my opinion all three committed the same error in thinking that, if one could design a natural science that no longer required an appeal to God, then God himself was no longer needed either. This is a logical fallacy, because there are many *more* questions than scientific ones to which God might be the answer. However, interestingly, neither Darwin himself, nor Huxley, nor Spencer wanted to call themselves atheists. They no longer *needed* God, which is not the same as claiming he did not exist. For various reasons each of them left open the issue of God's existence.[70]

Ernst Haeckel had been an orthodox Christian in his youth, but after he had been converted to Darwinism he became a fierce opponent of Christianity. At the same time he remained religious in the sense that he preached Darwinism as a religion of materialism. British historian of science Charles J. Singer wrote that Haeckel founded a movement that

---

[65] Watson (1929).

[66] Coyne (2009, 34).

[67] Cf. Stewart (2011).

[68] Huxley (1992; orig.: 1889).

[69] Haeckel (2015; orig.: 1879).

[70] Cf. Voltaire's remark: "If God did not exist, one would have to invent him" (*Si Dieu n'existoit pas, il faudroit l'inventer*).

wore the cloak of a religion and of which he was both the high priest and the congregation.[71] This was the man who, according to some, influenced the German mind in his time more than any other thinker, together with German philosopher Friedrich Nietzsche. Daniel Gasman has plausibly argued that Haeckel's "Social Darwinism" played a crucial role in the formation of Nazi ideology in Germany, and of fascism in Italy and France.[72]

2.5 IDEOLOGICAL CONSEQUENCES

2.5.1 VERIFICATION

Because of the disturbing mixture of biology and ideology in any discussion about evolution, we are singularly unimpressed by the constantly repeated assertion that evolution is a "proven fact."[73] Geoffrey A. Harrison et al. wrote already in 1964 that the theory of evolution "nowadays" has been completely "verified" in every respect.[74] Apparently, they had not read (or they ignored) the books of the greatest philosopher of science of the twentieth century, Karl R. Popper.[75] Other important books by Popper appeared after the book by Harrison et al.; but this 1972 comment is noteworthy: it is "a little hard to understand why natural selection should have produced anything beyond a general increase in rates of reproduction, and the elimination of all but the most fertile breeds." To this Popper added:

> This is only one of the countless difficulties of Darwin's theory to which some Neo-Darwinists seem to be almost blind. Particularly difficult to understand from this point of view is the transition from uni-cellular to multi-cellular organisms, which have new and peculiar difficulties in reproducing and especially in surviving after reproduction, and which introduce into life something new, namely death; for all multi-cellular individuals die.[76]

Popper has convincingly shown that scientific theories can never be verified (proven true) in the real sense of the word; at most they can be falsified. If you want to *verify* the claim that "all ravens are black," you

[71] Singer (1950, 480).
[72] Gasman (2004).
[73] Cf. the critical comments by Collins (2011, 119) on "fact" and "inference."
[74] Harrison et al. (1964).
[75] Popper (1959; 1963).
[76] Popper (1972, 271).

would have to observe all the ravens in the universe. But in order to *falsify* this statement, observing one non-black raven suffices. Correspondingly, in order to verify a scientific theory you would have to make an infinite number of observations, whereas one deviating observation is in principle enough to falsify the theory. In practice the matter is a little more complicated, as Hungarian philosopher Imre Lakatos[77] has shown, because scientists tend to cling to their theories even when they encounter deviating observations. This is one of the problems of evolution*ism*: no matter how many parts—genetic parts, morphological parts, paleontological parts—of the theory of evolution have been falsified (or at least attacked), belief in general evolution is unassailable. This is precisely why it is an *-ism*: for ideological reasons, no possible alternative is even *considered*.

In summary, *to claim that the theory of evolution has been fully verified is nonsense*.[78] Perhaps people may think that the *notion* of evolution has been sufficiently substantiated; that is, they may believe that there can be no doubt that general evolution has occurred. But even then, this is no more than a belief. Apart from that, the evolutionary hypothesis in the strictly biological sense of the word—that is, Darwinism, or neo-Darwinism, or neo-neo-Darwinism—has not been verified, and *cannot* be verified, just as no other scientific theory can be verified.

One may insist that no other theory explains the available genetic data as well as (neo-)neo-Darwinism does. But, first, this explanation is not *proof*. Second, the claim presumably means that, at this moment in history, one simply cannot *conceive* of any valid alternative. This consensus prevails until, in a "scientific revolution" in the Kuhnian sense,[79] suddenly someone comes up with a valid alternative, one that not only explains what the previous paradigm could explain, but also leaves fewer problems unsolved.

The contribution of Popper was that he elevated falsifiability to the level of an essential criterion of good science. A good scientific theory is one that (a) is in principle falsifiable, that is, may give rise to risky predictions, which may or may not come true, and (b) has thus far splendidly withstood a number of attempts at falsification. Popper stated explicitly: "Darwinism is not a testable scientific theory, but a *metaphysical* research program—a possible framework for testable scientific theories."[80] It was his opinion that, in this respect, Darwinism had the same status as Marxism and Freudianism; these too are theories that, as a matter of principle, could not be falsified

---

[77] See Lakatos and Musgrave (1974).
[78] On this see also van den Brink (2017, 17).
[79] Kuhn (2012).
[80] Popper (1976, 151); italics added.

because they had an answer for every criticism that might be raised. It is *impossible* to invent an experiment through which Darwinism, Marxism, or Freudianism could be falsified—and therefore *strictly as scientific theories* they are inadequate.

Popper did think that Darwinism was invaluable because the principle of natural selection explained many natural phenomena, such as the way bacteria adapt to penicillin (which, by the way, is nothing but micro-evolution, which itself is nothing but a misleading word for biological variation). However, at the same time, Popper maintained that the theory as a whole was unfalsifiable.[81]

Naturally, Popper was smothered with heaps of scorn from the evolutionist church. He therefore later changed his mind and recognized that natural selection is testable (which is *not* the same as saying that the idea of general evolution is testable!). But he elucidated (as reported by David Miller) that, although he considered Darwinism to be "an immensely impressive and powerful theory," and argued that

> the claim that it completely explains evolution is...very far from being established. All scientific theories are conjectures, even those that have successfully passed many severe and varied tests. The Mendelian underpinning of modern Darwinism has been well tested, and so has the theory of evolution which says that all terrestrial life has evolved from a few primitive unicellular organisms, possibly even from one single organism. However, Darwin's own most important contribution to the theory of evolution, his theory of natural selection, is difficult to test. There are some tests, even some experimental tests; and in some cases, such as the famous phenomenon known as "industrial melanism,"[82] we can observe natural selection happening under our very eyes, as it were. Nevertheless, really severe tests of the theory of natural selection are hard to come by, much more so than tests of otherwise comparable theories in physics or chemistry.[83]

The theory of evolution is unfalsifiable. In 2005, Nobel Laureate in physics Robert Laughlin observed that this theory is in fact anti-science

---

[81] I realize that other philosophers of science have different insights concerning this matter; see Sneed (1971).

[82] "Industrial melanism" refers to the dark color acquired by certain animals living in an industrial region where the environment is soot-darkened; once again, this involved only biological variation, not evolution.

[83] Miller (1985).

because it involves explanations that have

> no implications and cannot be tested. I call such logical dead ends antitheories because they have exactly the opposite effect of real theories: they stop thinking rather than stimulate it. Evolution by natural selection, for instance, which Charles Darwin originally conceived as a great theory, has lately come to function more as an antitheory, called upon to cover up embarrassing experimental shortcomings and legitimize findings that are at best questionable and at worst not even wrong. Your protein defies the laws of mass action? Evolution did it! Your complicated mess of chemical reactions turns into a chicken? Evolution! The human brain works on logical principles no computer can emulate? Evolution is the cause!... Biology has plenty of theories [to explain origins]. They are just not discussed—or scrutinized—in public.[84]

2.5.2 LEWONTIN AND GOULD

Other advocates of evolutionism have made claims similar to those of Popper: the principle of natural selection is unfalsifiable. I mention American geneticist Richard Lewontin and American paleontologist Stephen J. Gould, both of whom have or had Marxist leanings and cannot be suspected of having creationist sympathies; quite the contrary. Both admitted that many individual claims about natural selection cannot be verified. Imagine, Lewontin said, that you find somewhere a case of a phenotypic change that makes the organism better adapted to its environment. The theory says that natural selection must have played a role. However, suppose that you cannot find any signs of natural selection having occurred. Does this mean that the theory has been falsified? No. Darwinians might argue that you simply did not search well enough.[85]

There is no way anyone can demonstrate that natural selection did *not* play any role. Even staunch evolutionist Lewontin therefore concluded that not just the belief in general evolution, but also the theory of natural selection was of a metaphysical nature rather than of a scientific character: "Natural selection explains nothing because it explains everything." Of course, since then other mechanisms than natural selection have functioned, such as genetic drift and neutral evolution (Motoo Kimura[86]), but

[84] Laughlin (2005, 168–69).
[85] Lewontin (1972, 59).
[86] See especially Kimura (1983).

this does not basically change the validity of Lewontin's remarks about verifiability.

The Muse cannot be doubted; she simply wraps herself each time in a different veil. For this is the point: critical questions about neo-Darwinism do not imply that Lewontin or Gould had problems with the notion of evolution itself. This is because evolution*ism* is a faith that its advocates cannot possibly abandon. On the contrary, in 1997 Richard Lewontin said,

> Our willingness to accept scientific claims that are against common sense is the key to an understanding of the real struggle between science and the supernatural. We take the side of science *in spite* of the patent absurdity of some of its constructs, *in spite* of its failure to fulfill many of its extravagant promises of health and life, *in spite* of the tolerance of the scientific community for unsubstantiated just-so stories, because we have a prior commitment, a *commitment to materialism*. It is not that the methods and institutions of science somehow compel us to accept a material explanation of the phenomenal world, but, on the contrary, that we are forced by our a priori adherence to material causes to create an apparatus of investigation and a set of concepts that produce material explanations, no matter how counter-intuitive, no matter how mystifying to the uninitiated. Moreover, *that materialism is an absolute, for we cannot allow a Divine Foot in the door*.[87]

This statement is valid: no scientific theory may start from the notion of divine intervention, even if we have no alternative. Evolutionism may be a strange approach for many reasons, but it is always better than any alternative that demands a Creator God. Now, we may indeed agree that the assumption of divine intervention is not allowed in science. But four other statements are equally true.

(a) By itself, this prohibition ("God is not welcome in science") does not at all enhance the trustworthiness of the scientific theories in question.

(b) This prohibition itself does not constitute an argument *for* materialism at all. Looking only for material explanations does not necessarily mean that matter is all there is.

(c) Nor does this prohibition constitute an argument *against* theism. As I argued earlier, excluding God from our scientific explanations does not imply that God does not exist.

(d) What may be true for common scientific theories—they must exclude

[87] Lewontin (1997); italics added.

God—may not be true for unique (unrepeatable) origins in an extremely distant past.

Incidentally, in 1997, the atheist and evolutionist Lewontin remarkably answered the challenges of some of the most renowned popularizers of evolutionist science in his time. This group included

> ...Edward O. Wilson, Lewis Thomas, and Richard Dawkins, each of whom has put unsubstantiated assertions or counterfactual claims at the very center of the stories they have retailed in the market. Wilson's [books] *Sociobiology* and *On Human Nature* rest on the surface of a quaking marsh of unsupported claims about the genetic determination of everything from altruism to xenophobia. Dawkins's vulgarizations of Darwinism speak of nothing in evolution but an inexorable ascendancy of genes that are selectively superior, while the entire body of technical advance in experimental and theoretical evolutionary genetics of the last fifty years has moved in the direction of emphasizing non-selective forces in evolution.[88]

2.5.3 PARADIGM SHIFTS

In 2008, the National Academies Press in the United States published a book entitled *Science, Evolution, and Creationism*.[89] The authors documented "the overwhelming evidence in support of biological evolution," especially over against the claims of various kinds of creationism. Such references to "overwhelming evidence" always remind me of the very interesting—and preferably forgotten—story concerning the theories of gravity advocated by Isaac Newton. By the end of the nineteenth century, physicists generally had the same feeling that many biologists seem to have today. They believed that Newton's laws were not a theory but established fact. They believed that their scientific inquiry was more or less finished. One of them asserted that the only thing that remained to be done was to enlarge the number of decimals of the physical constants. Another well-known physicist, the German Max Planck, was discouraged in his youth by physicist Philip von Jolly from studying physics because in that discipline there were no new perspectives.[90]

Only very few of the physicists living then realized that, within a very short time, the entire foundation underlying their work in physics would be removed. At the beginning of the twentieth century, Max Planck and

---

[88] Ibid.

[89] National Academy of Sciences (2008).

[90] http://www.chemistryexplained.com/Pl-Pr/Planck-Max.html.

especially the German-American physicist Albert Einstein pointed out that Newton's laws had been based on presuppositions about the nature of reality (such as absolute space and absolute time) that strictly speaking were false. Today, Newton's laws are considered to be applicable only to that little segment of time and space (the macro-world) in which we happen to live. But in the broader sense, as an explanation for physical reality as a whole, today these presuppositions are considered to be wrong, strictly speaking (although for ninety-nine percent of everyday phenomena with which ordinary people deal, Newton is still quite useful).

By the way, biased physicists had a hard time accepting Einstein's totally new approach. In 1921, Einstein received the Nobel Prize, not for his more important general theory of relativity but for his theory explaining the photoelectric effect.[91] When he was nominated for the Nobel Prize, his theory of relativity was mentioned almost by way of apology, in the hope that this theory would not be used against him. Apparently, nothing is as difficult for a scientist as shifting from one paradigm to another. When French chemist Antoine Lavoisier altered the chemical paradigm and laid the basis for modern chemistry, *none* of the older chemists, although these were highly reputable scientists, like Joseph Priestley and Henry Cavendish, accepted the new paradigm, because they kept viewing the older paradigm as an established fact.[92]

The German philosopher Georg W. F. Hegel once uttered the famous statement, "But what experience and history teach is this, that peoples and governments never have learned anything from history."[93] In short, we learn from history that we do not learn from history. This can show us the tremendous power of a reigning paradigm. Nevertheless, already in 1983 the French biologist, Nobel Laureate and atheist François Jacob made the same comparison with physics that I just made, and observed that within neo-Darwinism a person similar to Albert Einstein would have to come on the scene in order to achieve a breakthrough.[94] In 1991, British biologist Rupert Sheldrake wrote something similar.[95] Around that time, Mae-Wan Ho and Sidney W. Fox wrote that they and other biologists saw a new paradigm emerging that would replace neo-Darwinism.[96] Earlier, Ho had published a book together with Peter Saunders under the title *Beyond*

---

[91] Friedman (2001, ad loc.).
[92] Cf. Yount (2008).
[93] Hegel (2004, 6).
[94] Jacob (1979).
[95] Sheldrake (1991).
[96] Ho and Fox (1988).

Neo-Darwinism.[97] She pleaded for an entirely new approach to evolution, one that is strictly non-Darwinian.

Please note that Ho gave and gives no consideration to rejecting the notion of general evolution. The same is true for Danish-Swedish biologist Søren Løvtrup, who held to evolution but called Darwinism a myth.[98] Similarly, British journalist and non-creationist Richard Milton attempted what the title of his book proclaims: *Shattering the Myths of Darwinism*.[99] Other (non-creationist) critics of (neo)Darwinism include French mathematician Marcel-Paul Schützenberger, American biochemist Christian Schwabe,[100] Canadian mathematician and biologist Brian C. Goodwin, English biologist P. Patrick G. Bateson, and American biologist Norman I. Platnick. On one Internet website, more than a thousand natural scientists with doctorates, many of them (emeritus) professors, declare: "We are skeptical of claims for the ability of random mutation and natural selection to account for the complexity of life. Careful examination of the evidence for Darwinian theory should be encouraged."[101] One of them, Russell Carlson, says on that same site, "To limit teaching to only one idea is a disservice to students because it is unnecessarily restrictive, dishonest, and intellectually myopic."

In his *Mind and Cosmos*, American philosopher and atheist Thomas Nagel was making a significant claim in the subtitle of his book, "Why the Materialist Neo-Darwinian Conception of Nature is Almost Certainly False."[102] This does not necessarily mean that neo-Darwinism will be rejected in its entirety; but its validity might apply only for a borderline condition, as in the case of classical mechanics (Newtonianism). For instance, neo-Darwinism may turn out to be valid only for cases that can be directly observed in nature and in the laboratory. But as a biological explanation for supposed general evolution, it has become passé. The hope of some is now focused on genome biology, *although in a sense this is just a sophisticated form of neo-Darwinism*, because it still involves genome mutations, and a bit of natural selection (although, admittedly, it has invented mechanisms that are impossible to root in traditional neo-Darwinian thought).[103]

However, please note that even if every biological explanation of supposed general evolution were to fail, faith in evolution itself will remain

97 Ho and Saunders (1984).
98 Løvtrup (1987).
99 Milton (2000).
100 Schwabe (2001).
101 https://dissentfromdarwin.org.
102 Nagel (2012).
103 The basic problems are still the same; see G. Barnard in Nevin (2009, 166–86).

unshattered. This is because evolution*ism* is indeed a faith, an *-ism*. In this respect, there is a significant difference here with the revolutions in physics and chemistry that we mentioned. In the latter, personal egos were surely shattered, but the paradigms involved did not bear the tremendous ideological load that evolutionism carries. There was no Lewontin or Dawkins claiming that there *could* be no alternative to the prevailing paradigms because God must be excluded.

2.6 THE CURRENT SITUATION

2.6.1 EVOLUTION, YES; (NEO-)DARWINISM, NO

Speaking of paradigm shifts, the Chinese-Swiss geologist Kenneth J. Hsü wrote in 1989 that the situation of Darwinism at that time was comparable to that of Ptolemy's geocentric theory during the Middle Ages.[104] Ptolemy's theory was science during antiquity, but it became a dogma after his predictions had been falsified by Galileo's new observations. Similarly, Hsü thought Darwinism was once a scientific theory, but during the twentieth century it became a dogma.

The American biologist and materialist Lynn Margulis became famous through her theory about endosymbiosis (the incorporation of one organism into another organism as an explanation for the origin of mitochondria and chloroplasts). She once said that history would ultimately judge neo-Darwinism to be "a minor 20th-century religious sect within the sprawling religious persuasion of Anglo-Saxon Biology. Neo-Darwinism is...complete funk."[105] In the 1990s, she was quoted to the effect that neo-Darwinians "wallow in their zoological, capitalistic, competitive, cost-benefit interpretation of Darwin—having mistaken him.... Neo-Darwinism, which insists on [the slow accrual of mutations by gene-level natural selection], is in a complete funk."[106]

In 1995, John Brockmann quoted her conclusion:

> The neo-Darwinist population-genetics tradition is reminiscent of phrenology [the pseudoscience of the skull knobs], I think, and is a kind of science that can expect exactly the same fate. It will look ridiculous in retrospect, because it is ridiculous.... I've been critical

---

104 Hsü (1989).
105 Mann (1991).
106 Ibid.

> of mathematical neo-Darwinism for years; it never made much sense to me. We were all told that random mutations—most of which are known to be deleterious—are the main cause of evolutionary change. I remember waking up one day with an epiphanous revelation: I am not a neo-Darwinist![107]

Margulis hastened to add, though, that she remained an evolutionist. Apparently, evolutionism is a belief you hold even when all the evidence for it disappears. (This reminds me of Jesus' saying, "Happy are those who don't see and yet believe," John 20:29 CEB.)

Incidentally, evolutionists may rightly say the very same thing about creationism: "Creationism is a belief you hold even when all the evidence for it disappears." But at that point it has become crystal clear that the battle has an ideological—if not religious—nature, and *not* a strictly scientific nature.

Historian of biology and anti-creationist Peter J. Bowler, in his books *Evolution: The History of an Idea* and *Darwin Deleted: Imagining a World Without Darwin*, dealt extensively with the many forms of non-Darwinian evolution that are current today, including its newest varieties.[108] In these alternative theories, natural selection is accepted as playing some role in small observable evolution (micro-evolution) but is no longer considered to be the central principle to explain supposed general evolution (macro-evolution). Fodor and Piattelli-Palmarini recently wrote, "We think of natural selection as tuning the piano, not as composing the melodies. That's our story, and we think it's the story that modern biology tells when it's properly construed."[109]

Let me simply mention here in passing that there have *always* been alternatives to Darwinism or neo-Darwinism, even if these have often received little attention or have met heavy criticism (in alphabetical order):[110]

(a) *Evo-Devo* (Evolutionary Developmental Biology; Sean Carroll[111]).
(b) *Facilitated Evolution Theory* (Marc Kirschner and John Gerhart[112]).
(c) The *Four-Dimensional* evolution theory (genetic, epigenetic, behavioural, symbolic; Eva Jablonka and Marion Lamb[113]).

---

107 Brockman (1995, 133).
108 Bowler (2009; 2013).
109 Fodor and Piattelli-Palmarini (2010, 21).
110 Cf. Rupke (2015); Strauss (2015).
111 Carroll (2016).
112 Kirschner and Gerhart (2006).
113 Jablonka and Lamb (2006).

(d) *Frontloaded-facilitated theory of evolution* (Peter Borger[114]).
(e) *Holistic evolution* (Jan C. Smuts; Adolf Meyer-Abich[115]).
(f) *Mechanistic evolution* (Israel Eisenstein[116]).
(g) *Neo-Lamarckism* (Pierre-Paul Grassé; Edward J. Steele[117]).
(h) *Neovitalism* (Hans Driesch; Rainer Schubert-Soldern; Edmund W. Sinnott; Paul Overhage[118]).
(i) The theory of *neutral evolution* (Motoo Kimura; Masatoshi Nei et al.[119]).
(j) *Nomogenesis* ("Evolution determined by law," Lev S. Berg[120]).
(k) *Organismic biology* (Ludwig von Bertalanffy[121]).
(l) *Pan-psychism* (Pierre Teilhard de Chardin; Bernard Rensch[122]).
(m) The *panspermia* theory (Fred Hoyle and Chandra Wickramasinghe[123]).
(n) The *post-Darwinian evolutionary approach* (my term; Eugene V. Koonin).[124]
(o) The *Prescribed Theory of evolution* (John A. Davison[125]).
(p) The *Third Way of Evolution* (a collection of various new approaches; Denis Noble; James A. Shapiro; Raju Pookottil; Suzan Mazur[126]).
(q) *Typostrophism* (Otto H. Schindewolf[127]).

It is important to distinguish between all these theories, as well as neo-Darwinism, as being more or less strictly biological theories on the one hand, and belief in general evolution on the other hand. If neo-Darwinism were to be shown to be wrong, this does not necessarily mean, as I said, that evolutionists would surrender belief in general evolution. *This is because they have no choice*. It would simply become that much clearer that evolutionism is a belief, for which people do not yet have a sufficient and satisfactory scientific explanation. Indeed, strict neo-Darwinism—evolution through a gradual accumulation of favourable mutations and natural

114 Borger (2009).
115 Smuts (2006); Meyer-Abich (1940).
116 Eisenstein (1975).
117 Grassé (2013); Steele (1981).
118 Driesch (2010); Schubert-Soldern (1962); Sinnott (1966); Overhage (1968).
119 Kimura (1983); Nei et al. (2010).
120 Berg (1969).
121 See, e.g., Von Bertalanffy (1955).
122 Teilhard de Chardin (1961); Rensch (1985).
123 Hoyle and Wickramasinghe (1984).
124 Koonin (2011).
125 Davison (2005).
126 Noble (2008); Shapiro (2011); Pookottil (2013); Mazur (2015).
127 Schindewolf (1994).

selection—has been largely abandoned. Today geneticists are focusing much more on duplication and recombination of already existing genetic material, and on the study of DNA sequences and of regulatory genes, than on new—almost always *un*favourable—gene mutations. This is what I have referred to as neo-neo-Darwinism. But this is merely a difference of degree from old-fashioned neo-Darwinism. It still leaves unanswered the question about where, throughout all those supposed 4.1 billion years, all the genetic material came from that was required for the supposed evolution from primitive unicellular organisms to the "Adam and Eve" of AEH advocates.

2.6.2 THE STATUS OF EVOLUTIONARY THEORY

I pose the question once more: Who would dare to assert that neo-Darwinism or neo-neo-Darwinism is an all-encompassing genetic theory that can sufficiently explain not only the changes we observe in nature and the laboratory, but also the supposed evolution from amoebas to humans? The premise of this critical question boldly claims, *if* there ever were general evolution, we have no adequate genetic theory to explain it. The premise boldly identifies the real nature of belief in evolution: it is just that—a belief. As Fodor and Piattelli-Palmarini recently stated, "'OK; if Darwin got it wrong, what do you guys think is the mechanism of evolution?' Short answer: we don't know what the mechanism of evolution is."[128] *But, of course, we still believe in general evolution.*

As natural scientist John A. Bloom put it,

> Many scientists view the theory of common ancestry as so obviously true ('the fact of evolution') that details about the exact mechanism of macroevolution (gradual change, punctuated equilibrium, or some yet unknown process) do not disturb them. Such a deep faith in this solely naturalistic model blinds many to any other possible explanations for this similarity, such as the theory of a common designer or engineer.[129]

Do all the people of BioLogos (identified in §1.6.2), and so many other Christian writers who seem to be more infatuated with evolution than some of their non-Christian colleagues, comprehend this message? The theory of evolution is a belief. You might say: it is a *doctrine* or *dogma* without a sufficient scientific foundation. A theory that, strictly scientifically speaking, is not necessarily false, but not necessarily right, either. Its nature is far more

[128] Fodor and Piattelli-Palmarini (2010, 153).
[129] http://www.asa3.org/ASA/education/origins/humans-jb.htm.

ideological, if not fundamentally religious, than it is strictly natural-scientific. In chapter 6, I will explain what I mean here by the term "religious."

It is interesting that initially Gijsbert van den Brink expressed himself cautiously regarding the theory of evolution. He says,

> I do not think that the theory of evolution has been "proved"—in my view, everyone stating that it is not proved is *a priori* right. "Proving" is a strong word, which in fact is relevant only in disciplines like logic and mathematics. It does seem to me, however, that the theory of evolution…when we try to look at it in as unbiased a way as possible in the light of all the empirical material that throughout the years has come to light, has good testimonials. At any rate, its testimonials are many times better than those of any other alternative.[130]

I have four comments to make about this.

First, even in logic and mathematics, a theorem can be considered proven only if one accepts certain axioms *a priori*. In our human world, proofs exist nowhere except within a framework that always demands faith (here, belief in the underlying axioms).

Second, it is naïve to think that one could look at the theory of evolution "in as unbiased a way as possible." I know of no natural science where ideological biases play as strong a role as they do in the dogma of evolution. I think I can study electromagnetism "in as unbiased a way as possible," because I see no ideological problems involved here. Whether light is more a wave phenomenon or more a particle phenomenon is scarcely an ideological matter. However, when a matter involves ideological dimensions, it is more difficult to approach that matter "in as unbiased a way as possible."[131] Again, among all the natural sciences no issue is more ideologically driven than the theory of evolution. What other field in the natural sciences has its own Dawkins to fiercely defend its own posture of atheism with respect to that field? There is none. Equally unparalleled in the natural sciences is the angry and sarcastic manner in which books authored by critics of evolutionism are reviewed (also on Wikipedia).

Third, van den Brink adds that the testimonials of the theory of evolution "are many times better than those of any other alternative." But what alternatives is he talking about? How many does he know? Does he mean

---

[130] van den Brink (2017, 17).

[131] With respect to evolution, Carlson and Longman (2010, 14) encourage Christians "to ground their theological and scientific beliefs in an *impartial* search for truth" (italics added).

alternatives for neo-neo-Darwinism? But did he indeed make a thorough study of, for instance, neo-Lamarckism, neo-vitalism, pan-psychism, or typostrophism, to mention just a few (see §2.6.1)? Or if he is talking about evolution in general, then yes, evolutionists have always emphasized that, on a strictly naturalist standpoint, *there are, and can be, no alternatives.* This was supposedly the great strength of evolutionism: for an atheist, evolution is the only option. And if you believe in general (particles-to-people) evolution on a strictly naturalist basis, you will automatically end up in atheism.

Fourth, in the rest of his book, and in interviews he has given, van den Brink's careful language disappears; he implicitly deals with general evolution as a fact that cannot be circumvented. He is prepared to revolutionize the theological approach toward Genesis 1–3 on behalf of a scientific theory concerning which he initially stated that it has not been proved. Not proved, yet a virtual certainty—that is his *de facto* attitude. And he is willing to pay a high theological price for this opinion. Was it worth it to make such a sacrifice for a theory that has not been proved?

2.6.3 FINAL REMARKS

Actually, *there is an alternative.* I do not know whether van den Brink is aware of it; after all, he has already told us that the theory of general evolution is much better than any other alternative. Notwithstanding his assertion, allow me to state the alternative. Before I do, I declare my great respect for some modern creationists, like the Australia-New Zealander physical chemist Jonathan Sarfati.[132] But my problem with his approach (and that of others) is that he sometimes gives the impression in his books of fighting the theory of evolution not primarily with scientific arguments, but with the Bible. This is not prohibited; I myself believe that a serious, historical approach to Genesis 1–3 does not allow for any theory of general evolution. Yet, methodologically I find it far more effective to combat evolutionism primarily with scientific arguments. Always begin fighting your opponent on common ground! Otherwise, he might reject you immediately because of your starting point.

This is the power of several scientists, of whom the British-Australian Michael Denton is one of my favourites. He earned his doctorate in biochemistry in London (UK). He is connected with the Discovery Institute, whose mission is this:

---

[132] See Sarfati (2004; 2008; 2010a; 2010b; 2015).

> Discovery Institute promotes thoughtful analysis and effective action on local, regional, national and international issues. The Institute is home to an inter-disciplinary community of scholars and policy advocates dedicated to the reinvigoration of traditional Western principles and institutions and the worldview from which they issued. Discovery Institute has a special concern for the role that science and technology play in our culture and how they can advance free markets, illuminate public policy and support the theistic foundations of the West.[133]

I don't know whether Denton is a religious person because in his books I find only scientific arguments.[134] And this is the way it should be; we should evaluate and answer science with science (as much as possible, because there are always ideological lions lurking in the shadows). In his first book, Denton strongly defended his case; his latest book is an update of that one. His thesis is that Darwinian evolution, in any form, is unable to explain the history of life, and he elucidates this with many arguments and examples. But he does not leave it there. Denton presents an alternative paradigm, one indebted to Darwin's great rival (*and* interlocutor), British anatomist, zoologist, and paleontologist Richard Owen. Denton's proposal is based on the idea of discrete biological forms, or types, which have the standing of natural laws.[135] Within types, endless variations are possible; but for crossing between types there is no decisive proof.

I wonder if non-biologist van den Brink is in a position to judge whether this alternative is scientifically better or worse than (neo-)neo-Darwinism. (Let him look again at the quotation from Laughlin in §2.5.1.) But no, he has already made up his mind. Evolution has not been proved, yet is a virtual certainty that we can hardly dismiss. So we have no other choice but to adapt our theology to it. This is a game I refuse to play. First, general evolution is a fact only for people who have ideological reasons for believing so (including those who are not aware of these), *and* for people who love belonging to the majority for fear of being excluded from the scientific community. Second, I will never allow my theological views to be co-determined by biological ideology. In this book, I hope to show the disastrous consequences of doing otherwise.

---

133 https://discovery.org/about.

134 See Denton (1986; 1998; 2016). Some recent scientific and/or theological books *contra* the theory of general evolution (in addition to those of Sarfati) are Collins (2003); Lennox (2007); Taylor (2008); Nevin (2009); Wood and Garner (2009); Richards (2010); Ashton (2012); Gauger et al. (2012); Brown and Stackpole (2014); Carter (2014); Poythress (2014); Sanford (2014); Snelling (2014); Brand and Chadwick (2017); Moreland et al. (2017); Paul (2017).

135 About Owen, see Rupke (2009).

I call the people of BioLogos as well as van den Brink "theistic evolutionists." They may not like this because they reject evolution*ism*. But first, as I have argued, in our post-Christian Western world no theory of evolution exists without evolution*ism*. Second, I wish to make clear through this expression that theistic evolutionism (just like evolutionary creationism) is neither fish nor fowl. Every Christian who accepts the idea of general evolution *and* believes in God's providence[136] is automatically a theistic evolutionist; behind the supposed process of general evolution there is God's guiding hand.

2.6.4 NEITHER FISH NOR FOWL

This is why theistic evolutionism is neither fish nor fowl. On the one hand, ordinary atheist or agnostic evolutionists abhor the notion of a God who is guiding things in the background because it always reminds them of the idea of a "God of the gaps" (Lat. *deus ex machina*).[137] Ordinary evolutionists believe in a purely naturalistic evolutionary process, whereas biblical Christians necessarily believe in a guided process. Theistic evolutionists often hardly seem to realize how wide a gap exists between them and ordinary evolutionists. They may not care, but the gap exists. Just listen to one of America's best known evolutionists, Stephen J. Gould: "Evolution substituted a naturalistic explanation of cold comfort for our former conviction that a benevolent deity fashioned us directly in his own image."[138] Among leading evolutionary biologists, 98.7 percent rejected a traditional theistic worldview and became functional atheists.[139]

On the other hand, theistic evolutionists have a conflict with many biblical Christians as well, who believe that God created plants and animals in a great variety of forms (this is the meaning of the well-known expression in Gen. 1, "after its/their kind"[140]). These are Christians who do not need an unproven theory, which moreover destroys their age-old theological approach to Genesis 1–3. This is what I hope to show in the following chapters.

Theistic evolutionists end up with no friends on either side. Ordinary evolutionists cannot be their friends because these abhor the "God of the

---

[136] See van den Brink (2017, 266–89).

[137] van den Brink (2017, 74–75) is tempted to introduce God in an effort to solve some of the biggest problems of evolutionary thinking (abiogenesis, the Cambrian Explosion) but shrinks from the "God of the gaps" idea. This is typical for him: I have never read a book that uses the phrases "on the one hand" and "on the other hand" as often as this one.

[138] Gould (2001, xi).

[139] Bergman (2010, 149–50).

[140] See more extensively, §4.4.1.

gaps" idea. And biblical Christians cannot be their friends, either, because these do not wish to read Genesis 1–3 through the Darwinian lens.

Let me end this chapter with a quotation from the Foreword that American dogmatician Wayne Grudem wrote for a book edited by Irish geneticist Norman C. Nevin.

> Evolution is secular culture's grand explanation, the overriding "meta-narrative" that sinners accept with joy because it allows them to explain life without reference to God, with no accountability to any Creator, no moral standards to restrain their sin, "no fear of God before their eyes" (Rom. 3:18)—and now theistic evolutionists tell us that Christians can just surrender to this massive attack on the Christian faith and safely, inoffensively, tack on God, not as the omnipotent God who in his infinite wisdom directly created all living things, but as the invisible deity who makes absolutely no detectable difference in nature of living beings as they exist today. It will not take long for unbelievers to dismiss the idea of such a God who makes no difference at all. To put it in terms of an equation, when atheists assure us that *matter* + *evolution* + 0 = *all living things*, and then theistic evolutionists answer, no, that *matter* + *evolution* + *God* = *all living things*, it will not take long for unbelievers to conclude that, therefore, *God* = 0.
>
> I was previously aware that theistic evolution had serious difficulties, but I am now [i.e., after reading Nevin's book] more firmly convinced than ever that it is impossible to believe consistently in both the truthfulness of the Bible and Darwinian evolution. We have to choose one or the other.[141]

[141] W. Grudem in Nevin (2009, 10).

3

A HERMENEUTIC SHIFT

> You will say to me then, "Why does he still find fault? For who can resist his will?" But who are you, O man, to answer back to God? Will what is molded say to its molder, "Why have you made me like this?" Has the potter no right over the clay, to make out of the same lump one vessel for honorable use and another for dishonorable use? (Romans 9:19–21)

THESIS

In the past, many interpretations of Genesis 1–3 operated within the same hermeneutical framework, one in which Scripture was basically allowed to speak for itself. However, modern attempts to reconcile Genesis with modern science demand an essentially new hermeneutic, which is fundamentally unacceptable, both to modern science itself and to biblical Christians.

3.1 A THEORY OF INTERPRETATION

3.1.1 THE PRE-CONDITIONS OF EXEGESIS

Not too long ago, American theologian William Brown, who was but one of many modern theologians like him, tried to find connections between the Old Testament creation texts and the findings of "modern science."[1]

[1] I put this phrase in quotation marks only this time to indicate to the reader that I am not always convinced that "modern science" is really modern, or really science (cf. chapter 2). Carter (2017, 41) says of Venema and McKnight (2017): "I discovered a key to deciphering the arguments of both authors: every time they say 'science,' one can replace that word with

Applying those findings to his understanding of these texts, he spoke of "a hermeneutical feedback loop" between the biblical texts and scientific theories.[2] This is one of few publications where I find some awareness of the "hermeneutical loops" involved in trying to read the Bible with the spectacles of modern science.

This chapter is about hermeneutics. This may sound dry, but in my view it is here where the Gordian knot is cut. Let me explain. The word *hermeneutics* comes from the Greek words *hermēneia* or *diermēneia*, "interpretation/translation" (see 1 Cor. 12:10), *hermēneutēs* or *diermēneutēs*, "interpreter/translator" (14:28), and the verbs *hermēneuō* or *diermēneuō* (Acts 9:36), in the sense of "to translate," and especially "to interpret." Now in theological language, hermeneutics is not about interpretation alone. To identify this activity, theologians use the term *exegesis*, which also comes from the Greek (from the Gk. verb *exēgeisthai*, lit., "to lead out," ie., to lead the meaning out of the text). Hermeneutics takes a step back, and asks: *How* do we interpret Scripture? According to what rules? Who decides these rules? From where do we get these rules? What justifies them? How are they applied? Do they change over time, or do they remain the same? Could they depend on, or be influenced by, prevailing ideologies and philosophies?

Hermeneutics is what we may call the *theory* of interpretation. This discipline is present not only in theology but in all forms of literary scholarship that investigates texts, especially ancient texts. This is no surprise, since when it comes to interpretation, the Bible is an ancient text just like other ancient texts, and must be interpreted carefully, according to well-defined rules. This explains why theological methodology has sometimes been treated as belonging to literary theory. The reason so many people study the Bible—many more than those who study Homer, Herodotus, or Horace—is that they believe the Bible is the Word of God, a special, inspired, and authoritative revelation from God. This is a presupposition of faith; it is a presupposition not only for exegesis, but also for hermeneutics. However, this does not alter the fact—but rather establishes it—that the Bible must be interpreted according to strict rules, whereby the interpreter demonstrates respect for the text.

Nowadays, hermeneutics has been extended to *all* forms of interpretation, including verbal and non-verbal communication. Everything is a "text" that must be "interpreted." What is the other person saying (meaning,

---

'philosophical naturalism' to see that it is not science (as an exploration of what we can directly test) to which they are referring."

[2] Brown (2010, 9).

intending) through this written word or that spoken word, through this gesture or that grimace? Yet, for the great majority of people, hermeneutics is first of all a theory about properly understanding the Bible. Centuries ago, Augustine delved into this important field of study.[3]

Among these basic rules are the following. First, a literal interpretation must always be preferred to a figurative interpretation. Second, "No prophecy of the Scripture is of any private interpretation" (2 Pet. 1:20 NKJV), which may mean: no one, not even the prophet himself, can interpret Scripture on his own, as he pleases. This has to do with a third rule: compare Scripture with Scripture. Each verse must be interpreted in the light of the rest of Scripture. This is what Jesus expressed in his statement: "Scripture cannot be broken" (John 10:35). From this follows a fourth rule: more difficult passages must be interpreted in light of more easily understandable passages. A fifth rule is: try to read the text not through the lens of your own culture, but through the lens of the culture of the author and of the first recipients. Try to stand in the shoes of the text's human author and its original readers.

3.1.2 PHILOSOPHICAL CONSIDERATIONS

Because orthodox Christians believe the Bible to be the inspired and divinely authoritative Word of God, not only literary criteria but also philosophical and religious criteria play a role. As with all sciences and humanities, hermeneutics is a discipline with its own philosophical foundations. These are ultimately based upon a pre-theoretical but no less rational *worldview*. Such a worldview is ultimately based upon a pre-theoretical, and even pre-rational (not *ir*rational but rather *supra*-rational) *faith* (i.e., faith is not just a matter of the immanent *ratio*, but of the transcendent human heart). I have analyzed this matter extensively in other works.[4] This matter has great practical significance. If we accept the Bible as the Word of God, we will scarcely be inclined to allow the natural sciences to tell us how we must read the opening chapters of Genesis.

It would be quite naïve to equate the natural sciences with God's so-called "general revelation." This egregious mistake was committed, for instance, by the Haarsmas.[5] Denis Lamoureux distinguishes between the "Divine Book of Works" ("offers overwhelming evidence for the evolution of humans")

---

[3] *De doctrina Christiana*; see http://www.ntslibrary.com/PDF%20Books/Augustine%20doctrine.pdf.

[4] Ouweneel (2014a; 2015); in these works, I also define a term like "religion." For more about this, see §§6.4 and 6.5.

[5] Haarsma and Haarsma (2011, chapter 4).

and the "Divine Book of Words."[6] Gijsbert van den Brink calls the theory of evolution "simply what the *book of nature* shows us at this moment."[7] This italicized phrase can hardly be anything other than an allusion to the Belgic Confession (Art. 2).[8] But if with this expression he is indeed referring to God's "general revelation," how could this "book of nature" show us picture A (evolutionary theory) "at this moment," and picture B or picture C at some other moments? What van den Brink apparently means to say is that the natural sciences give us a certain picture "at this moment," and, as is common in science, at other moments it will certainly give us different pictures. But this is something very different from the "book of nature" as intended in the Belgic Confession.

I find this identification of the natural sciences with the "book of nature" very disturbing. At most, the natural sciences may offer us a vague, weak, deficient, and certainly preliminary *approximation* of certain natural states of affairs. However, the more heavily such an approach is burdened with ideological biases that are by their nature anti-theist, and even anti-religious (see chapter 2), the more suspicious we must be toward such an approximation. Unfortunately, our world lacks any spokesperson who defends with holy fervour any strictly natural scientific theory at all—but the world is treated to numerous atheist and anti-religious spokespersons (people like British author Richard Dawkins[9]) who fiercely and ardently defend evolutionism. This unique state of affairs underscores the absolutely exceptional status of the theory of evolution (see §2.6.2). Never in history has there been such a conflict between, on the one hand, Scripture and orthodox Christianity, and on the other hand, atheist and anti-religious attackers who hide themselves behind supposed scientific "truths."[10]

This explains why we must introduce philosophy into our discussion. The hermeneutical problem is foundational not only to the natural sciences and to theology, but to the philosophy of science as well. On the one hand, what are the philosophical presuppositions involved in the natural sciences? On the other hand, what are the philosophical presuppositions involved in

---

6 In Barrett and Caneday (2013, 63–64).

7 van den Brink (2017, 255); italics added.

8 "We know [God] by two means: First, by the creation, preservation, and government of the universe; which is before our eyes as a most elegant book, wherein all creatures, great and small, are as so many characters leading us to "see clearly the invisible things of God, . . ." (Dennison [2008, 2:425]).

9 See Dawkins (1986; 2009).

10 Unless, of course, someone wishes to bring up the Galileo affair—but that conflict involved a struggle between newer scientific insights and Aristotelianism; see, e.g., Blackwell (1991); Petterson et al. (2014); Paul (2017, 116–17).

hermeneutics, and in theology more generally? In this respect, both natural scientists and theologians are often quite naïve. On the one hand, stating that (general) evolution (from amoebas to humans) is "a fact"[11] betrays a basic misunderstanding about what the natural sciences can do and are allowed to claim. On the other hand, the naiveté with which theologians adopt "what science has shown" and has supposedly proven is quite shocking, too. In both cases, some philosophical homework must be assigned and completed before moving on.

3.2 IS CHRISTIAN THEOLOGY POSSIBLE?

3.2.1 THE SCHOLAR'S ULTIMATE COMMITMENT

Before entering more deeply into the problem of hermeneutics, let me ask an introductory question: What is the difference between biology and theology when it comes to their academic character? Is biology a "neutral" enterprise from which we can freely adopt certain insights, whereas theology is obviously a "Christian" enterprise? Or is this difference not so obvious? Or let's ask: Is Christian theology possible? If so, what kind of enterprise is it? And if Christian theology is possible, then why not Christian biology? Or to ask the same thing from the opposite vantage point: Is a "neutral" biology possible? If so, what kind of enterprise is it? And if a neutral biology is possible, why not a neutral theology?

I know many theologians who are fervent Christians and who believe that the Bible is the Word of God. However, they seldom seem to ask themselves whether what they are doing is *Christian* theology, and whether such a Christian theology is possible. It seems to be self-evident to them that there is no such thing as a Christian biology or a Christian geology. Why do they assume that what they themselves are doing is *Christian* theology? Because they are Christians? But Christians doing biology or geology is not a sufficient criterion for their disciplines to be called Christian biology or Christian geology. Or is it Christian theology because the Bible, or at least the New Testament, is a Christian book? But there is no *a priori* reason why an orthodox Jew such as New Testament scholar Pinchas Lapide,[12] or even a Muslim or a Hindu, or even an atheist, could not be a successful theologian in the academic disciplines of literary theory or comparative religions (also termed "science of religion").

---

[11] As stated here: http://www.discovery.org/a/6401.

[12] See Lapide (2002).

Any academic can investigate Christianity within the field of the science of religion, or conduct a theological investigation of the Bible. As long as he or she observes the rules of hermeneutics, everything is fine. Such people need not be *Christians* personally; the question becomes whether being a Christian matters at all, as long as these scholars are academically competent. However, we should notice here that this word "competent" is quite ambiguous. In several recent books introducing readers to various fields of study, I have dealt extensively with the important distinction between what has been called *structure* and *direction* (or *directedness*).[13] Viewed structurally—I am referring to the horizontal dimension of academic standards—a Muslim or an atheist can be a "competent" theologian as long as he or she correctly plays the "language game" of theology (to use a term of Austrian philosopher Ludwig Wittgenstein[14]). One must correctly apply the hermeneutical rules and theological methods; one must obey the rules implied in the structure of theology itself. Speaking strictly structurally, "good" theologians are simply good academicians in that field; they do their job well.[15]

However, when it comes to the *direction*—this is the vertical dimension of faith or unbelief—the Muslim and the atheist, as well as the liberal theologian, are conducting apostate theology because they are working from an apostate ground motive.[16] This may sound harsh, but it must be said. The New Testament is very clear about the apostasy of the last days (2 Thess. 2:3; 1 Tim. 4:1 CJB; see also 2 Tim. 3:1–5; 4:1–4; 2 Pet. 3:17). We must expect to encounter it even at the heart of orthodox Christianity. Not only every scientist or scholar, but every theologian as well, is driven at their deepest level by apostate or by believing (i.e., biblical) ground motives. Seen from this viewpoint, an apostate theologian is a "bad" theologian—irrespective of one's academic standards, and irrespective of the orthodoxy of one's denomination—because one is working from an anti-biblical starting point. As such, these theological achievements are definitely and continually "incompetent." To be a horizontally "good" theologian one must have sufficient academic competence. To be a vertically "good" theologian one

---

[13] Ouweneel (2014a; 2015); this distinction was explained extensively first by Albert Wolters (1985).

[14] Wittgenstein (2009).

[15] A good example is Rudolf Bultmann (1971), who was a very liberal theologian, but from a purely academic viewpoint, his commentary on John is such an expert product that no academic commentator on John can ignore it.

[16] For an extensive explanation of the important term "ground motive," see Ouweneel (2014a; 2015).

must have an unshaken faith in the God of the Bible, and in the Bible as the inspired and authoritative Word of this God.

3.2.2 "NEUTRAL" SCIENCE?

I have joined others in emphasizing in many publications that neutral, objective, unbiased science (including theology) does not exist (see also chapter 2). In general, the scientist's ultimate commitment—one's deepest faith convictions—plays a much smaller role in the natural sciences than it does in the humanities. This is obvious from the reality that different natural scientists, with very different philosophical or religious backgrounds, often come to fundamentally the same results when carrying out the same scientific investigation. However, in the humanities, certainly in ethics and theology, this is definitely not the case. Here, Christian ethics and Christian theology are often radically opposed to non-Christian (to say nothing of an anti-Christian) ethics and non-Christian theology.

In his latest work, van den Brink apparently tries to reconcile certain (scholarly, yet Christian? or also neutral?) theological views with certain (scientific, and thus neutral?) biological insights. No wonder—belief in the "neutrality" of all science and scholarship reigns supreme. Not only the Dutch Christian philosopher Herman Dooyeweerd,[17] but also Hungarian philosopher Michael Polanyi[18] and Austrian philosopher Karl R. Popper[19] broke radically with the notion of "neutral" science.[20] Interestingly, van den Brink himself makes a fundamental error with respect to Dooyeweerd's arguments when he asserts: "They attempt to prove in a neutral-philosophical way that neutral philosophy is not possible!"[21] This suggests that Dooyeweerd, *the* great thinker who has comprehensively identified the religious biases of all human thinking, would have been unaware of the religious biases of his own thinking. Supposedly, Dooyeweerd was unaware of just how firmly he himself was trapped in the well-known "hermeneutic circle" (see the next section). This judgment, however, shows a basic misunderstanding of Dooyeweerd's thinking. Dooyeweerd was the first to point out that the claim that all thought is *a priori* religiously determined is itself necessarily *a priori* religiously determined.

---

[17] Dooyeweerd (1960; 1984).

[18] Polanyi (1973).

[19] Popper (1959; 1972).

[20] Dutch philosopher C. A. van Peursen (1995, 79) pointed out that, since the time of Dooyeweerd, awareness of such *a priori* ideological (and religious) commitment has become very widespread.

[21] van den Brink (2000, 335).

I find it significant that van den Brink apparently does not grasp this vital point of Dooyeweerd's thinking. His criticism of Dooyeweerd suggests that he himself is still not clear about the "neutrality" claim of modern science. At a minimum, he has no trouble approving such a radical paradigm like the theory of evolution as if it were a "neutral" piece of furniture with which one might comfortably decorate one's theological living room if desired.

This point is of the utmost significance for our present study. The less religiously coloured a natural-scientific hypothesis is—and the great majority of such hypotheses are such—the less will theologians be interested in them. The more religiously coloured a scientific hypothesis is—as is the case with the theory of evolution—the more that theologians seem to be interested. This is why I am seeking to explain what is an essentially new hermeneutic.[22] In their (rather "neutral"?) academic work, theologians feel pressured to incorporate into their work a scientific hypothesis that appears to them to be a fact proven by neutral, objective, unbiased natural science, instead of what it really is: *the most religiously coloured element within all of natural science.* This element is the Trojan horse they are unwittingly bringing within the walls of their theological city. When the same thing happened to the Trojans, the citizens of Troy shouted for joy—only to be killed by the Greeks soon thereafter.

It would be revolutionary enough if theologians were to choose a drastically new approach to Genesis 1–3 on the basis of some "almost neutral" scientific hypothesis. It would be revolutionary because *never before in history* had theologians allowed their approach to Scripture to be determined to such an extent by extra-biblical sources. However, what makes it even more revolutionary is that this shift involves not some "almost neutral" scientific hypothesis. Instead, theologians today are bowing before a science that, since the nineteenth century, has been conducted under the heavy pressure of ideological—and in the broad sense of the term even religious—biases (see chapter 2).

3.3 THE HERMENEUTIC CIRCLE

3.3.1 THE "NEUTRALITY" LIE

There is no such thing as "neutral" theology conducted in the light of "neutral" presuppositions. Similarly, there is no such thing as a "neutral" biology conducted in the light of "neutral" presuppositions. In this sense,

[22] Cf. VanDoodewaard (2016, chapter 1).

there is a clear similarity between theology and biology, a feature that characterizes all sciences and humanities. If we do not analyze this matter accurately and thoroughly, we may be easily confused by the claims that Christian theology is self-evident, just as self-evident as a non-Christian (religiously neutral) biology. Both claims are untrue.

Renowned Christian biologist Louis Pasteur, father of modern microbiology, reportedly wrote, "In each one of us there are two men, the scientist and the man of faith or of doubt. These two spheres are separate, and woe to those who want to make them encroach upon one another in the present state of our knowledge!"[23] Pasteur openly testified of his faith, but when he entered his laboratory, he left his faith outside the door. That is fine with me; I have been active in biological research for ten years, and cannot remember that my faith ever influenced the choice of my experiments, or the choice between alternative interpretations of their results. However, this is not essentially different from my theological work. When I frame a theological hypothesis, my faith presuppositions should not direct my academic work either. I must do justice to the biblical text, whatever my confessional biases may be; otherwise, I am not a good academician.[24]

This is why so many biologists can live very easily with the idea of a "neutral" biology. It is because the overwhelming majority of them are specialists, working in a very narrow field of investigation. By itself this is not a problem. But we should not expect such biologists to have a clear grasp of the philosophical presuppositions of their field (the "special philosophy of biology"), nor of the worldviews and faiths in which these presuppositions are embedded. The situation is no different with theologians. They may take "Christian theology" for granted, simply because they work with the Bible and the large majority of them are Christians. But then, they too are usually specialists. We should not expect such theologians to have a clear grasp of the philosophical presuppositions of their field (the "special philosophy of theology"[25]), nor of the worldviews and faith in which these presuppositions are embedded.

Whatever the situation may be, not only biologists but also theologians have some idea of "neutral" biology, from which you can freely adopt neutral, objective, unbiased insights. Usually this involves entirely innocuous questions, like: Does Job 39 give us an accurate picture of the ostrich? What is

---

[23] http://www.pasteurbrewing.com/louis-pasteur-a-religious-man/.

[24] See again Ouweneel (2015).

[25] Cf. Ouweneel (2014a; 2015). Some theologians scarcely seem to realize that such a thing as a "special philosophy of theology" exists at all. Whatever one may think of the views of Dutch theologian Harry Kuitert, his book on this subject (1996) is very lucid.

the identity of the hare chewing the cud (Lev. 11:6)? What does it mean that the eagle "renews its youth" (Ps. 103:5)? What does it mean that the grain of wheat "dies" in the earth (John 12:24)?[26] Biologists may shed light on some of these issues, but this never affects the interpretation of Scripture itself, let alone that it would affect any Christian doctrine. However, the situation is very different with issues that are not at all neutral, but ideologically freighted, issues which are nevertheless hailed by some theologians who deal with them as neutral biological knowledge on which the vast majority of biologists seem to agree. In addition, these theologians allow basic Christian insights on creation, the Fall, original sin, and thus even redemption, to be thoroughly affected by such "neutral" biological insights.

This is the new hermeneutic. For the first time in the history of Jewish and Christian thought, theologians are allowing their conclusions to be drastically modified by extra-biblical ideas, which are, moreover, ideologically not at all neutral. Please note that this is a double trespass. First, it is a major hermeneutical error to permit your theology to be influenced from the outside to such an extent that foundational Jewish and Christian teachings are heavily affected. Second, these influences from the outside world are not congenial at all; if they were, this feature might provide some small excuse. These influences are not neutral. They are ideologically driven, and in their roots and fruits they are anti-Christian, from Darwin to Dawkins. I say this despite all the Christian biologists who apparently saw no other option than to thoroughly embrace evolutionary thought. Among them are some very fine people, but I cannot agree with them.

3.3.2 NO ESCAPE

To me, the hermeneutic circle entails that whatever I say about a certain science, or about reality, is determined *a priori* by my deepest ideological presuppositions. (Actually, I prefer the phrase "deepest religious presuppositions," taken in its broadest sense: my deepest existential convictions, or better: my pre-rational ultimate commitment, or better: the supra-rational certainties that my heart possesses concerning the Ultimate Grounds of my very existence; see §6.3.1 and 6.4.) Where is the circle, then? In my view, if I claim that all science, at the deepest level, is religiously determined, this claim itself is determined by my ultimate religious commitment. And if someone claims that all science, at the deepest level, is *not* religiously determined but exclusively rational, and thus neutral, this claim itself is again determined by that person's ultimate religious commitment.

---

[26] For the latter question, see van den Brink (2017, 122n29).

Usually, the phrase "hermeneutic circle" is employed to make clear that a text as a whole can only be understood by grasping its parts, and that the parts can be understood only by grasping the whole.[27] This is what I call a *horizontal* description of the phrase. In addition to this, there is what I call a *vertical* understanding of the phrase: all my (immanent) scientific and scholarly investigations are determined by my (transcendent) ultimate commitment (which I call "religious commitment"), but this ultimate commitment is not timeless and unshakable: it is being continually influenced by everything I keep learning in the immanent-empirical world.

Once more, theology is not Christian because it is about Christianity; it is Christian when it is conducted by Christians, that is, by people who are regenerated in their heart and who have received, and live by, the Holy Spirit. By the same token, biology is Christian when it is conducted by Christians who fulfill these same conditions. But this is not enough. Very godly Christians may still be quite naïve concerning the philosophical foundations and worldviews underlying their discipline. This is (a) because they are specialists and have no idea about the philosophical foundations of their entire discipline in the first place, and (b) because the power of positivistic and scientistic thinking—here, the belief in a neutral, objective, unbiased science—is still overwhelming (despite postmodernism, and even post-postmodernism). So, theology or biology is Christian when such disciplines are practiced by Christians who have a basic awareness of how Christian or non-Christian convictions permeate the foundations of their discipline.

All four groups, Christian and non-Christian theologians and biologists, often have little idea about what actually comprises their "science" (in the broad sense of the term, including the humanities).[28] Some may think that "good" theologians are those who are led by the Spirit; but we have seen that this is a confusion of structure and direction. Conversely, if "good" theologians are merely theologians with high academic qualifications, then we have no reason at all to speak of Christian theologians, just as most scientists would not speak of Christian biologists. However, as soon as it is realized that all scientific disciplines are built on philosophical presuppositions (studied in the respective "special philosophies" of these disciplines), and that all philosophy is inevitably rooted in a pre-theoretical but rational worldview (no matter how inarticulate it may be), and that all worldviews are rooted in a supra-rational faith, and that all faith is either apostatic *or* anastatic (directed away from [think: apostasy] or toward God and his

[27] See more extensively Stegmüller (1996).
[28] Ouweneel (2014a).

Word, respectively)—things look very different indeed. Then there are such things as Christian and non-Christian and anti-Christian theology, as well as things such as Christian and non-Christian and anti-Christian biology.

As soon as we begin to understand these relationships, it will begin to dawn upon us how devastating it is when *Christian* theology allows itself to be drastically influenced, or even modified, by *anti-Christian* biology (more specifically, evolutionism).

We can never escape from our ultimate commitment, that is, from our faith concerning the Ultimate Ground(s) of our thinking. If we believe in a neutral theology or biology, it is because of our ultimate commitment (in this case, commitment to what we call scientism, the positivistic myth of a neutral, objective, unbiased science). If we subscribe to an ideologically coloured theology or biology, this too is because of our ultimate commitment, and so on. The *theoretical* question whether a Christian theology or biology is possible or desirable is determined *a priori* by an ultimate commitment, which is either congenial or uncongenial to God's Word.

To put it bluntly: whether you believe in the possibility of a Christian science depends on your Christian or non-Christian prolegomena. Such a hermeneutic circle is inevitable. I do not suppose that Richard Dawkins will flatly deny that his atheism and his fervent defense of the theory of evolution are closely related.[29] Conversely, I claim that Christian scientists and scholars who begin to deny the possibility of a Christian theology or biology do so under the influence of an ultimate commitment that is heavily permeated by humanistic influences in the broadest sense of the phrase.[30] Of course, Christian thinkers can come under the influence of non-Christian or anti-Christian thought, just as much as any other thinkers can. And this is where the problem begins: *Christian* theologians have come under the influence of anti-Christian thought (in this case: evolutionism). Explicitly choosing such a path demands a revolution in one's hermeneutic; this is the core message of the present chapter.

3.3.3 CHRISTIAN PLUMBING

I am under the impression that North American thought has been strongly influenced by British philosophical movements such as empiricism and analytical philosophy, whereas European thought is rooted more deeply in metaphysical movements (led by Immanuel Kant and Georg W. F. Hegel). No matter what the case may be, the "neutrality" postulate has become

---

[29] Cf. Dawkins (2009).

[30] A helpful introduction to this subject is Dooyeweerd (2003, especially chapter 3).

overwhelmingly influential in the Anglo-Saxon world. I encountered this problem when writing my recent book, *The World Is Christ's*.[31] In this work, I combat the two-kingdoms theology that has become so powerful in the North American Reformed and Presbyterian world. This theology claims that the world is divided into two kingdoms (realms, spheres): the sacred and the profane (common, secular). Theology belongs to the former, biology to the latter realm. Intuitively, I assume that among two-kingdom thinkers one will find few creationists but more theistic evolutionists. Their argument is that if the natural sciences are secular (which also implies that they are religiously neutral), Christians within the sacred realm can freely make use of (supposed) newer biological insights like those offered by theories of evolution.

Advocates of this two-kingdom theology love to make statements like these: neither the Fall nor redemption in Christ has changed the laws of logic or the laws of physics. They argue that an essential characteristic of the supposed "common (profane, secular, non-sacred, neutral) kingdom" is that the laws in this realm are valid for both believers and unbelievers. At a conference, one of the advocates of this view argued that there is no Christian way to change diapers or to pilot a plane.[32] A favourite dismissive argument in these circles is what I call the "plumbing argument." A Christian plumber is not called to install "Christian pipes" or to employ "Christian plumbing techniques." So since there is no Christian plumbing, neither is there anything like Christian politics, or Christian art, or Christian literature, or Christian biology. According to this type of irrational *non sequitur*, when consistently applied—that is, without the two-kingdom dichotomy—neither can there be a Christian church or Christian theology, any more than there can be Christian plumbing. This is like saying: there are no Christian chimpanzees, so there can be no Christian humans, either. It is this kind of thinkers who are the great proponents of "neutral" science.

Such superficial arguments may perhaps impress the general public, but they are obviously invalid. Plumbing is more Christian *or* non-Christian than some people think. Plumbing is a technical activity that is embedded in a certain view of technology as a cultural phenomenon. If you are a specialist in the field of plumbing, you may well never think about the foundational questions that precede and surround your activities. You just plumb for a living. But let us be thankful that there are people who ponder

---

[31] Ouweneel (2017, especially §11.4.2).

[32] For references, see previous note.

the basic questions of culture and technology.[33] As soon as we begin to consider such and many similar questions, we discover that there *are* no "neutral" activities—not even plumbing. How much more true is this in the case of science. Certainly, the many specialists in science may well never learn to look above the small horizon of their own narrow field. But at least *some* generalists should.

I remember that, in 1976, I could engage with perhaps only five people in the entire world who were working in the same field I was (homeotic mutations in *Drosophila melanogaster*, the fruit fly[34]); I remember a fellow in Switzerland, one in Spain, and one in the United States. I hated the situation. This kind of specialization was not in my nature. A specialist is someone who knows practically everything about practically nothing—but by nature I am more of a generalist: someone who knows practically nothing about practically everything (as I write this, my tongue lies firmly planted in my cheek). This is one reason—there were more—why I turned my back on the laboratory, and began studying, first, the special philosophy of biology, second, general philosophy (earning my PhD in 1986). Since then, I hope never to make such a foolish remark as this one: there is no such thing as Christian biology just as there is no such thing as Christian plumbing.

Ask people who say such things whether they would also be prepared to say: there is no such thing as Christian theology just as there is no such thing as Christian biology. Their answer will presumably be this: theology belongs to the sacred realm, whereas biology belongs to the secular realm. This is the answer of today's two-kingdom advocates. Perhaps in their heart of hearts, possibly without being aware of it, AEH advocates are also two-kingdom advocates. Within the *sacred* realm, to which theology belongs, they can freely borrow from the "neutral" or secular realm, to which belong biology, geology, paleontology, and so on.

3.4 WHEN IS A HERMENEUTIC REALLY NEW?

3.4.1 WOMEN IN OFFICE: A DIFFERENT HERMENEUTIC?

On July 3, 2017, the General Council of the World Communion of Reformed Churches (WCRC) met in Leipzig (Germany). On this day, the WCRC adopted a declaration approving the admission of women to all church offices.[35]

---

[33] See Ellul (1990); Schuurman (2003).

[34] Cf. Ouweneel (1975b).

[35] http://wcrc.ch/news/general-council-approves-statement-on-ordination-of-women.

Shortly thereafter, another, more conservative associate of Reformed denominations met in Jordan (Ontario): the International Conference of Reformed Churches (ICRC). To this conference belonged a denomination that had taken the initiative to form the ICRC, and was deeply involved in the organization: the Reformed Churches in the Netherlands (Liberated) (RCN). On June 15 and 16, 2017, this denomination did the same thing as the WCRC had done: all church offices were opened to women.[36] The local churches could determine for themselves whether they would implement this decision in practice, but in principle the door had been opened.

This decision of the RCN did not enjoy the favour of the majority of the members of the ICRC.[37] I happened to be in the area, and one of the local pastors delegated to attend the meeting told me that he and others would protest the presence of the RCN representatives. His argument was this: the RCN had adopted a different hermeneutic, one that no longer took Scripture seriously in all its teaching.[38] I gently asked him whether it was not possible that the RCN still took the Bible very seriously, but interpreted it differently on certain points. For instance, how would he, as a preacher, interpret 1 Corinthians 11:5, "Every woman who prays or prophesies with [her] head uncovered dishonors her head, for that is one and the same as if her head were shaved" (NKJV)? Does not Paul say here that women are allowed to pray and prophesy in church? And when they do so, must they cover their heads? My interlocutor did not back down. Paul had spoken "clearly, evidently, and obviously" in 1 Corinthians 14 and 1 Timothy 2, and the RCN had simply pushed aside this clear teaching.

Now my point at this moment is not at all to discuss the issue of women in ecclesiastical office. My point is to ask whether the RCN had indeed adopted a different hermeneutic. The claim that they have indeed adopted a different hermeneutic triggers me because I myself am leveling a similar accusation against AEH advocates. Some might suggest that the Canadian pastor and I are basically doing the same thing. In writing this book, am I merely identifying others with whom I disagree, arguing that they are following a new hermeneutic? Or is there an essential difference between the pastor's accusation and mine? Can I defend why the pastor's accusation is misplaced, and mine is warranted? I think I can.

You may or may not agree with the conclusion of the RCN; that is not my point right now. The point is that these RCN brothers—I know them—

---

[36] http://www.gkv.nl/historisch-moment-ambten-open-vrouwen/.

[37] https://yinkahdinay.wordpress.com/2017/07/18/rcn-suspended-from-icrc/.

[38] The same argument has been used before; cf. Johnston (1978).

remain deeply convinced that the Bible is the inspired and authoritative Word of God on all matters. They would deny that extra-biblical sources—such as Western society's view of women, modern feminism, the women's liberation movement—could ever change their view of Scripture, and could ever lead them to allow the Bible to say the opposite of what it actually says. They would argue that they honestly believe that their present view of women in office does *more justice* to the biblical data than their previous view. I repeat, you do not have to agree with their conclusions, but I fail to see what different hermeneutic could be at work here. A different hermeneutic operates with an altogether different starting point, different rules and values with respect to the exegesis of Scripture. In my view, a different hermeneutic as so described is not in play here.

3.4.2 OTHER EXAMPLES

Let me give a few examples of Bible passages on which Christians often differ fundamentally. Each time I will ask whether a different hermeneutic might be at stake.

Consider these verses: "For in one Spirit we were all baptized into one body—Jews or Greeks, slaves or free" (1 Cor. 12:13a), and: "In him [i.e., Christ] also you were circumcised with a circumcision made without hands, by putting off the body of the flesh, by the circumcision of Christ, having been buried with him in baptism, in which you were also raised with him through faith in the powerful working of God, who raised him from the dead" (Col. 2:11–12). These passages on baptism have generated a lot of controversy. Does the former passage say that through baptism a person becomes a member of the body of Christ, as Baptists claim, or does the verse speak of Spirit baptism, as Pentecostals say?[39] Does the latter passage (Col. 2) suggest that baptism has been substituted for circumcision as the covenant sign, and therefore must be administered to infants, as Reformed and Presbyterians claim?[40] Or does the passage speak of believers' baptism as an outward sign, of which (spiritual) "circumcision" (regeneration) is the inward counterpart?[41]

As far as I am concerned, I do not mind speaking of different *paradigms* functioning here.[42] But a paradigm is very different from a hermeneutic.

---

[39] Cf. Ouweneel (2007a, §8.4.1); see also Fee (1987) and Thiselton (2000) for the various views.

[40] Cf. Heidelberg Catechism, Lord's Day 28, Q/A 74 (Dennison, [2008, 2:795]); see Ouweneel (2016, ad loc.; 2018c, Appendix 1).

[41] Ouweneel (2018c, especially Appendix I).

[42] Unlike Lamoureux (2015, subtitle: *Is a Theological Paradigm Shift Inevitable?*), who speaks

In one paradigm, Bible passages are linked together and interpreted in a way that yields a certain "picture," such as a covenant picture. Others read the Bible as teaching a very different paradigm, and are convinced that believers' baptism is the "picture" given in the New Testament. The two parties have fiercely combated each other, and sometimes still do. Yet, aside from the most fanatic proponents on both sides, the parties are generally willing to acknowledge that, for the other party too, the Bible is the inspired and authoritative Word of God. That is, they are willing to accept that the other party is sincerely convinced that this is the most appropriate way to do justice to the biblical material. In other words, one party does not generally assert that the other party is employing a different hermeneutic. There may be different paradigms at work, but not different hermeneutics.

But now, imagine persons A and B both saying: I believe that women can become elders and pastors in the church. Person C answers: But that contradicts the statements of the apostle Paul. Person A replies to C: No, I sincerely think that you wrongly understand those passages by Paul; this is not at all what Paul wishes to say. Person B, however, replies to C: I know what Paul wrote, but that was his personal opinion. He lived in antiquity, we live in the twenty-first century. We look at women in a different way today.

Do we see what is happening here? Persons A and C have the *same* hermeneutic—even though they have opposing interpretations—but person B has a hermeneutic very *different* from that of A and C, even though A and B come to the same conclusion. The Baptists, Pentecostals, and Presbyterians whom I know usually have the same hermeneutic: they all accept the Bible as the authoritative Word of God; they simply explain certain passages differently. But liberal Christians have a different hermeneutic altogether: their personal evaluation of things determines what they wish to accept from the Bible, and what not.

Now I come to my point. If AEH advocates tell me: We also accept the Bible as the authoritative Word of God, but we merely explain Genesis 1–3 differently, my answer is: No, our hermeneutics are not the same. I will show you why. You may put a Baptist, or a Pentecostal, or a Reformed person in a room with only a New Testament; further, you may supply a list of all the passages where the words "baptize" and "baptism" occur; and then give them enough time to figure out what is the most biblical view of baptism. Some will come up with infant baptism—others will come up with believers' baptism. You may put a Baptist, or a Pentecostal, or a Reformed person in

---

of an inevitable "theological crisis" (47), in which, according to him, the doctrine of original sin will be discarded.

a room with only a New Testament; further, you may supply a list of all the passages where the words "woman" and "women" (and "wife/wives") occur; and then given them enough time to figure out what is the most biblical view of the woman's position in the church. Some will conclude that women may not exercise official authority in the church—others will conclude, that women may exercise official authority in the church.

So far, so good. But now suppose you put a Christian in a room with a complete Bible, you point out Genesis 1–3 in a special way, and ask what picture emerges of the origin of humanity. That believer may come up with various answers, *but one thing that believer will never say: Adam was an early hominid, or there was a population of advanced hominids and Adam was chosen from that larger population of hominids.* A Christian can discover and lay hold of such assertions only *outside the room*, namely, when one decides to be influenced by assumptions and conclusions that, by nature, belong to an entirely different sphere than the Bible, or are even hostile to it. Here, then, it is no longer the Bible alone that decides, but the Bible as read through the lens of modern science. At this point, a new hermeneutic is definitely at work.

3.5 PERSPECTIVISM

3.5.1 CONTRAST WITH CONCORDISM

Gijsbert van den Brink discusses extensively two ways of attempting to sail between the Scylla and the Charybdis, between, on the one hand, the data of modern science, and, on the other hand, the exegetical details of the Bible (cf. §1.5). According to him, the first of these two ways is *concordism*, which he defines as follows: "*Concordism is the hermeneutical view that biblical statements about the physical world correspond with scientific facts.*"[43] The second way is *perspectivism*, which he defines as follows: it is "*the hermeneutical view that, in interpreting the Bible, we must distinguish between the theological contents*[44] *and the worldview in which these contents lie embedded.*"[45] van den Brink tells us that he finds perspectivism an "attractive position,"[46] and understandably so. We must never read scientific facts into the Bible,

---

[43] van den Brink (2017, 114); cf. Collins (2011, 106–11).

[44] Strictly speaking, the term "theological contents" is a form of scientism here; "pistical contents" (or "faith contents") would be more correct. "Pistical" refers to matters of faith, "theological" to the *science* that deals with the matters of faith. Cf. chapter 4, note 56.

[45] van den Brink (2017, 122).

[46] Ibid., 128.

as did, for instance, the concordist Henry M. Morris.[47] I see this mistake in the Chicago Statement on Biblical Hermeneutics, which speaks of the "scientific accuracy of the early chapters of the Bible" (Art. XXII).[48]

However, the subject of the biblical worldview (see the second definition) is a tricky one. Are we just talking of the Bible writers' observational perspective, which was the same as ours, so that we, too, speak of the sun setting and rising?[49] Or does it go much further than this? I am afraid that in the most extreme sense, everything in the Bible that conflicts with the findings of modern science gets assigned to this (ancient, outdated) worldview, and thus may be quietly dismissed. Sometimes this ancient worldview is called ancient science by some of the authors in Barrett and Caneday,[50] who use the word "science" in an anachronistic way. Indeed, it is mistaken to call the Bible's references to nature "scientific facts," as does Morris; but it is equally mistaken to do the opposite, namely, refer to the Bible's references to nature as "ancient science," and thus dismiss them out of hand. Both views are equally *scientistic*; they entail an overestimation of science, as if all ordinary references to nature and the cosmos are, or ought to be, of a natural-scientific nature—and as such are either correct or outdated.[51]

In other words, van den Brink's definition of perspectivism does not tell us enough. Apparently, there is a moderate version and an extreme version of perspectivism. The latter dismisses the Bible's references to nature and early history as unreliable and outdated. The terminology is confusing. For instance, compare C. John Collins' distinction between, on the one hand, "literalistic" (scientistic) concordism, which he called "improper," and on the other hand, "historical concordism," which he called "proper."[52] My impression is that what he called historical concordism corresponds to what I just called moderate perspectivism. van den Brink recognizes only concordism and perspectivism; I (apparently with Collins) recognize an intermediate form (historical concordism or moderate perspectivism), one that does not deal with the Bible's references to nature and early history as

---

[47] Morris (2002).

[48] Cf. Geisler (1983) (http://www.bible-researcher.com/chicago2.html); cf. the Chicago Statement on Biblical Inerrancy (Art. XII), which speaks of "history and science" where it should speak of "history and nature" (http://www.bible-researcher.com/chicago1.html).

[49] Cf. van den Brink (2017, 111).

[50] Especially Denis O. Lamoureux in Barrett and Caneday (2013, 50, 179, 235); he also spoke of "ancient biology" (62), "ancient geography," and "ancient astronomy" (122, 235).

[51] N. L. Geisler in Radmacher and Preus (1984, 314) claimed implicitly that the language of Genesis is "scientific"; he appears to mean simply "correct, accurate."

[52] C. J. Collins in Barrett and Caneday (2013, 76–77, 192; cf. 244).

(correct or outdated) scientific facts, but definitely as historically truthful statements (see Appendix 2).

If we do not distinguish these *three* categories, we can easily imagine that extreme perspectivists (in Appendix 2, this is category 3) accuse the *middle* group (Appendix 2, category 2: historical concordists or moderate perspectivists) of scientific concordism (Appendix 2, category 1), as does Denis Lamoureux with respect to both John Collins and William D. Barrick.[53] Lamoureux's argument is rather simplistic: the Bible contains "ancient science" (such as the 3-tiered universe; see §§10.4.1 and 10.6.1), and teaches an historical Adam; in his view, Collins and Barrick are inconsistent because they reject the former but accept the latter. My reply is that Lamoureux himself is inconsistent: the Bible supposedly contains "ancient science" and teaches eternal salvation; Lamoureux rejects the former but accepts the latter. Where lies the basic difference between him, and Collins and Barrick? (For more on this see §10.4.1.)

In my view, Collins' reply was fully justified: "I have given literary reasons for rejecting Lamoureux's approach to Genesis; quite simply, to see it as 'science,' ancient or otherwise, is a colossal mistake."[54] Astonishingly, both "literalistic" (scientistic) concordists, such as Morris, and extreme perspectivists, such as Lamoureux, are making the identical mistake in this respect: they read science into Genesis 1–11. The former scientists do so to underscore how very accurate the Bible is, "even scientifically"; the latter scientists do so to reject all "ancient science," and thus to undermine the historical reliability of Scripture. Group 2 (historical concordists or moderate perspectivists) emphatically replies: there is no science in Genesis 1–11 but there are many (non-scientific) references to nature and the cosmos here (within the framework of what Genesis wishes to tell us about God, humanity, sin, and redemption). *And these references are always basically correct* (with some nuances that I will explain in the next chapter). To call references to nature "scientific" (as if only science has the prerogative to speak about nature) is one of the basic mistakes of Lamoureux and so many others.

3.5.2 THE SHIFT

Where are the precise boundaries between the three categories mentioned in the previous section? This is a bit like the Word of God, which is "piercing to the division of soul and of spirit, of joints and of marrow" (Heb. 4:12).

---

[53] In Barrett and Caneday (2013, 177–179, 230); notice Barrick's reply (253): "I do not believe that the Bible is a book of science."

[54] In Barrett and Caneday (2013, 193).

What the Word can finely distinguish and separate, we cannot. In line with this, I ask: Who is the scholar who ventures to disentangle the "theological contents" of certain parts of the Bible *from* the ("ancient, outdated") "worldview in which these contents lie embedded"?[55] As Kevin DeYoung observed:

> The Bible does not put an artificial wedge between history and theology. Of course, Genesis is not a textbook for the historical sciences or a science textbook, but that is far different than saying we ought to separate the theological wheat from the historical chaff. Such a division owes to the Enlightenment more than to the Bible.[56]

About one point van den Brink is right: this is a *hermeneutical* matter.[57] I endorse a perspectivism that does not treat biblical data as scientific facts but always analyzes them from the observational perspective. I would put it this way: the Bible does not contain scientific facts (nor should they be read into the Bible), but when the Bible speaks about nature, the cosmos, and history, it always does so correctly[58]—not according to the criteria of the natural or the historical sciences, but according to the criteria of ordinary, everyday, pre-scientific observation. It does so not only correctly, but also always from a *faith* perspective. That is, the Bible never intends to give us information about nature, the cosmos, and history as an end in itself—it always consistently teaches us matters of *God*, *faith*, and *salvation*. But if, in this religious context, it touches upon matters of nature, the cosmos, and history, it does so correctly from a practical, observational point of view.[59]

We must be precise here. John H. Walton is basically correct when he says, "The Bible is not revealing science, it is revealing God."[60] But what

---

55 van den Brink himself is clearly aware of this problem (2017, 132).

56 http://www.alliancenet.org/mos/1517/why-it-is-wise-to-believe-in-the-historical-adam#.WaaZAq2iGMJ.

57 Cf. Enns (2012, 126–27).

58 This phrase comes from the Statutes of the Evangelische Hogeschool (Evangelical College), of which I was one of the founders and lecturers. The thought goes back to Francis Schaeffer (1982, 2/1: *Genesis in Space and Time*). See also W. J. Ouweneel in Dekker et al. (2007, 146–50).

59 Cf. the Chicago Statement of Biblical Inerrancy: "Being wholly and verbally God-given, Scripture is without error or fault in all its teaching, no less in what it states about God's acts in creation, about the events of world history, and about its own literary origins under God, than in its witness to God's saving grace in individual lives" (http://www.bible-researcher.com/ chicago1.html).

60 J. H. Walton in Barrett and Caneday (2013, 116).

exactly does he mean? Are all biblical references to nature and cosmos examples of "ancient science" (see previous section), and therefore worthy of rejecting? I would put it this way: the Bible is not revealing the nature of the natural world as such—it is revealing God. But *within this framework* it also necessarily makes statements about nature, the cosmos, and history, and it does so in a reliable way, not from a modern scientific viewpoint but from an observational point of view (again, with literary nuances that I will explain in the next chapter).[61]

When I describe in this chapter the hermeneutical shift that has occurred in the position of AEH advocates, I am not dealing with the claim that biblical data may never be treated as scientific facts. I take this claim for granted, as I have explained many times.[62] The shift instead lies here: apparently, *everything* in the Bible that does not correspond with the findings of modern science—especially with the theory of evolution—is assigned to the ancient, outdated worldview of the Bible, which can therefore be quietly dismissed. It is quite confusing that van den Brink assigns these very different positions to the single position of perspectivism. I see a *cleft* between these two positions. I have always defended the moderate position, but I have always rejected the second, extreme position. The hermeneutical shift occurs when expositors move from the former to the latter position, falsely creating the impression that they are remaining within the framework of ordinary perspectivism.

In the words of John A. Bloom:

> The Bible focuses on salvation history and only fills in cultural and background details as they are relevant to the immediate context, not for the curiosity of readers thousands of years later. Nevertheless, the events that are described should be understood as real and historical, especially in light of their implications for mankind. For example, the institution of monogamous marriage, the origin of sin and the resultant corruption and mortality of mankind, are all grounded in God's interaction with a single male-female pair of human beings, from whom all mankind are descended. If anthropology forces us to replace Adam and Eve with a global *homo erectus* population that collectively evolved into modern man, then the theological foundation for the

[61] I am not talking—*contra* van den Brink (2017, 122n28)—about such trivial matters as whether Jesus was right when he called the mustard seed "the smallest of all seeds" (Matt. 13:32); presumably, Jesus was referring merely to familiar garden plants, as many expositors have pointed out.

[62] See Ouweneel (1987).

nuclear family, sin and death appears to be eroded. The credibility of the Bible when it speaks on these issues seems to be damaged: If it does not correctly explain the origin of a problem, why should one trust its solutions?[63]

3.5.3 FALK'S PERSPECTIVISM

Van den Brink was not the first to make the mistake of classifying two very different positions under the single heading of perspectivism. I found a striking earlier example with Darrel R. Falk. Just as van den Brink did later, Falk wished only to distinguish between two positions, thus obscuring the real picture. On the one hand, he refers to the view of "reducing the Bible to a scientific textbook";[64] this is what van den Brink and others call concordism. They reject this, and so do I. On the other hand, Falk's own view downplays the historical meaning of Genesis 1–3 by stressing "the use of figurative language, with its rich symbolism and deep meaning."[65]

There you have it: the language of Genesis 1–3 is either scientific language or figurative language. According to Falk, there is no third way (Lat. *tertium non datur*). Let us call Falk's two options the *scientistic* and the *figurative* views. Like so many other AEH advocates, Falk fails to see the third option, which could be called the *historical* view. This option has three important characteristics.

(a) This view rejects the absurd notion that Genesis 1–3 speaks scientific language, or aims to teach us scientific models (*contra* the scientistic view, which in fact is also historical, but in a wrong way).

(b) This view allows for metaphors, anthropomorphisms, and other figurative language in these chapters (*contra* any form of biblicistic literalism; see the next chapter).

(c) Yet the historical view maintains the fundamental historical character of Genesis 1–3 (*contra* the figurative view). Once again, if these chapters touch upon matters of nature, the cosmos, and history, they do so correctly from a practical, observational point of view (this is concordistic)—never as an end in itself (*contra* scientism), but to teach us spiritual truths (this is perspectivistic).

In (rightly) opposing the scientistic view, both Falk and van den Brink fail to distinguish between what I am calling the figurative and the historical views. A serious consequence of this erroneous approach is that Falk

---

63 http://www.asa3.org/ASA/education/origins/humans-jb.htm.

64 Falk (2004, 36).

65 Ibid., 32.

quotes some great teachers from church history, whose views he adduces as evidence for his own view: Augustine, John Calvin, and John Wesley.[66] It is my conviction that what these men were defending was the same historical view that I am defending: Genesis 1–3 are basically historical, but not in any scientistic sense. Because Falk does not distinguish between the figurative and the historical views, he thinks that Augustine, Calvin, and Wesley support his view, whereas in reality they support the historical view as I have described it. That is, they are against what we today call scientism, and they freely allow for metaphors and anthropomorphisms—yet, they maintain the historical character of Genesis 1–3.

3.6 CONSEQUENCES OF THE NEW HERMENEUTIC

3.6.1 A HIGH PRICE

The false presentation of things that we have just described explains why we are dealing here with a different hermeneutic. Many Christian denominations exist today largely because they read or apply the Bible in different ways. If you ask them to explain why they interpret the Bible as they do, they may send you letters, brochures, books, and website addresses, all of which explain their reasoning. Such writings and websites contain numerous *biblical* arguments. I know of few denominations that appeal to extra-biblical authorities, with one exception: those who appeal to dreams, visions, prophecies, and the like, which seem at times to be more authoritative for them than God's written Word. But where is the denomination that tells us that God has now revealed to them, either through a new reading of certain Bible passages, or through visions and prophecies, that Adam was an early hominid? You get my point. This is why I say that *a different hermeneutic* is operating here. This approach to the Bible argues this way: We always thought that we should read Genesis 1–3 in a certain way, but thanks to modern science we now realize that we should read these chapters in a different way.

Again, somebody might object that this is like saying: I used to read many Bible passages in a way that seemed to suggest that the earth was flat and was the centre of the universe, and that the sky was an enormous dome. But through modern science I now realize that in reality the earth is spherical, and that there is no dome over the earth.[67] You may wonder whether this is

[66] Ibid., 34–35. Cf. Augustine (1982).

[67] This argument is used by AEH advocates Falk (2004, 26–31) and van den Brink (2017, 113–14, 138–39).

correct. Does the Bible indeed teach a flat earth with a celestial dome over it? Opinions differ. But this is not the point right now. The point is, first, if you indeed believed that the earth is flat,[68] what difference would this make for your understanding of any biblical subject, any topic in systematic theology? None whatsoever. Second, what do you wish to conclude? That the Bible was wrong about such cosmological matters, and that through science we now know the proper view? No. The Bible was not written to teach us any form of scientific cosmology. Perhaps the ordinary experience of the Bible writers was indeed that the earth is flat, and the sky is a dome. *This is precisely what the earth and the sky still are in our ordinary experience*. A person who says, "The sun rises (or sets) at such-and-such a time," or "The Netherlands is a very flat country," is still using the ordinary worldview of the Bible writers. This is not at all surprising.

However, nowadays I encounter people who are saying, "We used to believe that at the beginning of the world there was one single pair of humans who were not born from other humans but formed directly by God, and from whom all humanity has descended—but through modern science we now know that this is not a correct understanding. Such a pair never existed. The animal beings we call humans descended over millions of years from other, less rational, less developed, and less moral hominids. Judged by reliable historical standards, the Bible is simply mistaken."

Please note, at this point we are not discussing whether the Bible does or does not teach science, or some ancient, ordinary, observation-based worldview. Rather, at this point the matter is far more serious because it involves important biblical subjects. Now the question has become whether we can maintain not only the historical significance of Genesis 1–3, but also the many doctrinal teachings as stated in the New Testament, like the following.

(a) Adam did not at all live at the "beginning of creation" (Mark 10:6; cf. Luke 11:30–51; John 8:44). Jesus was wrong.

(b) Adam (whoever he, or it [a population?], was) was not at all "formed" (1 Tim. 2:13)—which is a technical term referring back to Genesis 2:7, "God formed the man [Heb. *ha-adam*] of dust from the ground"—but was born from earlier hominids. Paul is wrong.

(c) Adam was not at all the "first man," more specifically a "man of dust" (1 Cor. 15:45–49), but the product of millions of years of hominid evolution. No fashioning from dust was involved, and certainly not a "first man" (unless one would speculate that, through a divine miracle, two *animal*

[68] Cf. https://theflatearthsociety.org/home/.

hominids produced the first real *human*). Paul is wrong again.

(d) Paul tells us that Adam was not born of a woman, unlike every person *after* Adam was (1 Cor. 11:8–12), but he is wrong again: whatever or whoever Adam was, it or he was born of a female, like the rest of us.

(e) Paul tells us that, at the Fall, Adam was not deceived, but the woman was deceived and became a transgressor (1 Tim. 2:14; cf. 2 Cor. 11:3). Paul is wrong again. Adam and Eve[69] (whoever they were) both fell, and this over a longer period of time. There was no question at all of the hominid "Eves" being deceived, and the hominid "Adams" not.

(f) Paul tells us that death came into the world through Adam (Rom. 5:12–14), but again he is totally mistaken. Let us assume that he means human death, not physical death in general. But even then, death among hominids had existed already for millions of years, and it was present before any hominid existed. There was no Adam (no matter how understood) through whom death entered the world; when the first man (no matter how understood) sinned, death already existed in the world.

The person who views Adam as a hominid, or as someone who belonged to a population of hominids, must pay a high price: that person must argue away every one of these five statements by the apostle Paul, plus a number of sayings by Jesus (§4.1.1). In what follows, we will see how advocates of this new hermeneutic try to rescue all these statements (something that in my view is an impossible task for AEH advocates). But at this point, I would declare that I cannot accept a theological theory that, first, leans so heavily for its conclusions on modern science, and second, pays such a high theological price for doing so.

Incidentally, let me summarize here what, apparently, is the Pauline understanding of the historical Adam. Adam was the first man on earth, formed by God (not from an animal but from dust), not born of a woman (although all women in the world sprang from him), not deceived (his wife was deceived) but consciously following his wife in the Fall. He was the man through whom human death entered the world. Between the historical Adam identified by Jesus and Paul (and by Genesis 1–3, of course) and the historical Adam designed by AEH advocates there is a deep, dark cleft.

Did Jesus and Paul refer merely to some literary figure, and not to a historical person (cf. §10.2.3)? If so, why did they not clearly tell us? Was it so difficult to tell us who Adam and Eve really were? And were Jesus and Paul

---

[69] Some may insist that we should not use the name "Eve" when describing events before Gen. 3:20 (which tells us that she was formed from Adam's side); but notice that in 2 Cor. 11:3 Paul does so.

themselves aware of what they were claiming? Were they themselves deluded, or were they trying to delude us? If we cannot trust them when they speak about Adam, how can we trust them when they speak about God?[70]

3.6.2 A VERY STRANGE POSITION

There is something very strange about this entire enterprise being pursued by AEH advocates. Why are they so eager to do something nobody has ever tried before: to "reconcile" (to use the term employed by James K. A. Smith; see §1.4.2) modern science and Scripture?[71] On whose behalf are they pursuing this endeavor? Certainly not to impress atheist evolutionists, who haven't the slightest desire for such a "reconciliation," such "coming to peace," such a "search for common ground," such a "rapprochement," such a "synthesis," and would probably denigrate its result (as Dawkins indeed did; cf. §1.4.3). In their view, Genesis 1–3, together with all of Genesis and much of the Bible, is mythical anyway. How could anyone with intellectual credibility try to amalgamate such myths and the results of modern science?

Instead of "reconciliation," "rapprochement" or "synthesis," James P. Moreland offers an interesting alternative: a "truce."

> Theistic Evolution is intellectual pacifism that lulls people to sleep while the barbarians are at the gates. In my experience theistic evolutionists are trying to create a safe truce with science so Christians can be left alone to practice their privatised religion while retaining the respect of the dominant intellectual culture.[72]

A second group that will be unimpressed by the attempts of AEH advocates consists of liberal theologians. They surrendered any historical significance of Genesis 1–3 (or 1–11) long ago. Karl Barth, Emil Brunner, and other like-minded theologians were called "neo-orthodox," but to many critics their views simply consisted of a new, more sophisticated form of liberalism. These theologians had no problem with a historical Adam, for under the influence of modern science, they relegated the stories about

[70] See Enns (2012, 93–117) on "Paul as an ancient interpreter of the Old Testament."

[71] For this terminology, cf. Falk (2004, title: *Coming to Peace with Science*); Miller (2007, subtitle: *A Scientist's Search for Common Ground Between God and Evolution*); Carlson and Longman (2010, title: *Science, Creation, and the Bible: Reconciling Rival Theories of Origins*; cf. 137: "*There is now rapprochement between science and Christian faith*"). Cf. Enns' term "synthesis" (chapter 1, note 37).

[72] Moreland (2007, 46); cf. biologist Rossiter (2015), who analyzes the theology of some theistic evolutionists such as Francis Collins (2006), Karl Giberson (2009), Kenneth Miller (2007; 2009), and John Polkinghorne (2011).

Adam's creation and the Fall to the realm of the supra-historical. These stories were not about historical events, that is, events that occurred in time and space. Adam is Everyman, and the Fall is Everyman's fall into sin; we all have fallen, and still fall.[73] Therefore, British Christian evolutionist Arthur Peacocke openly and unashamedly wrote, "The traditional interpretation of the third chapter of Genesis that there was a historical 'Fall,' an action by our human progenitors that is the explanation of biological death, has to be rejected."[74]

A third group that will be unimpressed by the efforts of AEH advocates are Christians who continue to maintain the traditional Jewish and Christian approach to Genesis 1–3, and do not replace this approach with an amalgamation of biblical exegesis and natural science. Reformed theologian Edward J. Young refuted the ideas like those defended by, for instance, Peacocke:

> When Adam sinned, he fell from an estate of being good into an estate of being evil. He was created by God as a creature of whom it could be said that he was "very good" [Gen. 1:31]. From this estate in which he was created by God he fell into an estate of sin and misery and by his disobedience plunged all men into that same estate of sin and misery. Furthermore, by my sin I did not fall from an estate of being "very good" into an estate of evil. I and all men like me were born into that miserable estate of sin, and when we sinned we simply showed that we were in such an estate. By sinning Adam became a sinner; by sinning we do not become sinners, we are already sinners. Sin does not cause us to fall from the estate wherein we were created, for we were born into a fallen estate. With Adam, however, the case was quite different. His sin brought him into a fallen estate. By disobedience he fell; by disobedience we simply show that we are already fallen. Hence, the experience of Adam was unique; it is his experience alone and not that of myself or of every man.[75]

To be sure, the position of AEH advocates does not go quite as far as that of the neo-orthodox theologians; in some way or another, AEH advocates

---

[73] Cf. Versteeg (2012, 22–23): "Just as [in AEH] the figure of Adam becomes the idea of 'man,' so Christ too, according to [Rudolf] Bultmann, appears to become an idea. When Christ is called the 'true man' by [Karl] Barth, he is still not the concrete historical man but the idea of the 'true' man."

[74] Peacocke (1993, 222).

[75] Young (1966, 60–61).

are attempting to maintain some notion of a historical Adam, but one that would be acceptable to modern science.[76] This is very strange; again, the AEH model is neither fish nor fowl. *Either* you believe in the historical Adam, as thousands of years of Jewish and Christian scholars have done—with endless variations, I know, but in this basic form: they accepted the Adam whom we know from Genesis 1–3 *and* (as far as Christians were concerned) from the Gospels and from Paul's letters. *Or* you say, along with honest liberals and nuanced neo-orthodox theologians: in light of modern science, forget about the historical Adam of theological tradition; Adam is Everyman, and his fall is Everyman's fall. But by way of contrast, this is the stratagem of AEH: they leave room for the theory of evolution *and* they claim to accept the historical Adam! Everybody should be happy. This Adam is not exactly the Adam with whom we are familiar, but he is instead an evolved hominid. Meanwhile, you can still tell the general public that you are continuing to uphold Genesis 1–3!

This is like claiming that I still believe in both capitalism *and* socialism. To be sure, neither my capitalism nor my socialism resembles capitalism and socialism as we have commonly defined them; yet I believe in both. As Denis Alexander asks, "Creation or Evolution: Do We Have to Choose?"[77] Of course, in his view you do not have to—as long as you do not understand by "creation" what Jewish and Christian expositors have believed for thousands of years ("For he spoke, and it came to be; he commanded, and it stood firm," Ps. 33:9). But neither do you understand by "evolution" what ordinary scientists believe about it, for the latter haven't the slightest desire to amalgamate their scientific notion of evolution with your religious notion of creation.

In my view, the reality is this: the evolved hominid defended by AEH advocates has nothing to do with Genesis 1–3 and with the Adam of Paul's epistles. Advocates of AEH have sadly fallen into the crevasse between two, or even three, positions: the position of common (agnostic or atheist) evolutionism, the position of liberal theology, and the position of traditional theology. They do not have modern scientists on their side, for these would jeer at their amalgamation. They do not have liberal theologians on their side because they tell AEH advocates, "Forget about the historical Adam; it's all myth." And they do not have those Jewish and Christian theologians on their side who, generally speaking, look at Genesis 1–3 *and* Jesus' and Paul's

---

[76] Cf. Reformed theologian Bruce Waltke (2007, 203). Waltke adheres more closely to the biblical text than does van den Brink, but he still accepts the theory of evolution.

[77] Alexander (2014, title).

declarations in the same way they have been understood for thousands of years. These theologians may be wrong, but if so, then Jesus and Paul were wrong as well.

In the coming chapters, I wish to elaborate further on this very strange situation.

3.6.3 A SLIPPERY SLOPE?

I am not among those who want no changes out of fear that *everything* will end up changing. In the context of this discussion, people often speak of a slippery slope or the thin edge of the wedge. In the creation–evolution controversy I have encountered the phrase "slippery slope scare tactics."[78] One of the pillar convictions of all forms of conservatism is: do not move, for then you cannot possibly go wrong. Yet, despite this fallacy of some conservatives, the question must occasionally be asked about where a movement might lead. This is a biblical principle: "When they sow the wind, they will reap a storm" (Hos. 8:7 GNT). We are responsible for the effect of our words and deeds. Like the man who was sitting with his dog near a hot lava stream, and casually threw a stick into the stream; his dog instinctively jumped after the stick—and was burned to death. The dog was not responsible for its death, but its owner was.

As soon as a theologian wishes to adapt to modern science, we have the right to ask: Why? And where will this lead? If I were to engage in personal conversation with an AEH advocate, I would soon be asking what my discussion partner believes about miracles. I would not begin with the bodily resurrection of Christ, because that is probably the last dogma that a once-orthodox Christian will surrender. But I might certainly ask about the three resurrections of ordinary people recorded in the New Testament Gospels (Luke 7:11–17; 8:40–56; John 11:1–44), and then about three resurrections recorded in the Old Testament (1 Kings 17:17–24; 2 Kings 4:18–37; 13:20–21).

If the reader dismisses my questions as pure suspicion, I readily admit this. I am suspicious when a person begins to make room for modern science in one's theological thinking. I cannot imagine how a person could invite modern science to change one's understanding of Genesis 1–3 so drastically, and suppose one will continue believing in Joshua's lengthened day (Josh. 10:12–14), in Elisha's floating axe head (2 Kings 6:5–7), or in the regressing shadow on the dial of Ahaz (Isa. 38:7–8), *all events that are in*

---

[78] J. H. Walton in Barrett and Caneday (2013, 238); in that context, he correctly insisted, "It is an overstatement to claim that if you believe X, you will soon reject the resurrection."

utter conflict with modern science. Or how could such a person continue to believe that Jesus multiplied bread and walked on water (Matt. 14:13–32)? Is it so unfair to ask questions like these? Modern science is a false friend when it comes to understanding the Bible. You may be enjoying the company of evolutionists as you discuss Genesis 1–3, but they will still mock you when you disclose your belief in biblical miracles.[79] If the historical Adam and the historical Fall (both in the biblical sense) are scientifically impossible, then so too are not only Joshua's lengthened day but also the resurrection of Jesus.

Such embracing of modern science reminds me of the old story about an Arab, who allowed his camel to stick its nose into his tent ("because it was so cold outside"), followed by other parts of its body, until the camel was entirely inside the tent and refused to leave.[80] The slippery slope argument has also been called the "camel's nose" argument. Don't allow the nose in because the rest of the camel will inevitably follow.

Of course, the resurrection of Christ is closer to the heart and scope of Scripture than the resurrection of the dead man whose corpse was thrown on Elisha's bones. But is Jesus' walking on the water also closer to this scope than Elisha's axe head floating on the water? Are Jesus' fishing miracles closer to this scope than Jonah's fish miracle? Are the miracles of the apostles closer to this scope than the miracles of Moses in Egypt? Who determines these answers? Again, it is someone's hermeneutic that determines the answer(s) to such questions. The Bible is *always* about God, salvation, and faith, in its broadest scope. This scope is *always* Jesus Christ, and the realization of God's counsels in him. However, *within this scope*, the Bible also frequently speaks of history, the cosmos, and nature. Such statements are never intended to teach us historical science or cosmology or natural science—this claim is such a platitude. However, *wherever* the Bible, within the framework of its scope, speaks about history, the cosmos, and nature, it always does so in a truthful, trustworthy way.

We take the relevant Bible passages seriously, also in terms of their historical dimension. And we do not allow any extra-biblical source to make these passages say something different from what Jews and Christians have always believed they said. Bible interpreters may have differed on many details throughout history. For instance, AEH advocates love to quote the church father Augustine, who made some remarks on Genesis that seem to fit their position. However, it will be very hard to find many rabbis or

---

[79] About biblical miracles, see Berkouwer (1952, chapter 7); Lockyer (1988); Brown (2006).
[80] Nunberg (2009, 118).

theologians living before the time of Enlightenment who did not believe in the historical Adam as Paul understood him.

Let me end this chapter with what American systematic theologian Lewis Sperry Chafer had to say about Paul's passages.

> Not one of the [Pauline] passages [on Adam and Eve] presents a rhetorical allusion. They are rather the basis of sound reasoning and the ground of far-reaching doctrine which is altogether sacrificed if the events recorded early in Genesis are not more than fable [including the creation and lapsarian myths of AEH]. The only motive that promotes argument against the historicity of these Mosaic records is that they seem absurd since, as it is claimed, they are unlike present human experience; but such reasoning not only assumes that God is restricted to those modes of operation which are current today, but that man is free to sit in judgment upon the Word of God"—which is in fact what happens if "modern science" is allowed to rule over Genesis 1–3.[81]

[81] Chafer (1983, 2:206).

4

THE PROBLEM OF HISTORICITY

For Adam was formed first, then Eve; and Adam was not deceived, but the woman was deceived and became a transgressor (1 Timothy 2:13–14).

THESIS

If Adam and Eve, as well as the Fall, were historical, they were so neither in the sense of journalistic practice nor of the historical sciences, nor of biblicistic literalism. Yet, Adam and Eve were really the first pair of humans in history, and the Fall was a genuine event that took place during their lives. This claim has essential importance for a proper theological understanding of the Bible as a whole.

4.1 IS GENESIS 1–11 HISTORICAL?[1]

4.1.1 JESUS' TESTIMONY

The narrative parts of the Bible are historically reliable, from beginning to end. This is because God's salvation, which begins in Genesis 3, does not hang suspended in some unhistorical or supra-historical vacuum, but lies embedded in biblical redemptive *history*. The apostles of Jesus Christ did not just proclaim some nice story of love and righteousness, but they proclaimed a *historical fact*: the bodily resurrection of Christ (see especially Acts 2–5).[2] Just as much as the Christian redemptive message

[1] Parts of this chapter have been adapted from two earlier publications: Ouweneel (2008; 2012).

[2] See extensively, Ouweneel (2007b, §12.3).

lies embedded in the historical facts of Jesus' life, atoning death, and bodily resurrection, so too Judaism lies embedded in the historical facts of the exodus from Egypt and the entrance into Canaan. The great Jewish and Christian festivals do not celebrate abstract principles but historical facts:[3] the exodus (Pesach), the lawgiving on Mount Sinai (Shavuot), the wilderness journey (Succoth), the deliverance from Haman (Purim), the rededication of the temple (Hanukkah), the birth of Christ (Christmas), his death (Good Friday), his resurrection (Easter), the coming of the Holy Spirit (Pentecost), and Christ's return to heaven (Ascension Day).

I mention these things to indicate the thoroughly historical character of Judaism and Christianity (very different from, e.g., animism, Hinduism, and Buddhism). Judaism and Christianity are based on history, on events that really happened.[4] The *description* of this history may be prosaic or poetic, objective or subjective, scholarly or non-scholarly, concise or detailed, but its basic purpose is to tell us about real persons and real events.

One consequence of the Bible's thoroughly historical character is that we cannot arbitrarily restrict historical reliability within the Bible, as if the historical reliability of the Gospels is highly important, but that of the Torah (the Pentateuch) is not, or is less so. Some would like to restrict historical reliability within the book of Genesis: from chapter 12 we are (mainly) on historical ground, before chapter 12 we are (mainly) on mythical, or stated more cautiously, on pre-historical (or even supra-historical) ground. This is not valid. To be sure, we believe *in Christ*, and not *in* Adam or *in* Noah. However, if we believe in Christ, we will also take seriously what he himself told us *about* Adam and Noah (see Genesis 1–11; more extensively §10.1).

(a) Jesus placed Adam and Eve at the "beginning of creation" (Mark 10:6; cf. Matt. 19:8). Also compare the word "beginning" in John 8:44, "your father the devil.... He was a murderer from the beginning," which refers to the time of the Fall—near the beginning of human history. Quite interesting statements! Was this merely a way of adapting or accommodating to simple souls who did not understand what had really happened? In other words, would Jesus today express himself very differently to us? For Jesus, the beginning of the world and the beginning of humanity more or less coincided; in Genesis 1, we read of merely five days between these events. And in Genesis 3 (which records events that occurred perhaps no later than the eighth day!), we are still at the "beginning," as Jesus explains.

---

3 Cf. C. J. Collins in Barrett and Caneday (2013, 159): "This is why 'history' matters: Biblical faith is a narrative of God's great works of creation and redemption, and not simply a list of 'timeless' principles."

4 Cf. Long (1994); C. J. Collins in Barrett and Caneday (2013, 146–48).

(b) We find a similar statement in Luke 11:50–51, where Jesus speaks of "the blood of all the prophets, *shed from the foundation of the world*," that "may be charged against this generation, from the blood of Abel to the blood of Zechariah." Jesus clearly declares that Abel's death virtually coincided with the "foundation of the world"; the events of Genesis 4 occurred relatively shortly after those of Genesis 1. Moreover, Jesus apparently thought Abel was just as historical as Zechariah, who is mentioned in 2 Chronicles 24:20–22.

(c) Let us look at Abel a little more closely. Jesus spoke of "the blood of the righteous Abel" (Matt. 23:35)—but who could this Abel possibly have been if Adam had been a chosen hominid individual or the member of a selected hominid population? "By faith Abel offered to God a more acceptable sacrifice than Cain, through which he was commended as righteous" (Heb. 11:4; cf. 12:24; Jude 11). "We should not be like Cain, who was of the evil one and murdered his brother. And why did he murder him? Because his own deeds were evil and his brother's righteous" (1 John 3:12). What could possibly be the historical sense of such statements if Adam had been a chosen hominid? What can AEH advocates tell us about these brothers, Cain and Abel, sons of Adam and Eve?

(d) Jesus spoke about Noah and his days as real historical events (Matt. 24:37–39; Luke 17:26–27).[5] The apostle Peter also spoke about Noah as a historical person and about the Flood as a historical event (1 Pet. 3:20; 2 Pet. 2:5; 3:5–6) "in the same historical way that he believed that Abraham's nephew Lot escaped the destruction of Sodom in Genesis 14 (2 Pet. 2:5–8)."[6] However, if our historical understanding of Genesis 1–3 must be adapted to the findings of modern science, what about the story of Noah's Flood?

4.1.2 COMMENTS

Based on the testimony of these opening chapters of the Bible, we have sufficient *a priori* reasons to accept the historicity of Genesis 1–11, but we are supported by the way Jesus and the apostles treated these chapters. The New Testament speaks with obvious acceptance of creation, the Fall, and the Flood as historical events. How could Jesus and the New Testament writers do this? I see only three options:[7]

---

5 Some have thought that Job 22:15–16 also points to the Noahic Flood: "Will you keep to the old [Heb. *olam*, ancient, possibly: pre-Flood] way that wicked men have trod? They were snatched away before their time; their foundation was washed away."

6 D. Anderson in Nevin (2009, 81).

7 These remind us of the three options that C. S. Lewis (1952, 54–56) furnished regarding the person of Jesus.

1. *Jesus accommodated.*[8] That is, he adapted to the ordinary way Genesis 1–11 was interpreted.

However, this seems to suggest that in his heart he knew better. But would this not make him a deceiver? How could we believe anything else he said, if we had to accept that Jesus occasionally told us things that he himself knew were actually very different? And why would he do so? Why could he not tell his listeners that Adam had been chosen from the human world at the time, just as afterward Noah and Abram had been chosen from among *their* respective generations?

2. *Jesus deluded himself.* He himself also accepted the traditional approach to Genesis 1–11, which modern science has now supposedly shown to be mistaken. Jesus did so because he himself, as Man and as an ordinary Jew, did not know any better.

Now I do know that, as a human person, Jesus did not know the day of his own return (Mark 13:32); but would he also not have known the truth about Adam and Eve, Abel and Seth, Enoch and Noah? In, through, and for him God created the world (John 1:1–3; Col. 1:15–17; Heb. 1:1–3); the Son of God existed when these people lived. He said, "Truly, truly, I say to you, before Abraham was, I am" (John 8:58). And would he, God the Son, not have known the truth about Adam and Eve, Abel and Seth, Enoch and Noah?

3. *Jesus spoke the truth.* He accepted the traditional approach to Genesis 1–11 because he knew these chapters to be historically accurate. I strongly prefer this third option. Jesus understood these chapters to be basically describing historical events. Therefore, we should never say that Genesis 1–11 are not really historical in the ordinary sense of the word, but that instead they contain good stories that Jesus and the apostles used for illustrating certain spiritual truths. Ostensibly, these Genesis stories would not be essentially different from Jesus' parables. However, Genesis 1–11 are not presented as didactic, ethical stories but as descriptions of real persons and events.[9]

Here is another example. We cannot accept the testimony of Hebrews 11 if we must assume that some of its heroes were historical figures (from Abraham on), whereas Abel, Enoch, and Noah were not.[10] The apostle Paul cannot place Adam in opposition to Christ as two representative heads

---

[8] This use of the term goes back especially to Calvin; see Huijgen (2011).

[9] This is why some expositors believe that the account of the rich man and Lazarus (Luke 16:19–31) is not a parable—the text does not use this word—because real persons (Lazarus, Abraham, Moses) are mentioned; thus A. MacLaren (http://biblehub.com/commentaries/luke/16-19.htm).

[10] Cf. Collins (2011, 90–91).

the way he does (Rom. 5:12–21; 1 Cor. 15:45–49)—the first Adam and the last Adam, the first man and the second man—if Adam is not at all the historical first ancestor of all humans that ever lived (see chapter 10). As David Anderson wrote,

> The supposed "gap"...between the non-historical theological stories of Genesis 1–3 and the true stories of Abraham, David and Jesus simply does not exist. The Bible interweaves them all into a seamless whole, and they must remain together or we must reject or ruin the whole garment.[11]

4.1.3 THE ORIGIN OF THE PENTATEUCH

It is important to notice that the claims in the previous sections about how Jesus and the apostles treated the early chapters of Genesis may be fully maintained, irrespective of one's views concerning the origin of the Pentateuch, and of Genesis in particular. Some AEH advocates, like Gijsbert van den Brink, pay scant attention to theories about the Pentateuch's origin. But BioLogos member Peter Enns gave the matter a lot of attention, especially to discredit the historical reliability of the books.[12] If Genesis was written after the Babylonian exile, its contents are less historically trustworthy than if it had been written by Moses. Yet, Jesus and the apostles assumed the historical accuracy of Genesis.

Jesus also accepted that Moses had written (at least part of) the Pentateuch: "Do not think that I will accuse you to the Father. There is one who accuses you: Moses, on whom you have set your hope. For if you believed Moses, you would believe me; for he wrote of me.[13] But if you do not believe his writings, how will you believe my words?" (John 5:45–47).

I am fully aware of the so-called *post-Mosaica* (portions of text that must have been written [long] after Moses lived)[14] but these do not contradict the claim that the bulk of the Pentateuch comes from Moses.[15] Philip said to Nathanael, "We have found him of whom Moses in the Law [i.e., the

---

[11] In Nevin (2009, 81).

[12] Enns (2012, 9–34).

[13] That is to say, Moses wrote of the Messiah, and I am the Messiah (see Gen. 3:15; 22:18 [cf. Gal. 3:8]; 49:10; Num. 24:17; Deut. 18:15).

[14] Gen. 11:31 (Chaldeans); 14:14 (Dan); 19:37 ("to this day"); 36:31–39 (Edomite kings); Num. 22:1 ("beyond the Jordan"); Deut. 4:38; 8:18 ("as it is this day").

[15] See Schaeffer (1982, 2:123–30); Walvoord and Zuck (1985); McDowell (1999); and especially Cassuto (1983). See also the careful analyses by Gispen (1974, 11–19), Hamilton (1990, 11–38), Collins (2011, 167–70), Paul (2017, 135–140), and other conservative expositors. See also Arnold and Beyer (1999); Alexander and Baker (2003).

Pentateuch] and also the prophets *wrote*, Jesus of Nazareth, the son of Joseph" (John 1:45), and Paul says, "Moses *writes* about the righteousness that is based on the law" (Rom. 10:5). For our present purpose, however, the question whether Moses wrote the bulk of the Pentateuch is less essential than that Jesus and the apostles viewed the historical parts of the Pentateuch as fully trustworthy. They would never have come up with the idea that Genesis 1:1–2:3 and in the rest of Genesis 2 contain two different creation stories, which supposedly partially contradict each other.[16] For them, Genesis 2:4–25 was undoubtedly nothing but an elaboration of God's work on the sixth day of creation (Gen. 1:26–31). Thus, Jesus referred in Matthew 19:3–9 to both Genesis 1:27 and 2:24 as parts of one and the same story. Conservative expositors have always believed in this unity.[17]

One way of downplaying the historical significance and accuracy of Genesis 1–11 has been to compare it with ancient pagan accounts of creation and some kind of a fall and a flood. This technique has been used since the rise of historical criticism, but it is used also by, among others, Peter Enns.[18] For instance, we hear about the Babylonian Enuma Elish, the Gilgamesh epic, and the Atrahasis story. However, even if all these accounts were older than Moses and the beginnings of Israel, they do not prove that Israel *adopted* its own creation and flood story from these pagan sources. On the contrary, because I believe in the divine inspiration of Scripture, I firmly believe that, in these pagan accounts, we have the *distorted* versions of the memories of Noah and his descendants, whereas in the Bible we have the *divinely inspired* version (cf. Paul's argument in Rom. 1:21–25). Kevin DeYoung summarizes the argument this way:

> The biblical story of creation is meant to supplant other ancient creation stories more than imitate them. Moses wants to show God's people 'this is how things really happened.' [In contrast to what the pagans say.] The Pentateuch is full of warnings against compromise with the pagan culture. It would be surprising, then, for Genesis to

[16] Cf. recently, Carlson and Longman (2010, chapter 6).

[17] For a recent restatement of the (especially linguistic) arguments see Kiel (1997, ad loc.); C. J. Collins in Barrett and Caneday (2013, 127–29); W. D. Barrick in ibid. (208); Paul (2017, 157–58).

[18] Enns (2012, 35–60); see more extensively, Dalley (1989); Clifford (1994); cf. Garrett (2000) on the "sources" of the Pentateuch. See the negative comments by A. McKitterick in Nevin (2009, 30–33). Cf. also the comments on these ancient parallels by J. H. Walton and C. J. Collins in Barrett and Caneday (2013, 98–100, 148–155); Collins (155) drew a positive conclusion from this comparison: "Certainly the parallels between Genesis 1–11 and these Mesopotamian stories argue that we should read these eleven chapters together."

> start with one more mythical account of creation like the rest of the ANE [Ancient Near East].[19]

4.2 OBJECTIVE–SUBJECTIVE

4.2.1 THE OBJECTIVITY/SUBJECTIVITY OF THE HISTORIAN

There is no real problem concerning the historical reliability of Genesis 1–11, as long as we take seriously Jesus and the apostles. However, this does not mean that we have really touched the notion of historicity as such. There is a conservative notion of historicity that is Western-scientistic, and inspired mainly by fear of liberalism. This notion claims that a biblical narrative is historical only if it contains an exact description of events according to the criteria of modern journalism or the historical sciences.

British theologian Christopher J. H. Wright offers an antidote to this view of historicity:

> We might certainly allow the [Bible] writer "poetic freedom," which I intend to mean that we can accept that not every little detail in a narrative shows exactly "what precisely happened if you had been there." But it is possible to tell something that really happened as a good story, and a good story can certainly be founded in history. The one does not exclude the other.[20]

I would put the matter still more strongly: no fundamental distinction exists between a "good story (based on the facts)" and an "'objective' account of the facts." Every person's account of any facts is necessarily a story—*his* story (*his* selection of the material, *his* presentation of things). Belief in the so-called "objective facts" was one of the delusions of positivism, which, unfortunately, also permeated theology. Facts are *always* subjective, always contextual; they are embedded in, and determined by, the context (the religious milieu, the social-cultural thought climate, the epoch, the prevailing philosophical and ideological paradigms, the personal beliefs) of the observer, or in this case, the expositor. In other words, facts exist only within the framework of a story.

---

[19] http://www.alliancenet.org/mos/1517/why-it-is-wise-to-believe-in-the-historical-adam#.WaaZAq2iGMJ.

[20] Wright (2001, 14). One illustration of Wright's claim is the different chronologies in the Gospels; the writers, especially Luke, took some liberty in rearranging the historical material, according to their theological purposes.

Consider this example of a teenage girl. How does her mother view her? And her physician? And her teacher? And her pastor? And a police officer? And a shopkeeper? And a boy who is infatuated with her? And a pedophile? From these individuals, we get eight different pictures of the girl. Nobody sees the girl as she exists objectively (least of all the girl herself). To put it more strongly: there *is* no objective picture of the girl. Similarly, neither modern scientists nor the Bible give us an objective picture of origins because such a picture does not exist. The biblical account is not objective but it views the origins of the world and of humanity within a certain pistical framework, in light of the Creator and his glorification.[21]

A good example of non-objective, and yet scholarly historiography is *The Waning of the Middle Ages*, the masterpiece by Dutch historian Johan Huizinga.[22] This work is a "good story" in which the subjective interpretation of the late Middle Ages can nowhere be severed from the objective facts (insofar as such facts exist at all). Is it an objective account of the facts? No—such a thing is impossible. Is it a subjective distortion of the facts? If so, it would be very bad scholarship. It does justice to "the naked facts" (if these exist), and it does so as responsibly as possible, but in an interpretative context. This is inevitable. The same is true of each journalist report and documentary, every scientific account, every manual of dogmatics—and *every Bible story*. Every story in the Bible is subjective in a double sense: first, each story is a rendering of the facts as the narrator has selected and interpreted them; and second, each story is experienced in a subjective, selective, and interpretative way by the reader. This is the case with every Bible story.

4.2.2 THE MYTH OF OBJECTIVITY

Listen once again to Christopher Wright:

> In modernism, the [Bible] reader, like any scientist, was a neutral observer of a given reality (the text and its meaning), which lay outside him- or herself. The objective "real meaning" was viewed as existing, just as the "real world," and it was the task of the interpreter to bring this meaning out. The postmodern view is that, even in science, the subjective observer is part of the reality that is observed.... The difference between subjective and objective has faded. And thus, the myth

[21] For the term "pistical," see chapter 3n44, and further chapter 6.
[22] Huizinga (1999).

of the "neutral observer" has rather devaluated in the new forms of science, and has perished in hermeneutics as well.[23]

In addition to seeing various dangers, Christopher Wright appreciated much that was positive in this new approach, such as the insight that the modern Western[24] approach to Scripture (also in conservative theology) is only one out of many possible contextual approaches.[25] We may never give up our belief in "a universal truth, as it is preached in the Gospel to all people and cultures,"[26] but this tells us nothing about the proper way of reading Scripture. There is an absolute, perfect truth of God—but our (everyday or theological) *representation* and *interpretation* of this truth are always relative and defective.[27] How else could we explain the great difference of opinion on matters like baptism, the Lord's Supper, Israel, church offices, eschatology, the gifts of the Spirit, and so on?

This claim that historiography in Scripture is not objective is not at all novel. Allow me to quote three Reformed theologians from the Netherlands. Herman Bavinck argued that the biblical historiography has "its own character"; Scripture is "often incomplete, full of gaps and certainly not written by the rules of contemporary historical criticism…. Even in the case of historical reports, there is sometimes a distinction between the fact that has occurred and the form in which it is presented."[28] The views of Nicolaas H. Ridderbos (d. 1981) have also contributed to viewing the early chapters of Genesis as a theological composition rather than an exact historical documentary.[29]

Berend J. Oosterhoff (d. 1996) put it this way: "Genesis 2 and 3 narrate to us facts, but these are communicated in symbolic language. This is a manner of rendering that is foreign to many of us Westerners, and is therefore difficult to accept. Poets and painters still practice it."[30] He explained this extensively, and argued, among other things, that the many anthropomorphisms in

---

[23] Wright (2001, 16).

[24] About this Western aspect, cf. Arnold H. de Hartog (quoted in Harinck [2001, 170]), who located an important cause of confusion about Scripture's authority in ascribing to Scripture a Western character. In his view, the Paradise story should be approached "not according to the shallow spirit of Western precision and literalism but according to the profound spirit of Eastern intuition or empathy with the essence of human history."

[25] Wright (2001, 17). An excellent introduction to this subject is Vanhoozer (2009).

[26] Ibid., 18.

[27] See extensively, Ouweneel (2007a, §13.4).

[28] Bavinck (*RD* 1: 447–48).

[29] Ridderbos (1963).

[30] Oosterhoff (1972, 193).

Genesis 2 and 3 certainly do not allow any exact historical interpretation; he adds, "It may be that the many anthropomorphisms in Genesis 2 and 3 determine the entire character of these chapters."[31] However, this does *not* mean that Genesis 1–2 constitute an allegory, or are mythical (in the sense of non-historical). They certainly deal with *historical facts*, but then "in the language and with the material that was proper to the ancient Eastern world."[32] Earlier Abraham Kuyper had stated that God gave us his revelation in Scripture "wrapped in the special Eastern symbolic-aesthetic language."[33]

Berend J. Oosterhoff emphasized that "The symbolic language does not de-historicize history, but is exactly what gives it expression."[34] Genesis 1–3 do not give us history in any usual sense of the word but do witness to us about the *historicity* of creation and the Fall. Thus, the biblical history, also in the early chapters of Genesis, is definitely a *historical* testimony.[35] To be sure, "The symbolic description does not entail less about realities than the exact-historical description…[Genesis 2–3] are about 'history'…the chapters do present history in the sense of what really happened."[36]

4.3 LITERALISM

4.3.1 BIBLICISM

In opposition to the modernist way of reading Genesis 1–3 (to which I would, of course, assign AEH), we find a biblicist way of reading these chapters. In my view, we must reject both. "Biblicism" can include many things,[37] such as the doctrine of mechanical inspiration, scientism ("true science is found only in the Bible"), antipathy toward theology ("if you have the Holy Spirit, you don't need any theology"), antipathy toward confessions ("only the Bible"), imitating biblical customs and states of affairs ("we must go back to Acts 2–4"), and so on.

One form of biblicism is literalism, which consists of an exaggerated adherence to the exact letter of a text. Literalism ties the reliability of Scripture immediately to the manner in which Scripture is read, namely, literally. When Genesis 2:7 says that God "formed," then he must have

---

[31] Ibid., 195; see 193–96.

[32] Ibid., 203; see 199–206.

[33] Kuyper (1894, 3:168; cf. 2:546).

[34] Oosterhoff (1972, 205).

[35] Ibid., 206.

[36] Ibid., 219.

[37] See Ouweneel (2012b, §§4.5.2 and 9.4).

fashioned with his own hands. If verse 8 says that "God planted," then he must have put plants in the ground with his own hands. If verse 15 says that God "took" Adam and "put" him in the garden, then he must have literally done so, like a person putting a chess piece on a chessboard.

Literalism is that particular form of biblicism that takes the Bible in an exaggeratedly literal sense, without taking into consideration the grammatical-historical meaning or the redemptive- and revelation-historical context of a Bible passage. Neither does this view take into consideration the particular stylistic figures like hyperbole and metaphor, the difference between diverse literary genres, in short: the normal factors that play a role in reading any text whatsoever. In fact, literalism is a form of rationalism.[38]

One of the most striking characteristics of literalism is that nobody can defend it with even an appearance of consistency without falling into absurdity. I emphasize this point because thoroughgoing fundamentalists are always chattering about how they always and everywhere take the Bible literally, whereas that is absolutely not so, and *could* not be so. Steve Falkenberg tells us that he had never met anyone who actually believed the Bible to be literally true. He said he knew a horde of people who say they believe the Bible is literally true but nobody is actually a literalist.[39] No wonder: it is simply impossible.

4.3.2 LITERARY AWARENESS

In relation to our subject, a striking form of literalism is a lack of literary awareness. Do those who say that they take the entire Bible literally believe that the speeches of Job and his friends—elaborate literary and poetic works—were spoken literally this way, were composed spontaneously by those men, and were written down by a stenographer? That would constitute a full-fledged misunderstanding of (a) poetry, and of the literary composition that undergirds such a work; and (b) the deeply sad encounter between Job and his friends, making it inconceivable that he and his friends did their best to speak such poetic improvisations as if it were a poetry competition. Even if we agree that Job really lived and really experienced these sufferings (on this, see Ezek. 14:14, 20; James 5:11[40]), even then the story was cast independently by an author into a literary form. The term "literal" is not at all useful here—the term "literary" is.

---

38 Cf. Loonstra (1999, 39, 62).

39 Falkenberg (2002).

40 About the historicity of Job, cf. Hartley (1988, 65–67).

We find something similar in the narrative of Jonah, especially in connection with his three-night stay in the fish (Jonah 1:17; 2:1, 10). One can take the story with historical seriousness (on this, see 2 Kings 14:25; Matt. 12:39–41; 16:4), as long as one does not exaggerate it to the extent of asking to what species this fish belonged, in order thereby especially to demonstrate the truth, or at least the plausibility, of the story. After all, the Septuagint tells us plainly in Jonah 1:17, and 2:1, 10—and Jesus cites this in Matthew 12:40—that Jonah was in the belly of a "sea monster" (Gk. *kētos*) (AMPC, CJB, NASB).[41] The Jewish tradition has sensed this more clearly by seeing a parallel between Jonah and Israel, both swallowed by a "dragon," which in Israel's case was Babylon (Jer. 51:34, 44).[42] As to what kind of animal provided lodging for Jonah during those three days, there is only one correct answer: a sea dragon.[43]

In general, literalism means a misapprehension with regard to a text's literary aspects: genre, metaphor, parable, allegorizing, anthropomorphism. I have known a physicist with a PhD who was convinced that every parable of Jesus must have happened literally, because Jesus spoke about them this way. "A man was going down…" (Luke 10:30), such that there once actually was a man who travelled down the road, fell into the hands of robbers, and was rescued by a Samaritan. Otherwise, if there never were such a man, Jesus would be a liar.

Behind this misunderstanding lies a general problem: what room does fiction occupy in Scripture, and what does this fiction mean for the credibility of Scripture?[44] Biblical fables[45] and parables are fiction; but may we not see the stories of Esther, Job, and Jonah as partially fictitious (even though these persons really existed) without the divine truth contained in them being attacked? In connection with Daniel, John J. Collins commented that "fiction and truth are not mutually exclusive."[46] The reason is that truth involves more than historical literal occurrence.

Once again: nobody can seriously believe that Job and his friends delivered long extemporaneous poems to each other, and that someone was

---

[41] In the Shepherd of Hermas (*Visio* 4:1; 6:9), *kētos* is an apocalyptic animal, like a dragon (cf. the dragon in Rev.); cf. §7.6.1.

[42] Kleinert (1868, 16–18); Cheyne (1970, ad loc.).

[43] Cf. Heb. *tannin*, "[sea] monster / [sea] dragon / [sea] serpent," in Isa. 51:9. Heb. *dag*, "fish," is a broader concept than "fish" in the strictly biological meaning, and includes all swimming animals (cf. Gen. 1:21: "the great sea creatures and every living creature that moves, with which the waters swarm").

[44] Cf. Goldingay (1994, 67–76).

[45] Examples of genuine fables are Judg. 9:7–15 and 2 Kings 14:9.

[46] Collins (1990, 36).

sitting nearby to record these stenographically. Whereas Genesis is basically a historical book, Job is basically a poetic book. In it, we are dealing with a literary composition that—under divine inspiration—is the product of a creative author. In saying this, I am not saying that the core of the story could not be historical. What I am saying is that one may legitimately wonder to what extent this is essential for the divine truth of Job (this despite the mention of Job in Ezek. 14:14, 20 and James 5:11).

4.3.3 GENRES, METAPHORS, ANTHROPOMORPHISMS

Let the reader note carefully what I *am* saying, and what I *am not* saying. The problem with historical interpretation versus literalism is that we can easily go astray in two directions.[47] The problem with literalism is that it maintains that a text is reliable, historically trustworthy, only if it is exactly, factually true in some sense associated with objective journalism or historiography. However, if this were true, there is nothing in the Bible that we could take historically literally because no part of the Bible was written in the style associated with objective journalism or historiography (if such forms of objectivity exist).

An excellent example is the comparison of the four New Testament Gospels.[48] They are no doubt literary compositions, differing in many respects, more comparable to paintings of Jesus than to photographs of Jesus. In this sense, they not only tell us much about the subject of the painting but also about the painters.[49] Abraham Kuyper compared the Gospel authors' way of writing to impressionism.[50] Yet, we accept the Gospels as thoroughly historical. The painted portrait of Jesus, as we find it in the Gospels, must *resemble* him; otherwise it would be unacceptable. Similarly, the description of the events in Genesis 2–3 must *resemble* historical reality; otherwise it is unacceptable, too. Yet, the main goal of Genesis 1–3, as well as of the Gospels, is to communicate real history. As Donald Guthrie said with respect to Luke's Gospel, "No-one would deny that Luke's purpose is theological [more accurately: is to convey a faith message[51]]. But this is quite different from saying that the history has been conformed to

---

[47] Cf. van den Brink (2017, 135): we should not throw out the baby of history with the bathwater of concordism (a form of biblicistic literalism).

[48] See extensively Ouweneel (2012b, §13.5).

[49] Cf. Macquarrie (1981, 30); Pentecost (1981, 24); Verkuyl (1992, 192); Van de Beek (1998, 124); Loonstra (1999, 68).

[50] Quoted in Harinck (2001, 113).

[51] For this distinction between "theological" and "faith message," see extensively, Ouweneel (2015). See also chapter 3, note 44 above.

the theology [i.e., the faith message].... It is truer to say that Luke brings out the theological [better: faith] significance of the history."[52]

But the second mistaken direction is this: we may abuse the frequent occurrence of genres, metaphors, parables, allegorizing, anthropomorphisms, and poetic parts in the Bible to argue that much, if not most, of the Bible is not historical at all. This would be a major mistake. I believe that the books of Esther, Job, and Jonah, as well as Genesis 1–11, are thoroughly historical despite their use of various literary genres, metaphors, parables, allegorizing, anthropomorphisms, and poetic language. Job and his friends existed, even if they did not speak to each other in the form of polished poems. Jonah existed, and he really was in the belly of the "fish," even if this "fish" is described as a sea dragon. And Genesis 1–3 is historically trustworthy, even if in these chapters we see God fashioning, planting, anesthetizing, tanning, tailoring, and sewing. Such metaphors and anthropomorphisms do not in the least affect the essentially historical nature of these chapters.

4.4 FIGURATIVE LANGUAGE IN GENESIS 1–3

4.4.1 GENESIS 1: *EX NIHILO*

It is interesting to note the various verbs in Genesis 1:1–2:7 that describe God's creative work. First, God simply speaks (commands), and thus calls non-existing things into existence: "By faith we understand that the universe was created by the *word* of God, so that what is seen was not made out of things that are visible" (Heb. 11:3); God "who...calls into existence the things that do not exist" (Rom. 4:17). This is God speaking, commanding, and calling. But in addition to this, the following verbs imply divine activity: "to create" (Heb. *bara*, Gk. *ktizō*), "to make" (Heb. *asah*, Gk. *poieō*), and "to form (to fashion)" (Heb. *yatsar*, Gk. *plassō*).[53]

In a sense, we may say that the Hebrew verb *bara* always entails creating "out of nothing" (Lat. *ex nihilo*), or more correctly: not out of already existing materials.[54] This is clear in Genesis 1:1 ("In the beginning, God created the heavens and the earth"). This view is found most explicitly in 2 Maccabees 7:28 (GNT), "So I urge you, my child, to look at the sky and the earth. Consider everything you see there, and realize that God *made it all from nothing* [others: not out of things that existed], just as he made

[52] Guthrie (1990, 107).

[53] Cf. also Heb. *kun*, Gk. *themelioō*, as in Ps. 119:90.

[54] Cf. Geisler (2011, 620–23).

the human race." Incidentally, notice the last phrase: also the human race was created out of nothing, or was not made out of things that existed. The verse in Maccabees may not be inspired, but it does reflect common Jewish conviction. Of course, Jews knew that Adam's body came from dust (Gen. 2:7); but they believed that Adam's spiritual essence was not made out of things that existed.

In Genesis 1:21 we read, "God created the great sea creatures and every living creature that moves, with which the waters swarm." This may not literally be a creation *ex nihilo*, because the expression "Let the waters swarm with swarms of living creatures" (v. 20) seems comparable to verse 11, "Let the earth sprout vegetation, plants yielding seed, and fruit trees bearing fruit." Yet, the presence of the verb "to create" suggests that an entirely new order of existence is being called forth here: *animate life*, that is, the animal kingdom. It "deserves" the use of the Hebrew verb *bara* because it represents a mode of existence that cannot be reduced to any earlier form of life. The waters swarming with water creatures suggest that animals share important features with vegetative life. At the same time, they are essentially different, as the use of the verb indicates (see extensively, §6.2).

This interpretation is supported by the fact that the Hebrew verb *bara* reoccurs in verse 27 no fewer than three times: "God created man in his own image, in the image of God he created him; male and female he created them." Here again, we have to do with a totally new mode of existence, as is indicated by the Hebrew verb *bara* (cf. Gen. 5:1–2; 6:7; Deut. 4:32; Ps. 89:47; Isa. 45:12). At the same time, when it comes to their physical aspects, humans are made from the same materials as the entire physical world; hence the Hebrew verb *yatsar* in 2:7, "Then the LORD God formed the man of dust from the ground and breathed into his nostrils the breath of life, and the man became a living creature [Heb. *nephesh chayyah*, living soul]." Animals are also "living souls" (see 1:20 JUB), but humans became living souls in an exceptional way, by God breathing into their nostrils the "breath of life" (Heb. *nishmat chayyim*). Therefore, just as animals constitute a brand new mode of creaturely existence as compared to plants, so humans constitute a brand new mode of creaturely existence as compared to animals. Those who praise the Lord are those who have *neshamah* ("breath"),[55] not the breath of animals but the one God himself breathed into them (Ps. 150:6).

---

[55] The Heb. word *neshamah* is related to the Heb. word *nishmat* that appears a few lines earlier; *nishmat* is in the so-called construct state, i.e., the form of a noun when it is followed by a possessive noun (here, "the breath *of* life").

We must take these facts completely seriously, not only in their biotic and psychical[56] but also in their historical sense. Creating animals and humans involved a new creative act of God—in fact *ex nihilo*, in the sense that animate life did not exist before the fifth day, and human life not before the sixth day. If we take such biblical statements seriously, this excludes the notion of evolution: animals were *created* by God, and humans were too; that which specifically characterizes animals and humans did not develop from earlier forms of life. This will become much clearer, I trust, in chapter 6.

I have no problem identifying verbs such as "to make" (*asah*) and "to form (or to fashion)" (*yatsar*) as metaphors. But describing a portion of history in metaphorical language does not change the fact that it *is* history. As Reformed theologian Joel Beeke explains, "We recognize that the Bible contains metaphors, poetry, anthropomorphisms, types, allegories, symbolic numbers, parables, and the like. However, we also believe that the Bible contains historical accounts of real people and events."[57] This, nobody denies. But Beeke would probably agree if I were to sharpen his point this way: the Bible contains metaphors, poetry, anthropomorphisms, and so on, *even within historical accounts of real people and events*, which does not make what is reported to be less historical. When Luke 23:45 literally says, "the sun stopped shining" (NIV[58]), no reader takes this literally; but we do understand this historically to mean that for three hours no sunlight reached the place where Jesus was dying.

Or take this creational example. The Bible often uses the imagery of God acting as a potter, fashioning either individuals (Job 10:8–9; 33:6; Ps. 119:73; 139:13) or the entire nation of Israel (Isa. 29:16; 45:9; 64:8; Jer. 18:6). This is an obvious metaphor—but one that expresses a historical reality. God made/fashioned/called forth plants and animals "according to its/their kind(s)" (Gen. 1:11–12, 21, 24–25), that is, in a variety of forms/types (cf. 1:12 VOICE, "seed-bearing plants of all varieties and fruit-bearing trees of all sorts"; also see 6:20 VOICE, "all kinds of birds").[59] The verbs may

[56] It is rather scientistic to use the terms "biological" and "psychological" here; "biotic" and "psychical" refer to life and psyche, respectively; "biological" and "psychological" refer to the respective *sciences* of life and psyche (cf. chapter 3, note 44 above).

[57] J. Beeke in Phillips (2015, 18).

[58] Gk. *tou hēliou eklipontos*; ESV: "the sun's light failed."

[59] For the meaning of this expression, cf. Gen. 6:20; 7:14; 8:19; Ezek. 47:10: "according to its/their kinds" = "all kinds of" (VOICE). Remarkably, neither Lamoureux, nor Walton, nor Collins seems to properly understand this expression; see in Barrett and Caneday (2013, 55–56, 67, 73–74, 169). It has nothing to do with the supposed immutability of species. The *Pulpit Commentary* says, "After his kind . . . seems to indicate that the different species of plants were

be metaphorical, but the historical reality behind them is not: God called forth plants and animals of all sorts.

4.4.2 GENESIS 1: *CHAOS* AND *COSMOS*

The significance of the creation *ex nihilo* is that the world has an explicit beginning in time. Therefore, the Bible can use expressions like "since the foundation of the world" (Matt. 13:35; 25:34; Luke 11:50; Heb. 4:3; 9:26; Rev. 13:8; 17:8), and "before the foundation of the world" (John 17:24; Eph. 1:4; 1 Pet. 1:20), or "from the beginning [of the world, or of creation]" (Matt. 19:4, 8; 24:21; Mark 10:6; 13:19; John 8:44; 2 Pet. 3:4), or "since the creation of the world" (Rom. 1:20). All these expressions are in conflict with the pagan notion of eternally disordered matter (Gk. *chaos*).[60] Creation has a beginning; before this beginning, there was only God. He did not prepare a *cosmos* (ordered world) out of *chaos*, as the Greek *demiurge* (demigod, intermediate being between [the supreme] God and the cosmos) was deemed to have done. No, the God of the Bible called forth from nothing the things that exist today—a notion that was entirely foreign to the pagan world.

The Big Bang theory is the prevailing new paganism because the supposed Big Bang accomplished exactly this: it produced a chaos of elementary physical particles. The demiurge who produced out of this chaos the present cosmos is no longer some distinct demigod but it is the "god" of natural laws. Whatever may change, the most natural laws existed from the very beginning[61] and have never changed since then (a remarkable thing, because in the world of evolution everything seems to be changing all the time; see §2.2.3). "In the beginning"—shortly after the Big Bang—there may have been "heaven" if you like (taken in the sense of the still condensed universe) but certainly no earth (cf. Gen. 1:1); supposedly more than nine billion years had to pass before the earth appeared. The Bible is wrong: in the beginning, there was no earth at all. And when our globe finally appeared, it was not shrouded in darkness, as the Bible claims, but the earth was a fiery ball. The Bible has it all wrong. In the beginning there was chaos; or in more scientific terms, a singularity of a tremendously high density and a tremendously high temperature.

---

already fixed. The modern dogma of the origin of species by development would thus be declared to be un-biblical" (http://biblehub.com/genesis/1-11.htm).

[60] About this matter, see Niditch (1985); Anderson (1987); A. McKitterick in Nevin (2009, 33–34).

[61] Some would say, except at the very first picosecond of the Big Bang itself (cf. https://en.wikipedia.org/wiki/Chronology_of_the_universe).

Interestingly, the phrase "Big Bang" was coined by English astronomer Fred Hoyle, who himself did not believe in it, nor in the theory of the expanding universe. He considered the idea that the universe had a beginning to be pseudoscience, especially because this hypothesis seemed to suggest that, at the beginning of the process, there must have been a Creator, or at least some Intellect that had set the thing in motion.[62]

4.4.3 MORE ON THE BIG BANG

I know, the Big Bang theory is just a theory, a model—but most natural scientists seem to believe it.[63] The Bible says, "In the beginning, God created the heavens and the earth," that is, at the very beginning there was the cosmos (Gk. *kosmos*, [ordered] world), roughly as we know it today. Most scientists say (in my terms), "In the beginning,[64] there was the chaos—and it took billions of years before anything like the present cosmos originated from it." (Please note that in the Bible the expression "the heavens and the earth," or "heaven and earth," is the common way of describing the entire universe; we find this, for example, in the Torah, Gen. 2:1, 4; 14:19, 22; Exod. 20:11; 31:17; Deut. 4:26; 30:19; 31:28.) If I cannot take the Bible seriously on this point, or if the Bible has to be taken figuratively here, why take literally those passages that tell us that the present "heaven and earth" will be replaced by a new "heaven and earth" (2 Pet. 3:7–13; Rev. 20:11; 21:1–5)? Why accept, on the one hand, a figurative ktiseology, and on the other hand, a literal eschatology?

There is another important point to be mentioned here. I really fail to see how a Christian could accept the Big Bang theory, explaining the

[62] Cf. the denial of this by Quentin Smith (1992) (https://infidels.org/library/modern/quentin_smith/ bigbang.html).

[63] But see Williams and Hartnett (2014).

[64] Cf. this quotation from David Darling (astronomer, publisher of the Internet *Encyclopedia of Science*): "What is a big deal—the biggest of all—is how you get something from nothing. Don't let cosmologists try to kid you on this one. They have not got a clue either—despite the fact that they are doing a pretty good job of convincing themselves and others that this is really not a problem. 'In the beginning,' they will say, 'there was nothing—no time, space, matter or energy. Then there was a quantum fluctuation from which....' Whoa! Stop right there. You see what I mean? First there is nothing, then there is something. And the cosmologists try to bridge the two with a quantum flutter, a tremor of uncertainty that sparks it all off. Then they are away and before you know it, they have pulled a hundred billion galaxies out of their quantum hats.... There is a very real problem in explaining how it got started in the first place. You cannot fudge this by appealing to quantum mechanics. Either there is nothing to begin with, in which case there is no quantum vacuum, no pre-geometric dust, no time in which anything can happen, no physical laws that can effect a change from nothingness into somethingness; or there is something, in which case this needs explaining" (https://www.newscientist.com/article/mg15120475-000-forum-on-creating-something-from-nothing/).

past 13.8 billion years, while at the same time believing that only the last 6,000–10,000 years of these 13.8 billion years, or of the 4.1 billion years of the supposed history of life, constitutes redemptive history. Many Christians also believe that one day the present cosmos will come to an end, and God will create an entirely new cosmos: "Then I saw a new heaven and a new earth, for the first heaven and the first earth had passed away" (Rev. 21:1).[65] How can a person believe in the Big Bang theory, and at the same time believe that "the heavens and earth that now exist are stored up for fire, being kept until the day of judgment and destruction of the ungodly" (2 Pet. 3:7)?[66]

But then, if it is true that the beginning of the Bible must be taken in a figurative sense, and that the end of the Bible must be taken in a figurative sense, why should we not take most things in between in a figurative sense? If the beginning and the end are not to be taken literally, *why would anyone take the resurrection of Christ literally*? Let's face it: the idea of a resurrection of the human body is just as unacceptable to modern science as the story of the creation and the Fall in Genesis 1–3, and as the story of the new heavens and earth at the consummation of the ages.[67]

Some critics find it unfair to jump from Genesis 1–3 to the resurrection of Jesus.[68] I can imagine that. Yet, such people forget that Jesus himself said, "If they do not hear Moses and the Prophets, neither will they be convinced if someone should rise from the dead" (Luke 16:31). Everybody agrees that here "Moses" stands for the Torah, the Pentateuch, including Genesis 1–3. If the reader does not believe "Moses" in Genesis 1–3, why would the reader believe if someone rose from the dead? Not conservative theologians but Jesus himself posed this question. In John 5:46–47 we hear Jesus declare, "If you believed Moses, you would believe me; for he wrote of me. But if you do not believe his writings, how will you believe my words?"

Moses wrote about Jesus not only in Deuteronomy 18:15 (cf. Acts 3:22; 7:37), but also in the early chapters of Genesis. Think of 3:15 (about the "woman's offspring"), and even earlier: "God, who said, 'Let light shine out of darkness,' [Gen. 1:3] has shone in our hearts to give the light of the knowledge of the glory of God in the face of Jesus Christ" (2 Cor. 4:6). Or still earlier: "In the beginning, God created the heavens and the earth" (Gen. 1:1), that is, these things were created "through the Logos" (John 1:3), in, through and for the Son (Col. 1:13–16; cf. Heb. 1:2). When both Genesis

---

[65] See the quotation of Greg Haslam in §9.4.3.

[66] *Contra* van den Brink (2017, 334–35).

[67] W. Marxsen (1968) is one of many theologians who believe that we no longer need to take Paul seriously, *also* when he speaks of a bodily resurrection from a grave.

[68] See Falk (2004, 208–10).

and John's Gospel start with "in the beginning" (Gk. *en archēi*), then the New Testament is telling us that God's Son *is* that "beginning" (Gk. *archē*, Col. 1:18), the "beginning of God's creation" (Rev. 3:14; cf. 21:6).

If we do not believe *everything* that the books of Moses contain—and I add, believe it in the way Jesus and Paul believed it—including Genesis 1–3, how can we believe the rest of the Bible? It is my earnest claim that AEH advocates are no longer hearing Moses properly;[69] he vanishes from their sight because they are looking at him through the lens of Darwin. We know what happened to Alice when she peered "through the looking-glass": she landed in a world that did not really exist. As Alice became a pawn in a bizarre game of chess, so Adam has become a hominid pawn in the evolution game.[70]

4.5 GENESIS 2–3

4.5.1 DUST AND CLAY

Genesis 2:7 says, "Then the LORD God formed the man of dust from the ground and breathed into his nostrils the breath of life, and the man became a living creature." This verse clearly teaches us that the first man was not born of a female—there *were* no women yet—but came forth immediately from God's hand. This comports with, on the one hand, 1:26–27: "Let us make man.... So God created man," and on the other hand, Paul's teaching: Adam was the "first man...from the earth, a man of dust" (1 Cor. 15:47), and this first "man was not made from woman" (11:8). These are unambiguous statements: the *first* man, and *not* born of a woman (even Jesus, "born of woman" [Gal. 4:4], was in this respect a more ordinary man than Adam).

There may be figurative language in Genesis 2:7; we do not know what it exactly means that God acted here as a potter forming the clay, or as one who literally blew breath into a heap of dust. But this does not change the basic message: God made the first man in such a way that his body consists of the same essential materials as the earth—as is the case with all of us, but Adam was first—whereas what made this man alive, the "breath of life," came from God himself. Rejecting an exaggerated literalism does not give us the right to fall into the other extreme of dismissing the entire historical message of the passage.

No matter how many interpretations of Genesis 2:7 there may be—how much of it is literal, how much is figurative?—the historical kernel must be

[69] They might no longer believe that Moses had anything to do with Genesis 1–3; see Enns (2012, 9–34).

[70] Carroll (1999).

clear: the first member and representative of *Homo sapiens* was *not* born of a female, but was created by God out of "earth." The fact that Job said to God, "Remember that you have made me like clay" (Job 10:9), and that Elihu said to Job, "Behold, I am toward God as you are; I too was pinched off from a piece of clay" (33:6),[71] does *not* mean that Genesis 2:7 is a description of Everyman. It is true, each human being is "formed out of the clay" (KJV). However, the Bible clearly says that Elihu was "the son of Barachel the Buzite, of the family of Ram" (32:2), and Job said of himself, "Naked I came from my mother's womb" (1:21). They were men of clay, but they definitely had their respective fathers and mothers. Adam did not; just as Seth was [the son] of Adam, so Adam himself was "[the son] of God" (Luke 3:38).

4.5.2 THE GARMENTS AND THE TREES

Another example of what, to some extent, may be figurative language, and is nevertheless historical, is the two trees in Genesis 2:9. That God "planted" a garden (v. 8) is presumably figurative—not the garden but the planting—just as God "forming" the man (v. 7), and God making garments (3:21). John Calvin argued centuries ago,

> Moses here, in a homely style, declares that the Lord had undertaken the labor of making garments of skins for Adam and his wife. It is not indeed proper so to understand his words, as if God had been a furrier, or a servant to sew clothes. Now, it is not credible that skins should have been presented to them by chance; but, since animals had before been destined for their use, being now impelled by a new necessity, they put some to death, in order to cover themselves with their skins, having been divinely directed to adopt this counsel; therefore Moses calls God the Author of it.[72]

Thus, Calvin fully maintained the historical truth of these verses, although his solution for how God "made" those garments is rather artificial; in his view, Adam and Eve did it themselves.[73] If God "made" the earth and the heavens (Gen. 2:1), why could he not have "made" garments of skins (3:21) in any unspecified way? I have no visual picture of either how God

---

[71] See the parallelism between "clay" (*chomer*) and "dust" (*aphar*) in Job 4:19; both illustrate the "simple, obvious fact that the human body is made of the common elements of the soil," says Buswell (1962, 159).

[72] Calvin, *Comm. Gen.* (ad loc.; http://www.ccel.org/ccel/calvin/calcom01.ix.i.html).

[73] The Benson Commentary suggests that angels might have been involved (http://biblehub.com/ commentaries/genesis/3-21.htm).

"made" the earth, or how he "made" those garments—yet I believe in the historicity of these events.[74]

The two trees in Genesis 2 and 3, the "tree of life" and the "tree of the knowledge of good and evil," are so important that I will devote separate attention to them in chapter 7. At this point I wish only to mention this. As we observed earlier in this chapter, we allow for hyperboles, metaphors, and anthropomorphisms in Genesis 2–3. However, this does not affect the fundamentally historical character of these chapters. They tell us about the first humans that ever existed on earth (the first problem for AEH), from whom all present humans descended (the second problem for AEH), who had been created as "very good" (the third problem for AEH), and who at a given moment in time fell into sin, and thus entered a sinful condition (the fourth problem for AEH). This was not a process of gradual deterioration, but a real *fall*: from a status of being very good to a status of being very bad. "In the place where the tree falls, there it will lie" (Eccl. 11:3). This is the picture of Genesis 3 for any ordinary reader: at one moment the "tree" of humanity was standing, and the next moment it lay on the ground.

4.5.3 THE SERPENT

To use another example: no one believes literally that after being cursed, the serpent henceforth "eats dust" (Gen. 3:14; cf. Isa. 65:25, "dust shall be the serpent's food"). Snakes would soon become extinct if they were to eat only dust. The statement makes perfect sense if we understand more clearly what (or rather, who) the "serpent" really *is*: "the great dragon... that ancient serpent, who is called the devil and Satan, the deceiver of the whole world" (Rev. 12:9), "the dragon, that ancient serpent, who is the devil and Satan" (20:2); "the serpent deceived Eve by his cunning" (2 Cor. 11:3). From passages like these it is evident that the serpent represented the devil (see more extensively §§7.4 and 7.5). Genesis 3:14 is not a statement about snakes in general, but about this particular snake. And consider especially verse 15: "I will put enmity between you and the woman, and between your offspring and her offspring; he shall bruise your head, and you shall bruise his heel." God speaks of enmity, not just between snakes and human beings in general, but between *this particular* serpent (Satan) and *this particular*

[74] Strangely enough, some rabbis thought that the "skins" referred to Adam and Eve's *own* skins, as if they had no skin before; see Talmud: Niddah 25a. Cf. the Targum of Onkelos: "And the Lord God made for Adam and for his wife vestments of honor upon the skin of their flesh" (https://books.google.nl/books?id=JJwCAAAAQAAJandpg=PA1andhl=nlandsource=gbs_toc_randcad=4#v=onepageandqandf=false). The Targum of Jonathan (or, of Palestine) has a different explanation: "vestments of honor from the skin of the serpent, which he had cast from him, upon the skin of their flesh" (ibid., PA157).

woman, who would be the "mother of all living" (v. 20). It is the enmity between the spiritual world of evil, represented by Satan and his demons, and the world of human beings.

As we will see, the final battle that is depicted in Genesis 3:15 is not between the serpent's offspring and the woman's offspring but between the "you(r)" in verse 15b, who can only be the serpent himself, and "he/his"—not a "they" (all the woman's descendants) but a "he": the woman's descendant *par excellence*, the Messiah. In the final battle it will still be the same serpent as the one of Genesis 3: "the devil and Satan."

In conclusion, in dealing with the significance of Genesis 2–3, it is ineffective to combat AEH by claiming that we should read these chapters literally. We will not get very far using terms like "literal" and "figurative." The real question at issue in these chapters is not how much is *literal*, but how much is *historical*. Did the events described really happen in time and space? It is the testimony of both the Old and the New Testaments that these events really happened, no matter how much figurative language was used to describe them.

4.6 THE HISTORICAL ADAM

4.6.1 PRO ARGUMENTS

Historicity in the Bible is of enormous importance, certainly in cases where this historicity is under heavy attack, as is the case with the historical Adam. I am inspired here by some of the arguments of Joel Beeke,[75] but I will develop them in my own way.

(a) *Genesis 1–3 is just as historical as Genesis 4–50*. Every unbiased reader, reading through these chapters, would be unable to notice anywhere in the text any caesura, a break, a transition in the book of Genesis, as if one senses a move from the figurative (if you like, even the mythical) to real history. Some expositors have viewed the story of the patriarchs as largely or entirely legendary, but many have opined that, from Genesis 12 onward, we are clearly dealing with history.[76] But where is the caesura? The section of Genesis 6–11 is written in largely the same style as the subsequent chapters (cf. §1.3.1);[77] who would dare suggest that the latter chapters are

[75] J. Beeke in Phillips (2015, 19–26).

[76] On this point, I recommend Kitchen (1995; 2006).

[77] I can agree to some extent with J. H. Walton in Barrett and Caneday (2013, 67), who says that, *qua* genre, Gen. 1–11 "is without peer"—as long as this does not imply a special (semi-historical?) status for these chapters as compared with Gen. 12–50.

less historical, except on the basis of evolutionary biases? And if Genesis 6–11 are historical, why not Genesis 1–5? Where are the basic differences in style, in literary genre?[78]

In Genesis 13:10 we read, "And Lot lifted up his eyes and saw that the Jordan Valley was well watered everywhere *like the garden of the LORD*, like the land of Egypt, in the direction of Zoar." Here in this part of Genesis, we find a cross-reference to the first part of the book, whose historicity is apparently being taken seriously. What Lot saw looked like the Garden of Eden, about which people in those days apparently still had some memory. It is like seeing something that reminds us of a thing or event from our early youth. Here in Genesis, people could have been looking at a wonderful valley and have been reminded of that other beautiful place from the early youth of humanity. Apparently, Adam and Eve had told their children about the wonders of Eden. In the future, God's people would be reminded once again of Eden: "For the LORD comforts Zion; he comforts all her waste places and makes her wilderness like Eden, her desert like the garden of the Lord; joy and gladness will be found in her, thanksgiving and the voice of song" (Isa. 51:3).[79]

(b) *Adam was the historical (fore)father of historical descendants.* We cannot argue that everything in Genesis except the story of Adam is basically historical, for all the subsequent chapters deal with his very physical offspring. Without Adam, there was no Seth; without Seth, no Enoch (to make a little jump; Gen. 5:1–24). Jude 14 tells us about "Enoch, the seventh of Adam" (cf. Heb. 11:5–6). There are no gaps in the genealogy of Genesis 5: it is Adam, Seth, Enosh, Kenan, Mahalalel, Jared, and Enoch, for a total of seven historical men; compare Luke 3:37–38, Jesus was "the son of Enoch, the son of Jared, the son of Mahalaleel, the son of Cainan, the son of Enos, the son of Seth, the son of Adam, the son of God." Once again, the total is seven. The texts are precise. Jesus had a real historical pedigree, going back beyond his legal father Joseph, all the way to Adam. And notice, just as truly as Seth was the son of Adam, Adam was the son of God, and *not* the son of some unknown hominid. His father was none other than God himself.

Similarly, from the opening verses of 1 Chronicles we are on solid historical ground. We are told matter-of-factly: "Adam, Seth, Enosh; Kenan, Mahalalel, Jared; Enoch," and so on—of course, beginning with enumerating seven historical men. If Abraham existed, then so did his forefather Noah. If Noah existed, then so did his forefather Enoch. If Enoch existed,

---

[78] For more on this, see §5.2.1.

[79] For a discussion of that other remarkable reference to "Eden, the garden of God" (Ezek. 28:12–19), see Ouweneel (2018a, Appendix 10).

then so did his forefather Adam. We cannot play around with these things. Genealogies were important in ancient times, more important than today. After the Babylonian exile, some men of the priestly family wished to function as priests: "These sought their registration among those enrolled in the genealogies, but they were not found there, and so they were excluded from the priesthood as unclean" (Ezra 2:62). In the same way, the chronicler told the post-exilic Jews: You are God's people because you can trace your pedigree back all the way to the patriarchs. That would have served his purpose, you would think (as does a similar genealogy in Matt. 1:1–17). But no, says the chronicler, we can go all the way back to the very beginning, to Adam. There, at the beginning of humanity, lie your earliest origins.[80]

(c) *Jesus spoke of Adam and other early humans as historical figures.* I dealt with this point earlier at the beginning of this chapter. Jesus spoke of the first husband and wife, and of Abel and Noah, as historical figures. How could he have warned people against God's imminent judgment (Matt. 24:37–38; Luke 17:26–27), which he compared to the Flood, if neither Noah or the Flood were historical? How could he warn about the "blood of Abel" if Abel were not historical, and had never really become the first martyr in human history? And how could Abel have been historical if his supposed parents, Adam and Eve, had never existed as genuine historical persons?

4.6.2 A "VERY GOOD" BEGINNING[81]

If we make a case for the historical Adam, the question may arise: Why does it matter? Why make a fuss about a tiny part of the Bible? The answer is that this tiny part turns out to be the very opening chapters of the Bible, and the rest of the Bible, especially the New Testament, attaches great value to these opening chapters. They are the pivot around which the rest of redemptive history is turning.

(d) *Without the historical Adam there was no originally "very good" humanity.* If we do not begin with the designation "very good" from Genesis 1:31,[82] and consider the Fall as a subsequent event, we lose the foundation of

---

[80] Enns (2012, 141) suggests that the chronicler wished to present Adam as the "first Israelite." Literally, Israel is Jacob, a distant descendant of Adam; Israelites are descendants of Jacob, not his ancestors. I do understand Enns' intention, though: in 1 Chron. 1:1 Adam is, figuratively speaking, the "first Israelite," Israel's first ancestor. Yet, he was also the "first man" here; cf. Isa. 43:27, "your first father," who according to David Kimchi was Adam; see Slotki (1983, 211).

[81] For the outline of the arguments in §§4.5.2–4.5.5, cf. J. Beeke in Phillips (2015, 27–40); again, the elaboration of these arguments is my own.

[82] J. H. Walton tried to reduce this evaluation of humanity as being "very good" to being functionally, not morally good, but J. C. Collins rightly rejected this view; see Barrett and Caneday (2013, 115, 130).

Christian soteriology, which involves the restoration of fallen humanity. On the AEH standpoint, we begin with hominids that[83] were already killing, stealing, lying, and promiscuously mating before any individuals existed that AEH advocates might identify as "Adam and Eve." For AEH, there never was a time when beings who were fully humans were "very good," who subsequently fell to become very bad. For AEH, a time never existed when death had not entered the human world, after which came the Fall, which introduced human death. Gijsbert van den Brink, too, has to admit that bullying, deceit, infanticide, and cannibalism had occurred among earlier hominids—and today among chimpanzees and bonobos—and that, therefore, a paradisal time when humans adhered to a morally high standard is very unlikely.[84]

The Dutch philosopher Herman Dooyeweerd, like others before him, summarized Christian truth in these three keywords: creation—Fall—redemption, which he called the Christian "ground motive."[85] Often the word "restoration" is added as a fourth element.[86] This paradigm was not the invention simply of Dooyeweerd or his direct predecessors; it has always been a basic motif of (Western) theology.[87] The four elements must be understood in such a way that restoration presupposes redemption, redemption presupposes the Fall, and the Fall presupposes the originally good creation. It is one of the claims of this study that the AEH model cannot maintain this three- or fourfold representation of spiritual reality, and thus necessarily deals a heavy blow to Christian soteriology. Without the originally good creation, we lose the biblical view of the Fall, and without the Fall (in the strictly biblical sense, not in the imaginary sense defended by AEH advocates), we lose the biblical view of redemption.

4.6.3 PROGENITOR OF ALL HUMANITY

(e) *The fact that Adam was the progenitor of the entire human race is quite evident in Scripture.* The apostle Paul called Adam the "first man" (1 Cor. 15:45–49). Elsewhere he referred to Adam as the "one blood" from whom all humans descended (Acts 17:26),[88] and his wife became the "mother of all living" (Gen. 3:20). Adam's sin had consequences for all his progeny, and

[83] Is this "which" or "who"? When did hominids switch from the "which" to the "who," from animals to persons? See extensively chapter 6 on this vital anthropological question.

[84] van den Brink (2017, 211).

[85] See Dooyeweerd (2003, chapter 1, §§8 and 10); cf. Ouweneel (2014a).

[86] Cf. the model defended by Wax (2011).

[87] Cf. extensively, Ouweneel (2018e).

[88] But cf. chapter 1, note 9.

because he was the progenitor of all humanity, his sin had consequences for all humanity. This is an important element of Paul's argument in Romans 5:12–21, as I hope to show in chapter 10.

It is rather embarrassing to encounter the facile suggestion by AEH advocates that there were many other people alive on earth during the time of "Adam" and "Eve" (regardless how the latter are understood).[89] Just as facile is their suggestion that the Fall of Adam and Eve also had consequences for the other people supposedly alive at the time.[90] This is totally against the central line in Scripture, as elucidated especially by Paul in Romans 5 and 1 Corinthians 15. He taught us that "in" (not just "through") Adam all fell *because he was the ancestor of them all*. All people were "in Adam's loins," so to speak, when he fell, to use the imagery of Hebrews 7:10 (Levi was in the loins of his forefather Abraham when the latter encountered Melchizedek; cf. §§9.5.2 and 10.5.1).[91]

Human evolution, as AEH advocates defend it (or at least allow for it), presupposes a world that in many ways was not "very good" at all. As we saw, physical death had supposedly entered the hominid world long before the time that "Adam and Eve" (however they are understood) "fell" (however this is understood). Back then females had difficulties giving birth, males dominated females, through painful labour hominids/humans obtained their food, thorns and thistles were growing on the earth. In Genesis 3:16–19 these things are described as *consequences of the Fall*, but in AEH these phenomena existed long *before* the Fall (however this Fall is understood). As Joel Beeke observes,

> If death and disaster did not arise from the curse and judgment of God upon Adam's sin, then how did it come into God's creation? Did God create a world of evil? Is God perhaps not the all-powerful Creator of all things, but only one limited influence among others? The fall of Adam is the hinge upon which our doctrines of creation and God turn. If we break the hinge, the whole system of biblical doctrine collapses.[92]

4.6.4 FOUNDATION OF GENDER RELATIONSHIPS

(f) *Jesus makes clear that, if you wish to understand God's mind concerning gender relationships, specifically human marriage and sexuality, you must look at the first humans on earth, at the beginning of creation:*

---

[89] van den Brink (2017, 226).
[90] Haarsma and Haarsma (2011, 256), although they do not necessarily condone this view.
[91] Cf. Ouweneel (1982, ad loc.).
[92] J. Beeke in Phillips (2015, 33).

> Because of your hardness of heart he [i.e., Moses] wrote you this commandment [i.e., "When you send your wife away, give her a certificate of divorce"]. But from the beginning of creation, "God made them male and female" [Gen. 1:27]. "Therefore a man shall leave his father and mother and hold fast to his wife, and the two shall become one flesh" [Gen. 2:24]. So they are no longer two but one flesh. What therefore God has joined together, let not man separate (Mark 10:5–9; cf. Matt. 19:4–6).[93]

What would have been the use of establishing God's basic rule for marriage, if this were based on a myth? Paul does something similar to what Jesus did: when he writes about the creation order for men and women, he goes back to the first humans because God's mind becomes manifest in their creation:

> For a man ought not to cover his head, since he is the image and glory of God, but woman is the glory of man. For man was not made from woman, but woman from man. Neither was man created for woman, but woman for man. That is why a wife ought to have a symbol of authority on her head, because of the angels. Nevertheless, in the Lord woman is not independent of man nor man of woman; for as woman was made from man, so man is now born of woman. And all things are from God (1 Cor. 11:7–12).

And:

> Let a woman learn quietly with all submissiveness. I do not permit a woman to teach or to exercise authority over a man; rather, she is to remain quiet. For Adam was formed first, then Eve; and Adam was not deceived, but the woman was deceived and became a transgressor. Yet she will be saved through childbearing—if they continue in faith and love and holiness, with self-control (1 Tim. 2:11–15).

For expositors, these passages contain numerous difficulties (Why a head covering? In what sense is the woman the "glory of man"? Why may women not teach or exercise authority? In what sense was Eve deceived, and Adam was not? How are women "saved through childbearing"?). These problems do not concern us now. My point here is this: when Paul explained gender relationships, he did not invoke abstract principles ("creation ordinances,"

---

[93] Cf. VanDoodewaard (2016, chapter 3).

if you like[94]), and certainly not some primeval myth, but he appealed to God's order as it was concretely manifested in the first pair of humans who lived at the beginning of humanity's history, created directly by God (cf. 1 Cor. 11:8 NKJV, "man is not from woman," i.e., the first man, Adam, was not born of a woman).

4.6.5 ADAM AS TYPE OF CHRIST[95]

(g) *As surely as Jesus is a historical figure, so too was his "type" (Rom. 5:14), Adam, a historical figure.* The first Adam "prefigured" the last Adam (CJB); "Adam became a pattern of the one to come" (NIV). For the term "type" see also 1 Corinthians 10:6, 11, "Now these things [i.e., the events during Israel's wilderness journey] became types of us…all these things happened unto them as types" (JUB). Not in literary theory, but apparently in the Bible, an "allegory" has more or less the same meaning as a "type": "Now this may be interpreted allegorically [one Gk. word: *allēgoroumena*]: these women [Sarah and Hagar] are [i.e., represent, depict] two covenants" (Gal. 4:24). Similarly, Adam and Christ are two covenant heads, in such a way that the former is a "type" of the latter.

Look at the first one, Adam, to get an idea of the last One (or to get an idea of how the last One is the counterpart of the first one). Paul works this out in 1 Corinthians 15:45–49:

- The first man Adam became a living being; the last Adam became a life-giving spirit.
- The first man was from the earth, a man of dust; the second man is from heaven.
- As was the man of dust, so also are those who are of the dust, and as is the man of heaven, so also are those who are of heaven. Just as we have borne the image of the man of dust, we shall also bear the image of the man of heaven.

What are the value and validity of this comparison between Adam and Christ if a historical Adam never existed in the first place? I do not mean the supposed historical Adam posited by some AEH advocates; I mean the Adam who was literally the "first man," created directly by God. Paul believed in *this* historical Adam, as did all believing Jews, including Jesus. Just for

[94] This is an important Reformed/Presbyterian notion; see Murray (1957), especially chapter 2 on "Creation Ordinances."

[95] Cf. J. Beeke in Phillips (2015, chapter 8).

a change, let me quote some of the deuterocanonical books (GNT): "You created Adam and gave him his wife Eve to be his helper and support. They became the parents of the whole human race. You said, 'It is not good for man to live alone. I will make a suitable helper for him'" (Tobit 8:6). "Wisdom protected the father of the world, the first man that was ever formed, when he alone had been created" (Wisdom 10:1). Apparently, this represented the current Jewish view of Genesis 1–3.

There are other passages in the New Testament that presuppose Adam to be a type of Christ. When Luke calls Adam "son of God" (3:38), this ties in with God calling Jesus his "Son" in verse 22. Notice the flow of Luke's argument: Jesus, the beloved Son of the Father, was (as it was believed) a son of Joseph, his legal father, and through a long line of ancestors, he was the son of Adam, the son of God. Immediately after this, Jesus, the Son of God, goes into the wilderness and is tempted by Satan, just as Adam, the son of God, was tempted by Satan (Luke 4:1–13).

Another example of Adam as a type of Christ is encountered in the use of the expression "image of God." Adam was created in the image of God (Gen. 1:26–27), but through the Fall this image was marred. It is significant that, when Adam fathered his son Seth, it is said that he did so "in his *own* likeness, in *his* image" (5:3).[96] However, Paul described the appearance of Jesus in this world as "the light of the gospel of the glory of Christ, who is the image of God" (2 Cor. 4:4). And elsewhere he says that Christ "is the image of the invisible God, the firstborn of all creation. For by [more correctly, in] him all things were created, in heaven and on earth, visible and invisible, whether thrones or dominions or rulers or authorities—all things were created through him and for him. And he is before all things, and in him all things hold together" (Col. 1:15–17).

Incidentally, notice how Adam is here implicitly both type and antitype (as in 1 Cor. 15:45, "The first man Adam became a living being; the last Adam became a life-giving spirit"). That is, he prefigures Christ, but also forms a contrast to him: Adam was image of God as the first human, created directly by God—Christ was, and is, image of God as the One in, through, and for whom God created all things. Christ is the image of God as *Man*—but this Man was and is in his person identical to the eternal Son of the Father, in, through, and for whom God created the universe.

---

[96] Mathews (1996, 310): "Adam has endowed his image to Seth, including human sinfulness and its consequences."

5

THE GIST OF GENESIS 1

For the LORD, who created the heavens (He is God, who formed the earth and made it; He established it and did not create it to be a wasteland, but formed it to be inhabited) says this,
"I am the LORD, and there is no one else" (Isaiah 45:18 AMP)

THESIS
Due to the theory of human evolution, many theologians today are confused regarding the biblical teaching about creation, humanity, and sin. Genesis communicates fundamental principles and facts concerning the creation of the universe in general, and of humanity in particular, and thus prepares us for the fall of early humanity (Gen. 3). Without a biblical doctrine of creation, humanity, and sin, one can have no biblical doctrine of redemption.

5.1 CONFUSION

5.1.1 CONFUSION ABOUT CREATION

In theology, some of the most controversial subjects are *creation*, *humanity*, and *sin*. Confusion surrounding them is rather regrettable because these subjects are so fundamental. During the last hundred years, they have become even more problematic. The subject of creation has become problematic to many Christians because of the rise of evolutionism. The topic of humanity has become problematic because, since the decline of nineteenth-century cultural optimism, humans have become a problem to themselves. The subject of sin has become problematic through the rise of depth psychology (psychoanalysis) and related psychological schools, which have given psychological interpretations of matters that formerly had been theologically interpreted as plain sin(fulness). For instance, psychoanalysis reduced personal guilt (in the theological sense) to a feeling of guilt (in the psychological sense).[1]

To begin with the first point: natural scientists should not determine the agenda of theologians (and theologians should not determine the agenda of natural scientists for that matter). For decades, Old Testament scholars who wrote commentaries on Genesis felt obliged to delve into all kinds of natural scientific problems instead of emphasizing the faith[2] content of Genesis 1–3. I have met many Christians who, when speaking of Genesis 1, insisted that the main issue is whether you believe in creation-in-the-traditional-sense or in evolution, in a young earth or an old earth, in a sudden or a gradual origin of the world, in an animal ancestry of humanity, in creation in six twenty-four hour days or otherwise, and so on. Or, when speaking of Genesis 6–8, they insisted that those chapters are especially about how the earth's strata and fossils originated.

Where are the theologians who still have the courage to let Genesis 1 say what it says (as carefully interpreted in the light of the full revelation of God, of course)? What has remained of the authoritative *speaking* of God? "For he spoke, and it came to be; he *commanded*, and it stood firm" (Ps. 33:9).

> Praise him, sun and moon, praise him, all you shining stars! Praise him, you highest heavens, and you waters above the heavens! Let them praise the name of the LORD! For he *commanded* and they were created. And he established them forever and ever; he gave a *decree*, and it shall not pass away (Ps. 148:3–6).

Clearly the psalmist believed that things appeared suddenly and immediately simply at God's command. The apostle Paul spoke of God who "calls into existence the things that do not exist" (Rom. 4:17). So Paul, too, believed that things appeared simply by God's word calling them forth. The commanding God of Genesis 1 is the One who became flesh in the person of Jesus Christ. *He* was the One who, when he was here on earth, spoke, and it came to be—immediately. He said "Ephphatha" (i.e., "Be opened"),

---

[1] Cf. Ouweneel (1984a, 110–115).

[2] I avoid the term "theological" because, of course, the Bible does not contain theology in the strictly scholarly sense of the term; see chapter 3 note 44 and chapter 4 note 56, and Ouweneel (2015).

and deaf ears were immediately opened (Mark 7:34). Jesus said, "Get up, take up your bed, and walk," and the crippled man immediately began to walk (John 5:8–9). Jesus said, "Lazarus, come out," and Lazarus came out of his tomb immediately (11:43–44). (And, as someone has observed, if Jesus had just said, "Come out," *all* the dead in that cemetery would immediately have come out of their tombs.)[3]

Perhaps one of the best examples of comparable instantaneous power in both God's creational work and Jesus' miracles occurred in Jesus' creation of wine at the wedding at Cana (John 2:1–11). As William Barrick observed, "Note that the normal production of wine results from a natural process requiring time and fermentation. Jesus, however, instantly changed the water into the very best of aged[4] wine—without using the natural process."[5] Barrick rightly compared this with "instant creation in Genesis 1," and wondered how some people can accept such instant creation in John 2, and not in Genesis 1.

Could God not do in Genesis 1 what the God-Man Jesus Christ did when he was on this earth? God had said, "Let there be light," and there it was, immediately (Gen. 1:3; cf. 2 Cor. 4:6). Please note that this is not the light emanating from some Big Bang, for the light of Genesis 1 was immediately employed to differentiate day from night (vv. 4–5), which assumes that the earth was in place, and that the light was directional, that is, had a (more or less) point source, whereas the "light" from the Big Bang would have been diffused through all space (see v. 1).[6] In Genesis 1, God is the King who commands his subjects; yes, he calls into existence new subjects that had not existed. Even a simple Roman centurion could say, "[I have] soldiers under me. And I say to one, 'Go,' and he goes, and to another, 'Come,' and he comes, and to my servant, 'Do this,' and he does it" (Matt. 8:9). In the same way, he expected Jesus to "only say the word" (v. 8), and the miracle would take place. "Only say the word"—what a powerful reflection of Genesis 1!

Claim: We will never grasp the essence of creation if we do not grasp the notion of the speaking (commanding) God, who revealed himself in the Son

---

3 Only twice did a miracle occur in two steps: the healing of the blind man in Mark 8:22–26, and the deliverance of the possessed man in Luke 8:28–33.

4 Creationists speak here of "apparent age," which characterized Adam and Eve as well, who, immediately after their creation, must appeared to have been a few decades old; see Sarfati (2015).

5 Cf. W. D. Barrick in Barrett and Caneday (2013, 83–84).

6 It assumes that the sun was in place, too—a well-known problem in the exegesis of Genesis 1, because the sun appeared first on the fourth day. Augustine had written (*De Civitate Dei* xi.6–7): "It is very difficult for us to imagine what sort of days [i.e., in Gen. 1:5] these could be." I have seen several plausible solutions to this problem, which need not concern us now.

of God, who is the Logos (John 1:1–3, 14)—a term that could be rendered as "Speaker." This was the God-Man Jesus Christ who commanded here on earth in the same way the Triune God had done in Genesis 1. In the Bible, creating is never a process of development; creation is always the product of God's command, God's calling forth.[7] The notion, defended by AEH advocates, of theistic evolutionism or evolutionary creationism has nothing whatsoever to do with creation in the biblical sense. It confuses God's providence with God's creation (notice the difference between creating and upholding in Heb. 1:2–3).[8] Incidentally, God's providence in Christ is also exercised by the "*word* of his power" (v. 3). He was the Lord who "rebuked the wind and said to the sea, 'Peace! Be still!'" (Mark 4:39). We find the opposite command in Psalm 107:25, "He *commanded* and raised the stormy wind, which lifted up the waves of the sea." "He sends out his *command* to the earth; his word runs swiftly.... He sends out his *word*, and melts them; he makes his wind blow and the waters flow" (Ps. 147:15, 18).

5.1.2 CONFUSION ABOUT HUMANITY

Since the decline of nineteenth-century cultural optimism, humans have become a problem to themselves. This is the reason why philosophical anthropology became a separate discipline for the first time in the twentieth century. In a lecture given during a North American tour (1959), Reformed Dutch philosopher Herman Dooyeweerd explained what in his—as well as our—time is so unique about the question: "What is Man?"[9] After each period in the history of Western thought in which all interest was focused on knowledge of the cosmos, humans began to feel dissatisfied and began to turn their thinking toward the central mystery of human existence. The difference with earlier epochs is that, today, the question "What is Man?" is no longer being asked from a theoretical viewpoint only. The question has become of vital, existential importance due to the needs of Western society and the fundamental crisis of our civilization.[10]

Unfortunately, few Christians have followed this development, and in their anthropological views most Christians have remained bound to Hellenist-scholastic models. When Dooyeweerd, who taught at the Free

---

7 Other metaphors include bringing about (Isa. 4:5; 41:20; 45:7; 48:7), and begetting (43:1, 15).

8 Clearly van den Brink (2017) is aware of the difference; he devotes a significant amount of attention to the supposed relationship between evolution and providence (268–89).

9 Dooyeweerd (1960, chapter 8: "What Is Man?"); see also Dooyeweerd (1984). Cf. also Sherlock (1997).

10 For an analysis of this lecture, see Ouweneel (1986, 13–17).

University of Amsterdam, tried to break away from traditional Calvinist anthropological views that went back to Plato and Aristotle, he was criticized by Reformed theologian colleagues, especially Valentijn Hepp and Hendrik Steen. Remarkably, within a few decades, that same Free University was won over to the cause of evolutionism, especially through the writings of biologist Jan Lever and geologist J. R. van der Fliert, but also liberal theologians like H. M. Kuitert. Within one generation, a shift occurred from a scholastic to an evolutionary anthropology. A radical Christian anthropology scarcely gained a hearing within the Free University, except in the thinking of Herman Dooyeweerd and Dirk H. Th. Vollenhoven, who were more highly appreciated by people outside that University than those within.

It is amazing that today, Christian theologians and natural scientists can so easily accept an evolutionary anthropology without apparently considering the consequences (cf. the next chapter). One method to account for such an anthropology is the ancient Hellenist method; I have heard Protestants defending the view that possibly, at a given moment, tens or hundreds of thousands of years ago, God took some hominid couple, and planted a human soul or spirit within them! *Voilà*, the creation of humanity. In fact, this is the official view of the Roman Catholic Church; listen to what Pope John Paul II said at a conference in 1996: "It is by virtue of his [i.e., Christ's] eternal soul[11] that the whole [human] person, including his body, possesses such great dignity. Pius XII underlined the essential point:[12] if the origin of the human body comes through living matter which existed previously, the spiritual soul is created directly by God."[13] As far as these popes were concerned, the human body might be a product of human evolution, but not the human soul/spirit. Such a view is a horrible amalgamation of ancient scholastic dualism and modern evolutionary thinking, both of them equally objectionable.

Other confessing Christian scholars have defended the view that humans do not differ from animals essentially, but only gradually; all typically human characteristics can be reduced to animal features.[14] Here, the vital question

[11] What could be meant here by Christ's "eternal soul"? Either this refers to Christ's human soul, which was *not* eternal, or it sounds like the heresy of Apollinaris of Laodicea, who taught that the Logos was the mind (Gk. *nous*) of Jesus' human body (thus denying his human soul; cf. Matt. 26:38; John 12:27).

[12] In his encyclical *Humani Generis* 36 (http://w2.vatican.va/content/pius-xii/en/encyclicals/documents/ hf_p-xii_enc_12081950_humani-generis.html).

[13] https://www.ewtn.com/library/PAPALDOC/JP961022.HTM.

[14] The former Jesuit and abbot Denis Diderot was possibly the first who, in his *Encyclopédie* (1750–56), placed humans on the same line as the "other" animals (See http://encyclopédie.

becomes: In what sense, then, were the first humans created "in God's image, after God's likeness" (cf. Gen. 1:26; also see 5:3; 1 Cor. 11:7; James 3:9)?

What a challenge for AEH advocates! In what sense could humans, in their view, possibly be image-bearers of God? Please note, humans are divine image-bearers not by virtue of a covenant relationship that God subsequently established between himself and early humans,[15] but by virtue of *creation*: "God *created* man in his own image" (Gen. 1:28). What does this mean if the first humans were nothing but ennobled hominids? I can see only three possible answers.

(a) *The scholastic-dualistic answer*: God planted a human soul into the first real humans (a form of psycho-creationism). In such scholastic circles, arguing continues about whether humans are dichotomous (spirit/soul + body) or trichotomous (spirit + soul + body).[16] Especially in some Charismatic circles, it is popular to say that a person has a spirit, is a soul, and dwells in a body;[17] or that the soul is our inner life, and that, through their spirit, people are in contact with God and with other people; or that the soul is the bridge, the intermediary, between spirit and body.[18] These are nothing but varieties of the ancient Hellenistic-scholastic dualism.

(b) *The strictly evolutionist answer*: all essential human features are reducible to animal features. This is a real dilemma for AEH advocates: (a) dualism or (b) reductionism. In the former case, they end up in the arms of ancient pagan thinkers (preeminently Plato and Aristotle), in the latter case, they end up in the arms of modern pagan thinkers (preeminently Charles Darwin, and his successors, like Thomas Huxley in the nineteenth, and Richard Dawkins in the twentieth and twenty-first centuries). However, if this path is chosen, then AEH advocates have no answer as to how animal hominids (which cease to exist when they die) could evolve into humans, who are imagebearers of God and are destined for eternity (see §5.3.2).

(c) *A radical-Christian approach*, which shuns scholastic dualism as well as evolutionary reductionism, and highlights the unique character and position of humans (see chapter 6). For me, the most exquisite example of this is the philosophical anthropology developed by Dooyeweerd and Vollenhoven, and, in their line, elaborated by, among others, Andree

---

eu/index.php/ naturelle/845426839-zoologie/21975744-ANIMAL).

[15] *Contra* van den Brink (2017, 226, 237) and others.

[16] See Collins (2011, 95); cf. Haarsma and Haarsma (2011, 253).

[17] See Andrew Wommack (http://www.awmi.net/reading/teaching-articles/spirit-soul-and-body/).

[18] See extensively, Ouweneel (2008, chapters 6–8).

Troost,[19] Danie Strauss,[20] Gerrit Glas,[21] and myself.[22]

Claim: AEH advocates have made scarcely any attempt to find a middle path between anthropological dualism and anthropological reductionism. If they were to pursue such a path, they would discover that AEH itself could never satisfactorily solve this dilemma. And if they were to follow what I have described as the third option, a radical-Christian anthropology, they would soon discover that this would sever AEH at its roots. This I hope to show in the next chapter.

5.1.3 CONFUSION ABOUT SIN

The doctrine of original sin (in Dutch and German called "inherited sin" [*erfzonde*, *Erbsünde*]) has always played a great role in Christian theology (see more extensively chapter 9). What is it? Can we still believe in it? Some theologians say that they have great trouble with it. They argue that people are guilty because they sin personally, not because Adam sinned. They say that people cannot inherit sin. What we inherit from Adam is his imperfection and mortality, not his sinfulness. Other Christians argue that we speak too much, and too one-sidedly, about human sinfulness, and thereby humans are divested of their grandeur as creatures of God. They emphasize that this grandeur was not annulled by the Fall (whatever this Fall exactly entailed).

These ideas are related to the age-old battle about whether, also *after* the Fall, humans still are the image of God, or, if one so wishes, whether they still exhibit the image of God.[23] Whether this question is answered affirmatively or negatively, in both cases it appears we end up in theological quicksand. In the former case, the radical character of the Fall is supposedly imperiled; in the latter case, the question arises whether humans, if they are no longer images of God, can still be called human.

In any case, any soteriology (doctrine of divine salvation) stands or falls with the doctrine of the Fall. For many centuries, Roman Catholics and Protestants were convinced that human history began with *good* people ("And God saw everything that he had made, and behold, it was very good," Gen. 1:31), through the Fall they became *bad* people, and through divine salvation they become good people again (to put the matter in its most

---

[19] Troost (1989; 2005).
[20] Strauss (1991; 2009; 2014).
[21] Glas (1995; 1996; 2006).
[22] Ouweneel (1984a; 1986).
[23] See Ouweneel (2008, chapter 5 and §8.4).

elementary form). The Fall is the transition from good to bad people, from the state of purity and innocence to the state of evil and mischief. Here again, the AEH position faces a tremendous dilemma. It seems to have only two options:

(a) First, it might assume that such a transition from good to bad never occurred. In this case, the Fall is viewed as part of human development, part of a (sometimes painful) learning process from less mature (imperfect) to more mature (less imperfect) (a process of improvement, maturation).

(b) Second, AEH advocates might assume that a transition from good to bad did occur, but this transition was hardly a drastic one. In this view, whatever preceded the Fall was an evolutionary process in which hominids did not serve the true God, and *did* kill each other, steal from each other, lie to each other, commit unchastity, etc. Long before this supposed Fall, there were people like Cain in this world, so to speak. Long before this supposed Fall, there were people like Lamech who exclaimed: "I have killed a man for wounding me, a young man for striking me" (Gen. 4:23).

Any option that AEH advocates may choose *always* necessarily begins with humans who had hominid ancestors, who were not alone in the world, who were already acquainted with human death, as well as with human wickedness. Whatever the good intentions of some AEH advocates, they are trying to invent a square circle. *You cannot invent a moral rupture of some kind, occurring somewhere in the process of human evolution, and then call this the Fall.* People may invent all kinds of scenarios, but then they should not claim that their invented "Fall" has anything to do with the Fall depicted in both the Old and the New Testaments. Their "Fall" is nothing but a construction born of embarrassment.

Claim: the biblical Fall is a transition from a good to an evil state, in the lives of the first humans (and with consequences for the entire creation). Any "Fall" in AEH thinking, no matter how cleverly designed, can never correspond to the Fall as recorded in the Bible. This statement has tremendous consequences for soteriology. Remember the biblical picture: creation of "very good" people (and everything else), then the Fall, making these people (and their descendants) evil, then redemption, restoring people to goodness.[24] Compare the AEH story line: it begins with people who have a long history of evil and death behind them. Whatever "Fall" may have occurred, redemption can never be any form of restoration to the original state of goodness, since such a state of goodness never existed. Not only is soteriology greatly imperiled,

---

[24] Redemption is much *more* than the restoration of the original good (cf. Ouweneel, 2018e), but it certainly *includes* it.

but so is eschatology as well. In fact, every subject of systematic theology is threatened by the approach of AEH because all of systematic theology is built upon the foundation of ktiseology, anthropology, and hamartiology. This will become much clearer in subsequent sections and chapters.

5.2 THE AIM OF GENESIS 1[25]

5.2.1 EXEGETICAL QUESTIONS

One question that precedes any exegesis of Genesis 1 concerns the literary genre of this chapter. Is Genesis 1 narrative prose? Poetry? Something else? Do we find here prophetic language, or a special form of wisdom literature?[26] Such studies on genre have often been used as an escape: if the chapter is poetry, or prophecy, or wisdom literature, one need not expect it to provide precise historical descriptions. (Of course, this is not immediately obvious; there is no *a priori* reason why poetic or prophetic writings, like Ps. 104, should be less historically trustworthy.)

Indeed, a verse like Genesis 1:27—

> So God created man in his own image,
> in the image of God he created him;
> male and female he created them.

—has a typically poetic form, characterized by parallel lines, according to the standards of common Hebrew poetry. This is why the ESV prints the verse in the same form as all other biblical poetry. But this is the *only* verse in Genesis 1 where it does so, and rightly so.[27] The chapter itself is not poetry at all,[28] nor does it have the form of prophecy (as in the prophetic

[25] Commentaries on Genesis (or the first part of it) that I have found helpful (without condoning all the views presented in them) are Grant (1890; 1956); Coates (1920); Böhl (1923); Aalders (1981); Jacob (1974); Leupold (1942); Morant (1960); Cassuto (1961); Frey (1962); Kidner (1967); Mackintosh (1972); Gispen (1974); Rice (1975); Morris (1976); Young (1976); Cohen (1983); Blocher (1984); Hamilton (1990); Sailhamer (1990); Ross (1997); Westermann (1984); Collins (2006); and Payne (2015).

[26] About these questions, see Oosterhoff (1972, 207–19); Wenham (1987, ad loc.).

[27] In Genesis 2, the ESV prints vv. 4 and 23 as poetry, and in Genesis 3, vv. 14–19.

[28] Falk (2004, 32) erroneously speaks of "the possibility of the creation account in poetic terms"; cf. D. O. Lamoureux in Barrett and Caneday (2013, 232): "Genesis 1 is built on an ancient poetic framework, casting doubt on the belief that this chapter is 'an objective description of God's creative activities'," thus suggesting a contrast between the poetic and the historical. The term "objective" is meaningless and inappropriate here (cf. §4.2).

books), nor the form of wisdom literature (as in Job, Ps., Prov., and Eccl.).[29] As Kevin DeYoung notes: "The opening chapters of Genesis are stylized, but they show no signs of being poetry. Compare Genesis 1 with Psalm 104, for example, and you'll see how different these texts are. It's simply not accurate to call Genesis poetry. And even if it were, who says poetry has to be less historically accurate?"[30]

It is hardly possible to argue that Genesis 1 is of a different literary genre than Genesis 2–50, or, if one so wishes, than Genesis 12–50. At most, its language is loftier than that of later chapters.[31] Genesis 1 presents itself as history:[32] first, God did this, and then, the next day, he did that, and so on. In my very first book on the creation–evolution problem,[33] I pointed to the striking parallelisms between the days of Genesis 1 and the days of John 1–2 (1:29, 35, 43; 2:1; see Appendix 4).[34] Scripture does not supply us with any reason for viewing the days in John 1–2 as literal and those in Genesis 1 as non-literal.

It is pointless to argue that history is always human history, and that therefore there was no history before Genesis 2 (or 1:27). Genesis 1 is the description of God's creation of the world as the explicit *imminent domain of humanity*. Everything in the chapter opens up to this goal. First, there is the creation of the subjects, and finally the vice-gerents themselves appear.[35]

> Let us make man [*adam*, singular with both singular and plural meaning] in our image, after our likeness. And let them [plural] have *dominion* over the fish of the sea and over the birds of the heavens and over the livestock and over all the earth and over every creeping thing that creeps on the earth.... Be [plural] fruitful and multiply and fill the earth and *subdue* it, and have [plural] *dominion* over the fish of the sea and over the birds of the heavens and over every living thing that moves on the earth (Gen. 1:26, 28).

---

[29] See more extensively, R. D. Phillips in Phillips (2015, chapter 5).

[30] http://www.alliancenet.org/mos/1517/why-it-is-wise-to-believe-in-the-historical-adam#.WaaZAq2iGMJ.

[31] Collins (2006, 44); Collins in Barrett and Caneday (2013, 74, 248) speaks of "exalted prose narrative," van den Brink (2017, 130) speaks of "artistic prose."

[32] This is sometimes readily admitted by theologians who themselves do not believe in its historicity; cf. Barr (1993).

[33] Ouweneel (1974). Cf. also the interesting study by Stephen W. Boyd (https://www.icr.org/i/pdf/technical/Statistical-Determination-of-Genre-in-Biblical-Hebrew.pdf).

[34] Cf. Carson (1990, 168); Kinney (2013, on John 1 and 2).

[35] A Jewish tradition says, "The world was made for man, though he was the last-comer among its creatures. This was design. He was to find all things ready for him"; see Ginzberg (1969; https://philologos.org/__eb-lotj/vol1/two.htm#1).

There is no doubt that the remainder of the Bible treats Genesis 1 as history. Well known to any Reformed churchgoer are not just Moses' words but God's own words to Moses:

> Remember the Sabbath day, to keep it holy. Six days you shall labor, and do all your work, but the seventh day is a Sabbath to the LORD your God. On it you shall not do any work, you, or your son, or your daughter, your male servant, or your female servant, or your livestock, or the sojourner who is within your gates. For *in six days the Lord made heaven and earth, the sea, and all that is in them*, and rested on the seventh day. Therefore the LORD blessed the Sabbath day and made it holy (Exod. 20:8–11).

> ...the Sabbath...is a sign forever between me and the people of Israel that in six days the LORD made heaven and earth, and on the seventh day he rested and was refreshed (Exod. 31:16–17).

Nor is it valid to argue that Adam's name is the Hebrew word for "man" or "human," so that the person Adam was supposedly *any* (hu)man (Everyman). In general, translators render the Hebrew word *adam* as "man" when the word has the article (*ha-adam*),[36] and as a proper name when it has no article.[37] Usual exceptions are Genesis 1:26 (*adam*, "man" or "humans" or "humanity"); 2:5 ("no man"); 2:24 (here, "man" is not *adam* but *ish*, the male person, like in 4:1b, 23).[38] Anyone who studies these passages carefully will soon discover that, with the exceptions mentioned, *adam* and *ha-adam* refer to one and the same person.

This man is not Everyman—he is a particular man, without human father and mother, whose wife came forth from his side, who was innocent of any evil before the Fall, who was placed in a wonderful garden, and who fell because he followed his wife into evil after she had listened to the devil. This is not Everyman, because every other man/woman has a father and a mother, receives his wife/her husband in a very different way (cf. 2:24!), and *is sinful from his/her birth.* Everyman does not fall; we are fallen creatures from conception: "What is man [Heb. *ēnosh*, i.e., mortal, fragile man], that

[36] This appears in Gen. 1:27; 2:7–8, 15–16, 19 (3x), 20a, 21–23, 25; 3:8–9, 12, 20, 22, 24; 4:1; 5:1 (2x), 2–5. In the Septuagint, the name "Adam" appears for the first time in Gen. 2:16, 19–23; in the KJV in vv. 19–23.

[37] This appears in Gen. 2:20b; 3:17, 21; 4:25.

[38] Cf. the study by Hess (1990); P. G. Ryken in Barrett and Caneday (2013, 214) erroneously explains the names Adam and Eve as "Man" and "Woman"; *adam* is "human," *chawwah* is "life."

he can be pure? Or he who is born of a woman, that he can be righteous?" (Job 15:14). "How then can man [*ēnosh*] be in the right before God? How can he who is born of woman be pure?" (25:4).

The well-known dogmatician Wayne Grudem rightly remarked,

> While there are sincerely held differences on that question [i.e., what room is there for evolution?] among some Christians with respect to the plant and animal kingdoms, these texts [Gen. 2:7, 21–22] are so explicit that it would be very difficult for someone to hold to the complete truthfulness of Scripture and still hold that human beings are the result of a long evolutionary process.[39]

In other words, you can hardly be an AEH advocate, and have an orthodox view of Scripture.

5.2.2 GOD AS CREATOR-KING

God in his creational work is described in various ways.[40] Here are a few examples. In Job 38, Psalm 104, and Isaiah 40, God is especially the Creator-Builder: laying foundations, determining the earth's measurements, stretching the measuring line over it, laying the cornerstones, and so on. Obviously, we must take into account the poetic nature of the chapters mentioned. Another image is that of God as the Creator-Farmer or Creator-Gardener, which we find in Genesis 2:4–15. It is possible that Genesis 1:2 ("The earth was without form and void, and darkness was over the face of the deep") points to God as Creator-Warlord, if we take the Hebrew words *tohu wabohu* (cf. Isa. 45:18 *tohu*; Jer. 4:23 *tohu wabohu*) and "darkness" (Jer. 4:23, "I looked on the earth, and behold, it was without form and void; and to the heavens, and they had no light") as descriptions of evil (possibly due to the fall of Satan).[41]

In Genesis 1, God is clearly the Creator-King. The activities of building or fashioning are certainly present (see "making" in vv. 7, 16, 25–26),[42] but he is especially the commanding God; his voice is enough to bring into existence the content of his command. Of course, this is not the complete

---

[39] Grudem (1994, 265).

[40] See Ouweneel (2008, §2.4).

[41] See extensively, Kroeze (1962); Ouweneel (2018a); I dealt with this matter already in Ouweneel (1974). To the question how there could have been evil before the Fall, theologians have usually answered that the condition of Gen. 1:2 must have been due to the fall of Satan.

[42] God works with his hands in creation as a potter; cf. Job 10:8–9; 33:6; Ps. 119:73; 139:13; Isa. 29:16; 45:9; 64:8; Jer. 18:6.

picture; in Genesis 1:1–2:3 God is also the workman, who goes through his week of work, and rests on the Sabbath (2:2–3).[43] God's rest was really a refreshment (Exod. 31:17, Heb. *wayyinaphash*, from *nephesh*, "breath, soul"; cf. MSG: "on the seventh day he stopped and took a long, deep breath"; he heaved a sigh of relief, so to speak).

Yet, the picture of the commanding God is prominent in Genesis 1 (and why can commanders not get tired?). Ten times we hear the creative "and God said…" (Heb. *wayyomer Elohim*).[44] The psalmist said, "By the *word* of the LORD the heavens were made, and by the breath of his mouth all their host" (Ps. 33:6). "Let them [i.e., all the works of creation] praise the name of the LORD! For he *commanded* and they were created" (148:5). "By faith we understand that the universe was created by the word [Gk. *rhēma*] of God" (Heb. 11:3). This sheds a special light on the *Logos* (the "Word"), who in John 1:3 is the One through whom all things were made. If in Genesis 1 God created through his word, in John 1:1–3 (cf. vv. 14, 18), this Word received a face, namely, that of the Son of God, who became the Man Jesus Christ.

God's speaking is not a magical speaking, as if certain words had inherent power, independently of the One who utters them.[45] It is *God's* word; it is his power that they contain. Thus, Hebrews 1:3 uses the expression "the word of his *power*," which is a Hebraism: "his powerful word"; through this word God has created and upholds all things.[46]

In the background of God's speaking in Genesis 1 is, as many have pointed out, Israel's polemic with the surrounding nations, who served their own idols, especially the sun and moon (cf. Deut. 4:19; 17:3; Job 31:26–28; Rom. 1:22, 25).[47] In opposition to this, Genesis 1, where the words "sun" and "moon" do not even occur, shows that the "greater light" and the "lesser light" are nothing but servants of God, instituted "to separate the day from the night," and to let them "be for signs and for seasons, and for days and years" (v. 14). The great "sea creatures" (in Hebrew one word, *tanninim*, elsewhere translated "dragons," e.g., Ps. 74:13; Isa. 27:1; 51:9; Ezek. 29:3; 32:2) were

---

43 Cf. C. J. Collins in Barrett and Caneday (2013, 145).

44 A Jewish tradition says, "With ten Sayings God created the world, although a single Saying would have sufficed"; see Ginzberg (1969; https://philologos.org/__eb-lotj/vol1/two.htm#1).

45 Cf. Berkouwer (1971, 125): "But 'magic' is only apparent when we stand outside the fellowship of the Lord and conceive of the expulsion of the 'power of darkness' apart from his communion."

46 Here it is *rhēma*; in the Greek Ps. 33:6 (Septuagint: 32:6) it is *logos* ("By the word of the LORD the heavens were made"), the term of John 1:1–3. Cf. Ouweneel (1982, ad loc.).

47 The notion of this polemical character does not imply that Gen. 1 is a demythologized text (an already extant Eastern text, divested of its mythical features), as A. Donald in Nevin (2009, 35–38) seems to think.

very threatening to Israel, especially *Rahab* (Job 9:13; 26:12; Ps. 87:4; 89:10; Isa. 30:7; 51:9), *Behemoth* (Job 40:15–24), and *Leviathan* (Job 3:8; 41:1; Ps. 74:14; 104:26; Isa. 27:1). But in Genesis 1:14–18, these water monsters are nothing but creatures of God, which are called forth merely by his voice, and are totally under his control.[48]

Notice that such things can never be played off against the historical character of Genesis 1. The chapter does not teach us merely that God is "in control" when it comes to the sun, the moon, and the stars, or when it comes to the great invisible (spiritual) powers in "the air" (cf. Eph. 2:2). No, this chapter tells us also that God *called forth* these lights and these powers. And in Genesis 1 and 2, such calling forth never implies development. The sun and moon did not develop out of some original nebula, but they were called forth by God's voice. "Let there be lights"—and immediately there were lights: sun, moon, and stars. This is essentially identical to Jesus saying, "Let there be wine"—and immediately there was wine (John 2:1–11), or "bread," and there *was* bread (6:1–15). This is what God's commanding word entails: immediate actualization.

Incidentally, what can the theory of evolution tell us about the development of Rahab, Behemoth, and Leviathan, and the other *tanninim*? Nothing, I am afraid. But these are created realities, just as the sun and the moon. And they appeared at God's calling. "For he spoke, and it was" (Heb. *ki hu amar wayyehi*, Ps. 33:9).

The theistic evolutionist (or evolutionary creationist) says, as it were: God spoke, and behold, after some millions or billions of years, it was there; or God spoke extremely slowly, and things appeared extremely slowly. The Bible says, God spoke, and behold, it appeared immediately. The Son of God whom we see in the Gospels as One who is commanding, is the Second Person in the Godhead, that is, the very God who was commanding in Genesis 1. In the Gospels, the things and persons commanded obeyed the God-Man Jesus Christ immediately; I understand that we are being told in Genesis 1 that the things and persons commanded also obeyed God the Father, Son, and Holy Spirit immediately. The same person who rebuked the wind and commanded the sea—and they obeyed immediately (Mark 4:39)—is the One who, in the beginning, commanded the sea and the dry land, the plants and the animals, and finally human beings, to appear. The people asked, "What sort of man is this, that even winds and sea obey him?" (Matt. 8:27).

---

[48] Cf. Spykman (1992, 161–63); see further §7.6.1.

If people wish to be theistic evolutionists or evolutionary creationists—the title matters little—this is up to them. But let no one tell us that their position can be harmonized with Genesis 1–3. The Bible's God who was "calling into existence the things that do not exist" (Rom. 4:17) is not the same as the god who was guiding a development that took millions of years. What King would God be if he gave his command, and he would have to wait millions of years before his command could be considered fully obeyed? Evolutionary creationism confuses God's providence with God's creation. He providentially rules the world through the "word of his power" (Heb. 1:3). God commands his creatures all the time (Ps. 104). This is his providential rule. It is not *creation*.

Evolutionary *creationists* apparently have no proper idea of God's work of *creation*. Providence involves guiding what already exists. Creation involves calling forth what does *not* yet exist. In biblical terminology, if this happened throughout millions of years, by a process of development, this would be God's work of upholding. If something appears instantaneously in response to God's command, then this is creation. It is highly questionable how much room AEH has for *this kind* of (biblical) *creation*, except perhaps the creation of Genesis 1:1, at the very beginning.

5.2.3 *BERESHITH*

Genesis 1 begins with the Hebrew word *bereshith*, "in [the] beginning," a well-known term to indicate the beginning of a king's rule (Ezra 4:6; Jer. 26:1; 27:1; 28:1; 49:34; cf. Gen. 10:10, "The beginning of his kingdom was…").[49] In other words, "in the beginning," God created for himself a kingdom—and he had appointed his Son to be heir of this kingdom already *before* this beginning (cf. Heb. 1:2). No chaos power, no *tohu wabohu*, no darkness,[50] can threaten this kingdom, no matter how strongly Satan and his angels had attempted this. Indeed, Jesus speaks of the "kingdom of Satan" (Matt. 12:25–27), but he also shows that the power of the Holy Spirit is stronger than that of Satan: "But if it is by the Spirit of God that I cast out demons, then the kingdom of God has come upon you" (v. 28). "For the kingdom of God does not consist in [idle] talk but in *power*" (1 Cor. 4:20). "You will receive power when the Holy Spirit has come upon you" (Acts 1:8).

---

49 Notice the difference with John 1:1–2: no matter what "beginning" may be in mind (including that of Gen. 1:1), the Logos *was* already there. Cf. also the phrase "in/from the beginning" (Gk. *kat' archas*) in the quotation in Heb. 1:10, which in the original (Ps. 102:25) is "of old, long ago" (Heb. *lephanim*).

50 Notice that the kingdom of Satan is a realm of darkness (cf. Luke 22:53; Acts 26:18; Col. 1:12–13; 1 Pet. 2:9).

This (creative) *power* is manifested in Genesis 1, as is evident already in verse 3: "And the Spirit of God was hovering over the face of the waters." When God sends forth his "Spirit" (Heb. *ruach*, which can also mean "breath"), his creatures appear (Ps. 104:30; cf. 33:6). As Elihu said, "The Spirit [Heb. *ruach*] of God has made me, and the breath [Heb. *neshamah*] of the Almighty gives me life" (Job 33:4; cf. Isa. 42:5). Word and Spirit are intimately related here (cf. Eph. 6:17): the Spirit is the breath of God's mouth, and his word is carried, so to speak, by the breath of his mouth. Merely with his voice, the King commands each creature to come into existence, and to take the place assigned to it. If the "new birth" is through "water" (I take this to be an image of God's Word; cf. John 15:3; Eph. 5:26) and Spirit (John 3:3–6), so too was the birth of the first creation.

God's Kingship came to light, for instance, by his giving names to each of his creatures: to light and darkness (Gen. 1:5), to the firmament (v. 8), to the dry land and to the gathered waters (v. 10). In agreement with this, Adam as God's vice-gerent (vv. 26, 28) gave names to *his* subjects, the animals (2:19–20), and even to his wife (3:20; although she is more co-regent than subject). In this way, Adam did two things: he expressed his insight into the essence of all God's creatures,[51] and he exercised the authority that God had given him over them. No true kingship is conceivable without these two characteristics: wisdom and authority (see Prov. 20:26, "A wise king winnows the wicked").

Name-giving and name-changing are the prerogative of kings. Pharaoh Neco changed the name of Eliakim (2 Kings 23:34), and Nebuchadnezzar, through his chief eunuch, changed the names of Daniel, Hananiah, Mishael, and Azariah (Dan. 1:7). The "last Adam" did the same as the first Adam: he gave names to Simon (naming him Cephas or Peter, both meaning "rock"; Mark 3:16; John 1:42), as well as to James and John (naming them Boanerges, "sons of thunder"; Mark 3:17),[52] and he gives new names to the conquerors of Pergamum (Rev. 2:17). And above all, God gave Christ a new name— indeed, the name to which every knee shall bow (Phil. 2:9). Such name-giving belongs to the power of those who are in charge (cf. Gen. 17:5, 15). It is one of those features that express God's Kingship in Genesis 1.

---

[51] Cf. names given in the Bible to which the meaning is immediately *added* (in Gen. 3:20; 4:25; 5:29; 10:25; 11:9).

[52] Perhaps, Jesus gave Nathanael (John 1:45–51) the name Bartholomew (Matt. 10:3)—or gave Bartholomew the name of Nathanael.

5.3 ANIMALS AND HUMANS

5.3.1 RELATIONSHIP

Rabbi Benno Jacob was an expositor who pointed to a fine distinction between animals and humans in Genesis 1.[53] About the animals of the fifth day of creation we read, "And God blessed them, saying, 'Be fruitful and multiply and fill the waters in the seas, and let birds multiply on the earth'" (v. 22). Especially the latter phrase indicates that God did not really address the animals, just as he had not addressed the light and the light-bearers, the plants and the trees, mentioned earlier in the chapter. However, about the humans created on the sixth day we read, "And God blessed them. And God *said to them*, 'Be fruitful and multiply and fill the earth and subdue it, and have dominion over the fish of the sea and over the birds of the heavens and over every living thing that moves on the earth'" (v. 28). The correspondence is that in both cases God blessed the respective groups of creatures, and called upon them to be fruitful and multiply. The differences are not only that, in the second case, God called upon the first humans to have dominion over the animals, but that he spoke *to* them. In the former case, we read "and he blessed them, saying" (Heb. *wayyebarek otam... lemor*); but in the latter case, we read "and he blessed them, and he said to them" (Heb. *wayyebarek otam...wayyomer lahem*). In the former case, God pronounced blessing over the animals, so to speak; in the latter case, he explicitly addressed his blessing to the first humans.

God continued to speak to the first humans in verses 29–30, as well as in 2:16–17. He addressed them as beings created in his image and after his likeness in order to enable him to have a relationship (communion, fellowship) with them. I repeat, God did not choose from a multitude of hominids some people with whom he could *enter* into a (covenant) relationship, as some AEH advocates wish to suggest. No, this relationship was there from the beginning of human existence; this relationship belongs to humans *being* the very images of God. This relationship with God is not an element that is *added* to the identity of the first humans; no, from the outset this relationship belonged to their very essence. We will return to this important matter in the next chapter.

We may wonder whether Genesis 3:8 ("the sound [or, voice] of the Lord God walking in the garden in the cool [lit., wind[54]] of the day") points to

---

[53] Jacob (1974, ad loc.).

[54] Nachmanides rendered the phrase as "In the wind of day," suggesting that God's manifestation was signaled by a strong wind (cf. 1 Kings 19:11); see Cohen (1983, 14). God's *ruach* was working in creation (Gen. 1:2), on the *first* day (v. 5), and I wonder if there is not a parallel

God's *habit* before the Fall.

> They heard the voice of the LORD God walking in the garden—The divine Being appeared in the same manner as formerly—uttering the well-known tones of kindness, walking in some visible form (not running hastily, as one impelled by the influence of angry feelings). How beautifully expressive are these words of the familiar and condescending manner in which He had hitherto held intercourse with the first pair.[55]

5.3.2 FROM DUST TO DUST

AEH advocates look diligently for the possible *similarities* between Adam and the animal kingdom. The Bible tries to make the *differences* between humans and animals as clear as possible. A sombre man, who has decided to know nothing more than what is "under the sun," may conclude: "What happens to the children of man and what happens to the beasts is the same; as one dies, so dies the other. They all have the same breath [Heb. *ruach*], and man has no advantage over the beasts, for all is vanity" (Eccl. 3:19). Verse 21 asks: "Who knows whether the spirit [Heb. *ruach*, breath] of man goes upward and the spirit [Heb. *ruach*, breath] of the beast goes down into the earth?" The answer comes almost at the end of the book, though it is not always properly expressed in the translations: "the dust [i.e., the human body] returns to the earth as it was [cf. Gen. 2:7; 3:19], and the spirit [Heb. *ruach*] returns to God who gave it" (Eccl. 12:7; cf. Gen. 2:7).

One might put it this way: the dust returns to the "mother," the spirit to the "Father"; see how the mother's womb and Mother Earth (which in the Bible is never deified, unlike in paganism!) are identified in the following passages: "Naked I came from my mother's womb, and naked shall I *return*" (Job 1:21; cf. 10:9; 34:15). That is, coming from the womb and going to the earth are not essentially different, as the Latin words make clear: *mater* = mother, *materia* = matter; the earth is "mother substance." "For you formed my inward parts; you knitted me together in my *mother's womb*. . . . My frame was not hidden from you, when I was being made in secret, intricately woven in the depths of the *earth*" (Ps. 139:13, 15). The deuterocanonical Sirach 40:1 (RSV) says, "Much labor was created for every man, and a heavy yoke is upon

---

here: God's *ruach* is here again, *ruach hayyom*, the "wind/Spirit of the day of Adam's Fall," *or* the wind/Spirit of *every* day that God had walked with the first humans until this day (which might have been a total of not more than one or two days).

55 Jamieson-Fausset-Brown Bible commentary (http://biblehub.com/genesis/3-8.htm).

the sons of Adam, from the day they come forth from their mother's womb till the day they return to the mother of all."[56] We *are* "dust" (through our mothers), and we *return* to "dust" (i.e., the earth) (cf. Gen. 3:19).

5.3.3 ETERNITY BEINGS

Notice in particular the implication of Ecclesiastes 12:7: animals exist for a time; they disappear forever into the earth's dust, and their last breath evaporates in the air. The same happens to the human body, but in their essence human beings live on, for they are *beings of eternity*. Therefore, the same author who says that God has put "eternity" (Heb. *olam*) into the human heart (Eccl. 3:11),[57] also says that the human "spirit" (Heb. *ruach*) returns to its Giver, God (12:7; cf. Gen. 2:7). The spirit lasts forever, according to Wisdom 2:23 (RSV), "God created man for incorruption, and made him in the image of his own eternity." A person's *body* may be corrupted, but one's essence lives on in eternity—God's eternity.

Human beings are forever—so then let AEH advocates tell us at what point in supposed human evolution did hominids stop existing at their death but instead become "eternity beings"? What a dilemma! Why do we never hear AEH advocates explaining such essential things to us? When and how did the change happen? Or maybe this too was the kind of event where divine intervention was indispensable? Is it the case that, on the one hand, to reassure modern science, they advocate human evolution, while on the other hand, to reassure orthodox Christians, they posit some kind of divine intervention? There is no "hereafter" for animals—there is one for humans. Or what is perhaps more significant (and more in line with Old Testament revelation): there is no resurrection for non-human hominids, but there is one for humans (Isa. 26:19; 53:10; Dan. 12:2).[58] There is a "resurrection of the dead" (Matt. 22:31), more specifically, a "resurrection of life" and a "resurrection of judgment" (John 5:29), a "resurrection of both the just and the unjust" (Acts 24:15). Who could take AEH advocates seriously if they do not answer these questions for us:

(a) When and how did this tremendous transition occur, namely, that two hominid parents, that were still *animals* that ceased existing at

---

[56] Where languages have different genders for nouns, the noun "earth" seems always to be feminine: Heb. *erets*, Gk. *gē/gaia*, Lat./Span. *terra/tierra*, Dutch/Ger. *aarde/Erde*.

[57] Others render it as "a sense of eternity" (GW, NOG, VOICE), or "a sense of past and future" (NRSV).

[58] Other Old Testament passages (Hos. 6:2; 13:14; Ezek. 37:1–14) seem instead to refer to a spiritual "resurrection" (revival) of Israel; opinions differ.

death, gave birth to *human beings* who were designed for eternity? What changed temporary animals into eternity beings?

(b) If from the beginning, each of these humans were either just or unjust, either en route to eternal life or en route to eternal judgment (see the verses just quoted), then according to what criteria was this to be determined?

(c) What divine message—that is, something essentially more than mere "moral consciousness"—did these early humans possess so that they had a fair opportunity for choosing the right way, the way to eternal life, the way that was according to God's criteria?

The Bible underscores the suddenness of human appearance on the scene of this world, as God's representative, destined to live with God *forever*. In opposition to this, AEH advocates hold firmly to a *gradual* development of humans from early hominids. On the one hand, in order to underscore the special significance of humans as compared to animals, as well as the special significance of the Fall, AEH advocates must invent one or two major ruptures in hominid evolution of any kind, the ruptures that involved the origin of humans (in the biblical sense) and some kind of a Fall. On the other hand, to maintain membership in the community of other evolutionists, they must maintain the relative gradualness of human evolution. Thus they are confronted head-on with a very uncomfortable choice: either they do not belong (any longer) to the community of those who wish to maintain the historical character of Genesis 1–3, or they do not belong (any longer) to the community of ordinary evolutionists, who cannot account for these ruptures at all. The only community prepared to welcome AEH advocates is the company of liberal Christians, who no longer maintain the traditional approach to Genesis 1–3 as historical.

5.4 IMAGE OF GOD

5.4.1 LITTLE LOWER THAN *ELOHIM*

The exalted position of Adam is essentially different from that of even the highest animals: *the latter* were only subjects, *he* (Adam) was the ruler. Imagine again the dilemma confronting AEH advocates: in some way or another, one out of many hominids, or a selected hominid population, is no longer a subject (or group of subjects) of the great King, but is called to kingship himself/itself. Adam was destined by God to rule[59] the world on God's behalf. This is why I called him a viceroy or vice-gerent.[60] But the biblical expression is different: Adam was created "in the image of God," that is, as his representative on earth (see §§6.1.3 and 6.4). In the Egyptian and Mesopotamian society, the king or a person of high authority was sometimes called "image of God."[61] In Genesis 1, this expression is applied to Adam and Eve. Just as the pagan king represented his deity, which was the real king of the nation, so Adam and Eve, as vice-gerents, represented God the King.

Let me extend the parallel a bit further.[62] At the exodus from Egypt, God had a conflict, not just with the Pharaoh of Egypt, but with the "gods of Egypt" (Exod. 12:12; Num. 33:4). These gods were invisible spiritual powers, hostile toward the God of the Bible. Pharaoh was nothing less than the earthy representative, even the embodiment, of his god. Similarly, Nebuchadnezzar was the earthy representative, even the embodiment, of his god. This god was an angelic "prince"; for this word "prince" (Heb. *sar*), see Daniel 10:13 and 20–21, where we hear about the "princes" (angelic rulers) of Persia and Greece. In the case of the king of Babylon, the embodiment went so far that the description of this king, after he had died, merges into the description of his angelic prince (Isa. 14:12–15[63]). In Jeremiah 51:34, we read that the earthly king "devours" Judah, and in verse 44 the god Bel must spit out that same Judah; here, too, the king and his god are one.

These things may help us understand to what an exalted position Adam and Eve had been called: as "images" of God they were representatives, if not virtual embodiments, of God. This is the original sense of Psalm 8:5 (apart from its application in Heb. 2:4–5 to Jesus, the Son of Man): God made humans just a little lower than God himself, *crowning* them with glory and honour, as is fitting for vice-gerents: "You placed the son of man just beneath God *and honored him like royalty*, crowning him with glory and honor" (VOICE). As a healthy counter-balance, we notice two things, however:

---

[59] Heb. *radah*, lit. "to trample down" (cf. Num. 24:19 GNT), hence "to have dominion," often royal dominion (1 Kings 5:4; Ps. 72:8; 110:2).

[60] Another well-known term is "steward," someone managing the domain of an owner (e.g., Bingham, 1981; https://www.ewtn.com/library/PAPALDOC/JP2STWRD.HTM), honouring God's supreme Kingship.

[61] Hamilton (1990, 135).

[62] See Ouweneel (2018a, §3.1, and Appendix 10).

[63] The church fathers applied this passage, as well as Ezek. 28:1–10, to the fall of Satan; this is correct insofar as the angelic princes of Babylon and Tyre are nothing but angels of Satan; in Rev. 12:9 and 20:2, Satan, the "dragon," is the angelic prince of the restored Roman Empire of the end times as described in Rev. 13; see extensively Ouweneel (2018a).

(a) *Creation*: even in creation, Genesis 1 may underscore human grandeur, but Genesis 2 underscores human smallness. In 1:26, Adam is nothing less than "image of God," but in 2:7 he is nothing more than "dust from the ground."[64] Both are true: according to their bodily aspect, humans are but dust; according to their spiritual aspect, they are God's own image and likeness. A human being is glamour and clay, dignity and dust, eminence and earth, grandeur and ground at the same time (see further in chapter 6).

(b) *Fall*: whatever grandeur Adam and Eve may have possessed when created they largely lost in and through their Fall. No hominid was being exalted to become "image of God"—how should one imagine such an event?—but the opposite occurred: the first pair of humans, who had been created "in the image of God," became the most miserable of all creatures.[65] Notice again verses 1–3 in Genesis 5, "When God created man, he made him in the likeness *of God*.... When Adam had lived 130 years, he fathered a son in his own likeness, after his image, and named him Seth." Although some have argued that the two sentences express the same thing,[66] I cannot help seeing a contrast here (see further in the next chapter).

5.4.2 ADAM AND CHRIST

It is interesting to see the use Hebrews 2 makes of Psalm 8:

> For it was not to angels that God subjected the world to come, of which we are speaking. It has been testified somewhere,
>
> "What is man, that you are mindful of him,
> or the son of man, that you care for him?
> You made him for a little while lower than the angels;
> you have crowned him with glory and honor,
> putting everything in subjection under his feet." [Ps. 8:4–6]
>
> Now in putting everything in subjection to him, he left nothing outside his control. At present, we do not yet see everything in subjection to him. But we see him who for a little while was made lower than the angels, namely Jesus, crowned with glory and honor because of the

[64] Cf. Sailhamer (1990, 40–41).

[65] The French Christian philosopher Blaise Pascal (1623–1662) in his *Pensées* (1995; "Thoughts," sections 3 and 6) has beautifully described this double-sidedness: human *grandeur* and human *misère* (wretchedness).

[66] Rabbi Nachmanides; see Cohen (1983, 22).

> suffering of death, so that by the grace of God he might taste death for everyone (Heb. 2:5–9).

The common view of the Psalm has been well summarized by Abraham Cohen: "From one point of view man is so insignificant in comparison with the vastness of God's works that it is surprising He deigns to give him a thought. On the other hand, he is the human lord of the earth and endowed with powers which make him little less than divine."[67] Cohen did not view Psalm 8 as a Messianic psalm, and without Hebrews 2 we would not have done so either. I am not aware of any Jewish expositor who ever regarded the psalm this way. Hebrews 2 recognizes Jesus in this psalm because of the common New Testament knowledge that Adam is both type and antitype of Christ (cf. Rom. 5:12–21; 1 Cor. 15:45–49). "Type" implies similarity, "antitype" implies contrast (see §4.5.5).

On the one hand, the similarity is this: the first Adam (with Eve) was appointed to be ruler of God's creation (Gen. 1:26–28), the last Adam was appointed to be the ruler of the world to come. On the other hand, the remarkable contrast is as follows. Let us look first at Adam, who was "made a little lower than God [Heb. *elohim*, or heavenly beings, angels]" (Ps. 8:5) in the sense that this "man of dust" was highly exalted, so highly that, having been created in the image and after the likeness of God, he was just a little lower than God himself. Cohen again: "There is an obvious reference to Gen. 1:27, where man is declared to have been created *in His own image*. As compared with the rest of the animal kingdom, the human being is on a far higher plane owing to the Divine element which is in him. For all that, he is *less than God*, the Being Who is infinite."[68] Here, Cohen points to the essential difference between animals and humans in a way that may appeal to AEH advocates.

Let us now, by way of contrast, look at Christ. Adam was exalted above the animals in being (a little) "less than God." However, in the case of Christ it was the other way round. He was "in the form of God," but "made Himself of no reputation, taking the form of a bondservant, [and] coming in the likeness of men" (Phil. 2:6–7 NKJV). Expositors may differ about whether, according to Hebrews 2, Jesus was "made for a little while [or, made a little] lower than the angels" when he became Man, or only when he had died.[69] But in any case, he was in the form of God and took a lower

[67] Cohen (1985, 18).

[68] Ibid., 19.

[69] Cf. KJV, "who was made a little lower than the angels for the suffering of death"; see Ouweneel (1982, ad loc.).

position. In short: Adam was *exalted* to a position of being "a little lower than the angels [or, than God]," Jesus was *humbled* to a position of being "a little lower than the angels."

5.4.3 THE GOD OF GODS

No AEH advocate can ever do justice to the grandeur of Adam and Eve's position before the Fall. It is inconceivable how, during some supposed human evolution, a particular hominid pair was made vice-gerents of the world, including vice-gerents over all the other hominids. Reflect for a moment on the implausible notion of some AEH advocates that God would have chosen an entire *population* of hominids, in order (a) to promise them everlasting life (in contrast with all other hominids, who ceased and cease to exist at their death), (b) to elevate those who were chosen, to the grandeur of the image and the likeness of God, and (c) to make them vice-gerents over the other hominids and the rest of the world.

The calling of the first pair of humans to subdue the earth and to have kingly dominion over it and over all its inhabitants (Gen. 1:26, 28) relates to God's own Kingship. As the principal Ruler, God had the right to install his highest creatures in this office of vice-gerents. To this end, he *did not select* them from the animal world, but *created them entirely separately* from it, so that every confusion was excluded. It is as if God spoke to the first humans the words of Pharaoh: "You shall be over my house, and all my people shall order themselves as you command. Only as regards the throne will I be greater than you" (Gen. 41:40). After their redemption from Egypt, the Israelites said, "The LORD will rule as king[70] forever and ever" (Exod. 15:18 GW; after this, we hear similar words many more times). In a comforting way, God testified in Genesis 1 to all Israelites, who lived in fear of the other nations, or even more, of the spiritual powers behind these nations: the Almighty God is in control, all things are subjected to him, every creature is at his disposal.

The Fall did not change anything about this divine rule, as testified, for instance, by Psalm 104. Not only in Genesis 1, but also after Genesis 3, we may perceive in God's creation God's "eternal power and divine nature," as the apostle Paul puts it (Rom. 1:20). The One in and through and for whom God created the world is also the One in whom he holds it together (Col. 1:16–17). The Son through whom God created the world, is the One who "upholds the universe by the word of his power" (Heb. 1:2–3).

---

[70] The phrase "rule as king" is one Heb. word, of the root *mlk*, different from the root *rdh* used in Gen. 1:26, 28; cf. Num. 23:21 and Deut. 33:5, where God is referred to as *melek*, "King."

The gods are a reality, but they are nothing but demons, fallen angels, even "angels of Satan" (cf. Matt. 25:41; Rev. 12:7, 9). Those who serve God have nothing to fear, for God is the "God of gods" (Deut. 10:17; Josh. 22:22 NKJV; Ps. 136:2; Dan. 2:47; 11:36): "There is none like you among the gods, O LORD" (Ps. 86:8). "For who in the skies can be compared to the LORD? Who among the heavenly beings is like the LORD, a God greatly to be feared in the council of the holy ones, and awesome above all who are around him?" (89:6–7). "For the LORD is a great God, and a great King above all gods. In his hand are the depths of the earth; the heights of the mountains are his also. The sea is his, for he made it, and his hands formed the dry land" (95:3–5). "For great is the LORD, and greatly to be praised; he is to be feared above all gods. For all the gods of the peoples are worthless idols, but the LORD made the heavens. Splendor and majesty are before him; strength and beauty are in his sanctuary" (96:4–6; cf. 97:7, 9). "All worshipers of images are put to shame, who make their boast in worthless idols; worship him, all you gods!... For you, O LORD, are most high over all the earth; you are exalted far above all gods" (97:7, 9). "For I know that the LORD is great, and that our LORD is above all gods. Whatever the LORD pleases, he does, in heaven and on earth, in the seas and all deeps. He it is who makes the clouds rise at the end of the earth, who makes lightnings for the rain and brings forth the wind from his storehouses" (135:5–7; cf. 136:2).

I quote these verses to remind us of the grandeur of prelapsarian Adam and Eve.[71] As we read each of these passages from the Psalms, we remember: *the image of this God was reflected in Adam and Eve.* Their grandeur was nothing less than the creaturely reflection of God's own grandeur. No hominid could ever have satisfied this ideal according to which God *created*—not developed or selected—the first humans.

> I look up at your macro-skies, dark and enormous, your handmade sky-jewelry, moon and stars mounted in their settings. Then I look at my micro-self and wonder, Why do you bother with us? Why take a second look our way? Yet we've so narrowly missed being gods, bright with Eden's dawn light. You put us in charge of your handcrafted world, repeated to us your Genesis-charge, made us lords of sheep and cattle, even animals out in the wild, birds flying and fish swimming, whales singing in the ocean deeps" (Ps. 8:3–8 MSG).

---

[71] Some rabbis understood this "grandeur" very literally: "Rab Judah said in Rab's name: The first man reached from one end of the world to the other...[see Deut. 4:32 Heb.]. But when he sinned, the Holy One...laid His hand upon him and diminished him [see Ps. 139:5]" (Talmud: Sanhedrin 38b).

5.5 THE GOAL OF CREATION

5.5.1 AGAIN, THE HISTORICITY OF GENESIS 1–3

Genesis 1 intends to introduce to us the first humans, coming forth directly from God's hand into the freshly created world. More than that, the chapter implicitly anticipates the Fall, and the redemption that God has prepared before the foundation of the world. That is, Genesis 1 and 2 do not describe some "creation-in-itself," but a creation that we must understand in order to understand the Fall. Similarly, we must understand the Fall in order to understand salvation. As Frederick Grant put it, the story of the old creation is recapitulated in a wonderful way in the new creation (cf. 2 Cor. 5:17; Gal. 6:15).[72] He said this with reference to Paul's words: "God, who said, 'Let light shine out of darkness,' has shone in our hearts to give the light of the knowledge of the glory of God in the face of Jesus Christ" (2 Cor. 4:6). This is not some arbitrary typological application of Genesis 1:3, but gives the essence of it: just as Genesis 1 and 2 tell the story of the first creation, so Genesis 3 and the rest of the Bible tell the story of the renewed creation.[73] *This makes sense only if Genesis 1–2 is just as historical as the rest of the Bible.*

This juxtaposition, which is also an opposition, is stated more clearly elsewhere in Paul:

> Thus it is written, "The first man Adam became a living being" [Gen. 2:7]; the last Adam became a life-giving spirit. But it is not the spiritual that is first but the natural, and then the spiritual. The first man was from the earth, a man of dust; the second man is from heaven. As was the man of dust, so also are those who are of the dust, and as is the man of heaven, so also are those who are of heaven. Just as we have borne the image of the man of dust, we shall also bear the image of the man of heaven (1 Cor. 15:45–49).

This entire comparison is inconceivable unless the first Adam's story is just as historical as the last Adam's story. In other words: *if there never was a "man of dust" (a man who came instantaneously and directly from dust), how can we know there is a "man of heaven" (a man who came, and will come, directly from heaven)?*

---

[72] Grant (1901, 549).

[73] Please note, this renewed creation is *not* a different and brand new creation, but the exaltation of the first creation; see Ouweneel (2017, chapter 9). In *this* sense, the entire Bible is still about the first creation.

5.5.2 CREATION AND RESURRECTION

For the reasons just adduced, I believe that evolutionary creationists will inevitably and ultimately begin to doubt the bodily resurrection of Christ, and his bodily return from heaven. As soon as the natural sciences begin to govern the interpretation of the early chapters of the Bible, they eventually they will also govern our interpretation of the Gospels and of the last chapters of the Bible. What hermeneutical rule determines where in the Bible the natural sciences are allowed to exert their influence, and where not?

Consider (1) the origin of life (abiogenesis). This occurred a long time ago, and no human being was present. The manner of this origin can be explained fundamentally in only two possible ways:

(a) It happened spontaneously, purely according to the natural laws known to us. However, the natural laws known to us make it virtually impossible that life could originate spontaneously. Yet, it could not have occurred otherwise because appeals to divine intervention have no place in science.

(b) It is impossible that life could originate spontaneously according to the natural laws known to us. Therefore, this must have happened through divine intervention. In normal science such an assumption is inappropriate; however, here we are not dealing with normal science, but with a unique event, which no human being witnessed.

Now compare this with (2) Christ's bodily resurrection from the dead, in which all orthodox Christians believe. Again there are fundamentally only two possibilities for explaining how it happened:

(α) It happened spontaneously, entirely according to the natural laws known to us. However, the natural laws known to us make it virtually impossible that a really dead person could become alive again. Yet, it could not have occurred otherwise because appeals to divine intervention have no place in science.

(β) It is impossible that a really dead person could become alive again according to the natural laws known to us. Therefore, this must have happened through divine intervention. In normal science such an assumption is inappropriate; however, we are not dealing with normal science here, but with a unique event, to which no human being was a witness.

In my view, topics (1) and (2) are totally comparable. How, then, is it possible that, when it comes to topic (1), some Christians choose (a), whereas, but when it comes to topic (2), these same Christians choose (β)? Where is the logic here? Can they not comprehend the fear of orthodox Christians that when it comes to topic (2), although they still choose (β), eventually, in order to maintain logical consistency, they will choose (α), or rather – since nobody really believes in (α) – they will reject the bodily resurrection of Christ altogether? Of course there are Christians who believe both (b) and (β); at least they are consistent. However, the danger is that ultimately they will undertake two steps. First, they may gradually move from (b) to (a) with respect to topic (1), because of the supposedly powerful argument claiming that appeals to divine intervention are scientifically unacceptable, even when origins are concerned. From there it will not be such a big step to move from (β) to (α) with respect to topic (2).

1 Timothy 1:18–19 teaches an important principle: "This charge I entrust to you, Timothy, ...that...you may wage the good warfare, holding faith [Gk. *echōn pistin*] and a good conscience. By rejecting this, some have made shipwreck of their faith [Gk. *tēn pistin*]." Paul says here, If you don't hold "faith" (Gk. *pistis* without the article), you will also make shipwreck of your faith (Gk. *pistis* with the article). I take it that the first "faith" describes the condition of the heart, while the second "faith" is the Christian truth (as in Jude 3, "the faith that was once for all delivered to the saints").[74] Losing basic Christian truths begins, not simply by making theological or philosophical choices, but with making a *moral* choice of the heart. This is why the road of evolutionary creationism is so dangerous. Someone may freely choose to travel this road—but Paul clearly tells us where this road may end.

5.5.3 A TIME PROBLEM

What is inconceivable in evolutionary thinking is explicitly taught by Scripture: the entire cosmos, with its billions of galaxies, was created for humanity. "For the LORD, who created the heavens (He is God, who formed the earth and made it; He established it and *did not create it to be a wasteland* [Heb. *tohu*], *but formed it to be inhabited*) says this, 'I am the LORD, and there is no one else'" (Isa. 45:18 AMP). God created the sun, moon, and stars to serve humanity, something implied in Genesis 1:14–18 (cf. also Jer. 33:20, God's covenant with the day and the night).[75] And he created the earth not to remain a wasteland (as it was in Gen. 1:2, *tohu wabohu*) but to be inhabited by humans: "God, the LORD, who created the heavens and stretched them out, who spread out the earth and what comes from it, who gives breath to the people on it and spirit to those who walk in it" (Isa. 42:5).

All the plants and all the animals, all the mountains and all the seas, are there for humanity. As the crown and the head of creation, the first humans

---

[74] See various interpretations at http://biblehub.com/1_timothy/1-19.htm.

[75] People have tried to express this philosophically with the phrase *anthropic principle*: the fine-tuned universe looks as if it were designed for humanity; see Barrow and Tipler (1988).

receive God's creation as a gift from his hand. They were called to rule this creation on behalf of God and for the honour of God, but they were also allowed to enjoy it freely. This is still the case after the Fall: "foods that God created to be received with thanksgiving by those who believe and know the truth. For everything created by God is good, and nothing is to be rejected if it is received with thanksgiving, for it is made holy by the word of God and prayer" (1 Tim. 4:3–5).

One of the most remarkable purposes for which God created the world is implied in Hebrews 1:1–2, where God reveals himself in his "Son, whom he appointed the heir of all things, through whom also he created the world."[76] Please note that Christ is called here the "heir of all things" before it is said that these same "all things" were created "through him." Already before the foundation of the world, in eternity, God made his Son to be the heir of the things that still had to be created. This seems to imply that these things were not only created *through* Christ, but also in order to give them as an inheritance *to* Christ (cf. the intriguing "for" Christ in Col. 1:16; see §5.6.2). This means that, from the beginning, God's creation was Christocentric; it was (eschatologically) designed to be the domain of the *kingdom of God*,[77] as this was entrusted initially to the first Adam, and finally to the last Adam (1 Cor. 15:45–47).

Though my confession has no probative value, yet I must admit—as I did before—that I have always had trouble with the idea that the history of humanity (*Homo sapiens*) supposedly would have spanned between 100,000 and 200,000 years, and that redemptive history would only span 6,000 to at most 10,000 years (assuming that we are living in the last days; 2 Tim. 3:1; 2 Pet. 3:3). This proportion is merely between 3 percent and 10 percent. I find it hard to believe that this was God's plan when he created humanity. Believers believe they have been elected before the foundation of the world (Eph. 1:3–5), yet after creation, humans supposedly had to wait for 90–97 percent of cosmic history before they heard the first words of God's good news. I find this difficult to accept. Therefore, I fully understand why liberal theology wants simply to discard those 6,000–10,000 years, and to abandon the notions of a special creation of *Homo sapiens* and of a historical Fall, a particular event in time and space. I do not agree with their position, but I do understand it.

What I do not understand is that some people wish to appear "orthodox" in some sense (hence the quotation marks!), and therefore try somehow

---

[76] Cf. Ouweneel (1982, ad loc.).
[77] Cf. Verkuyl (1992, 104).

to retain the idea of human creation, a historical Adam and a historical Fall—all the while embracing evolutionary thinking *in toto*, including the supposed evolution of *Homo sapiens*. I do not understand this position. It simply will not work, because it will please neither full-blown evolutionists nor conservative Christians, and in the end this amalgamation cannot be maintained. On the contrary, such a view will inevitably, necessarily, and gradually move to a more and more liberal position.

5.6 THE WONDER OF A GOAL

5.6.1 EVOLUTION HAS NO GOAL

What if world history had no goal? What if our individual lives had no purpose? How should we then live? Should we *invent* a goal and a purpose? Of course, this is what evolutionists accuse religious people of doing: since people cannot live without a purpose, religion invents one for them. Some evolutionists were proud to announce their adult insight: *evolution has no goal*, so try to cope with this "scientific discovery." Evolution has no purpose; it simply happens (see Simpson in §2.3.3). In case we want to speak metaphorically about a "goal" of evolution, then the goal is simply the successful copying of genes. But that, too, does not happen in a strategically planned manner, for evolution is—metaphorically speaking—"blind and indifferent."[78] This is the point of evolutionary atheist Richard Dawkins; the subtitle of one of his books was: *Why the Evidence of Evolution Reveals a Universe Without Design*.[79] The main title of the book is *The Blind Watchmaker*, which hints at the famous watchmaker argument in William Paley's book *Natural Theology*.[80] Paley adduced it as an argument to prove the existence of God: a watch clearly points to a watchmaker. Yes, says Dawkins, there *is* a watchmaker, but he is blind. He is called evolution. Not only ordinary watchmakers can make watches; evolution can too, says Dawkins. Paley forgot that watches do not procreate, but living organisms do. The name of Dawkins' god is evolution.

Over against this naturalistic attitude, we find, for instance, the theory of Intelligent Design (ID).[81] Advocates of ID hold that certain features of

---

[78] http://crucialconsiderations.org/science-and-philosophy/evolution/why-evolution-has-no-goal/.

[79] Dawkins (1986).

[80] Paley (1802).

[81] Some recent publications on this topic include Berlinski (2009); Dembski and Witt (2010); Meyer (2010; 2014); Klinghoffer (2015); Axe (2017); and Van Bemmel (2017, 109–29);

the universe and of living things are best explained by an intelligent cause, not an undirected process such as natural selection. Through the study and analysis of a system's components, they believe to be able to determine whether natural structures are the product of chance, natural law, intelligent design, or some combination of these. Naturalist and evolutionist opponents have tried to dismiss ID as a disguised form of creationism,[82] but this is formally incorrect. ID advocates do not overtly subscribe to the idea of divine creation; they do not even hint at what the "intelligent cause" to which they refer might possibly be. However, the mere mention of a possible "intelligent cause" is enough to provoke the fury of naturalists.

Of course, all Christians (and Jews, and Muslims) believe in "intelligent design" in the sense that they hold the universe to be intelligently designed by God (see Ps. 139:13–16). The only point is whether ID advocates have really shown that the existence of ID can be scientifically proven. But this is not under discussion now. What is under discussion is that "our intuition that life is designed"[83] may possibly not be confirmed by biology but is surely confirmed by what we know from Scripture. There was an Intelligent Designer at the beginning of world history, who created the world for a purpose.

Perhaps this purpose can be summarized with this statement by the apostle Paul (1 Cor. 15:28): in the end, God will be "all in all" (Gk. *[ta] panta en pasin*, perhaps: "all things in all his own," cf. CJB). In Colossians 3:11 Paul speaks about Christ who "is all, and in all" (Gk. *[ta] panta kai en pasin*); in all who are his own Christ is everything. And in Ephesians 1:23, he speaks of Christ's body, the church, which is "the fullness of him who fills all in all" (Gk. *ta panta en pasin*).

5.6.2 CREATION'S CHRISTOCENTRICITY

I cannot stress enough the Christocentricity of creation: Christ

> is the image of the invisible God, the firstborn of all creation. For by [better: in[84]] him all things were created, in heaven and on earth, visible and invisible, whether thrones or dominions or rulers or

---

cf. the comments by Steve Fuller in Nevin (2009, 124–34).

[82] Cf. Forrest and Gross (2007, title: *Creationism's Trojan Horse: The Wedge of Intelligent Design*).

[83] See the subtitle of Axe (2017).

[84] Gk. *en* instrumentalis; in fact, any rendering of this preposition as "by" (cf. both John 1:3 KJV and Heb. 1:2 KJV, where the Gk. *dia* appears) is mistaken as a description of the Son's role in creation; the world was created "in" (*en*) or "through" (*dia*) him, not "by" (*hypo*) him.

authorities—all things were created through him and for him. And he is before all things, and in him all things hold together. And he is the head of the body, the church. He is the beginning, the firstborn from the dead, that in everything he might be preeminent. For in him all the fullness of God was pleased to dwell (Col. 1:15–19).

Notice especially the third preposition: all things were created not only "in him" (Gk. *en autōi*, by virtue of Christ), not only "through him" (Gk. *di' autou*), but also "for him, unto him" (Gk. *eis auton*). That is, Christ was not only the means but also the goal of creation. (In a broader sense, such a formula is true of the Triune God: "From him and through him and to him are all things," Rom. 11:36; therefore all his works praise him: Ps. 8:1; 19:1; 103:22; 145:10.) Again, let me repeat the earlier question. All things were created for (or, [un]to) Christ—but did he have to wait almost 13.8 billion years before the first humans appeared? It is rather mind-boggling to accept that this could ever be what the Father had in mind for his Son. Who am I to say that this would be *impossible*? All I am asking, as a counter-question, is this: Is it *credible*? Is it *plausible*? If somebody were to tell Moses or Paul about it, would they indeed nod their wise heads and say, We presumed it all along? Does this timetable fit the Bible's description of God and his creation?

You need not tell me that, for God, a thousand years are as one day (Ps. 90:4; 2 Pet. 3:8). I have heard these words abused several times, in quite different contexts, and they are irrelevant here. If the supposed 200,000 years of the evolution of *Homo sapiens* were followed by, say, 200,000 years of redemptive history, I would not be complaining. But 6,000–10,000 years? Even in this brief period of time we hear the faithful saints praying, "Return, O LORD! How long? Have pity on your servants!" (Ps. 90:13). It did not take long before the first heretics began to wonder: "Where is the promise of his coming? For ever since the fathers fell asleep, all things are continuing as they were from the beginning of creation" (2 Pet. 3:4). Indeed, it did not take long before leading Christians began to say, "My master is delayed," and began to beat their fellow servants and ate and drank with drunkards (Matt. 24:48–49). But what are these 200 or 2,000 years compared to the supposed 200,000 years of the evolution of *Homo sapiens*? The master returns "after a long time," says the parable (25:19)[85]—but what are 2,000 years compared to the 2–3 million years that the genus *Homo* has

---

[85] At any rate, these two parables in Lk. 25 show that Jesus did not necessarily expect his return at a very short term; *contra*, e.g., Albert Schweitzer (2005); cf. Erickson (2007, 1199–2000).

supposedly been walking around on this earth? Perhaps the Bible has it all wrong. But could it not be that the theistic evolutionists (or evolutionary creationists) have it all wrong?

5.6.3 THE BEGINNING

Let us listen again to what Jesus had to say about this. When asked about the possibility of divorce, his answer was, "What did Moses command you?" His interrogators said, "Moses allowed a man to write a certificate of divorce and to send her away." Now notice Jesus' reply:

> Because of your hardness of heart he wrote you this commandment. But *from the beginning of creation*, "God made them male and female." "Therefore a man shall leave his father and mother and hold fast to his wife, and the two shall become one flesh." So they are no longer two but one flesh. What therefore God has joined together, let not man separate (Mark 10:3–9).

In Jesus' reply, the beginning of creation coincides with the beginning of humanity, *in concreto*, the first pair of humans. From the beginning of light until the beginning of humanity, it was six days (according to Gen. 1); this warrants our claim that the two beginnings more or less coincided. But if almost 13.8 billion years separated the beginning of the cosmos and the beginning of *Homo sapiens*, then Jesus either did not know any better, or he deceived his listeners. *Or* he was speaking the simple truth, but then those 13.8 billion years never existed in the first place.

If Jesus were merely wishing to accommodate himself to the level of his listeners, he could easily have used a different vernacular, in order to avoid saying things that are literally untrue. I prefer another option: he simply took Genesis 1 at face value, accepting its historical character because he believed in the historicity of Genesis 1–3. He was in good company: Moses did too (Exod. 20:8–11; 31:16–17), and Paul did too (Rom. 5:12–21; 1 Cor. 11:8–12; 15:45–49). Or more correctly: Paul and Moses were in the good company of Jesus. The question that God asked Job could have been addressed to Moses and Paul as well: "Where were you when I laid the foundation of the earth?" (Job 38:4). The answer is: Nowhere. With the Son of God it is different. *He was there* when the earth was created. As we have seen, it was even in him and *through* him that all things were created, and even *for* him. He knows everything about creation, infinitely more than AEH advocates and I know, even more than Moses and Paul knew. Therefore, what he told us about creation is true and trustworthy. Thus, this entire controversy that we are discussing ends up being a Christological controversy.

I do not wish to read too much into Proverbs 30:4, but I like the question it asks: "Who has established all the ends of the earth? What is his name, and *what is his son's name*?" Literally, this may be merely a rhetorical question. But spiritually it is perfectly true to say that the One who established the world is named Father, Son, and Holy Spirit. As John Gill (eighteenth century) put it,

> Since it is the Lord alone and his own proper Son to whom these things can he ascribed, say, What is his name? That is, his nature and perfections which are incomprehensible and ineffable. Otherwise he is known by his name Jehovah [read: YHWH], and especially as his name is proclaimed in Christ and manifested by him and in his Gospel; and seeing he has a Son of the same nature with him, and possessed of the same perfections, co-essential, and co-existent, and every way equal to him, and a distinct person from him, say what is his nature and perfections also; declare his generation and the manner of it; his divine filiation, and in what class it is.[86]

[86] http://biblehub.com/proverbs/30-4.htm (I have edited the text lightly).

WHO IS MAN?

When I look at your heavens, the work of your fingers,
the moon and the stars, which you have set in place,
what is man that you are mindful of him,
and the son of man that you care for him?
Yet you have made him a little lower than the heavenly beings
and crowned him with glory and honor (Psalm 8:3–5).

THESIS

The belief in human evolution always leads to both anthropomorphizing (humanizing) animals and zoomorphizing (animalizing) humans, in order to relate them as closely together as possible. Theological views of the essence of a human being based on the idea of human evolution are inherently tempted to downplay and underestimate the true view of humanity as we find it in biblical thought. Once again: without a biblical anthropology there can be no biblical soteriology.

6.1 ELEMENTARY BIBLICAL DATA

6.1.1 ADAM THE HOMINID?

It is rather astonishing to discover how easily AEH advocates posit that the first humans arose from a supposed hominid population. N. T. Wright claims, "*God chose one pair from the rest of early hominids for a special, strange, demanding vocation*. This pair (call them Adam and Eve if you like) were to be the representatives of the whole human race."[1] James K. A. Smith writes:

> From out of this [evolutionary] process there emerges a population of hominids who have evolved as cultural animals with emerging social systems, and it is this early population (of, say, 10,000) that constitutes our early ancestors. When such a population has evolved to the point of exhibiting features of emergent consciousness, relational aptitude, and mechanisms of will—in short, when these hominids have evolved to the point of exhibiting moral capabilities—our creating God "elects" this population as his covenant people.[2]

Gijsbert van den Brink insists:

> It is rather obvious that we were anatomic-modern people before we began to develop the capacity of symbolic thought and acting, and thus—be it gradually, or more saltationally through "emergence"—became modern people cognitively and behaviorally as well.[3]

What is so incredible in such statements from people who consider themselves orthodox Christians? It is the ease with which such authors deliberately or inadvertently ignore what the Bible tells us about humans. Here are just a few things by way of introduction (a more detailed exposition will follow).

1. First and foremost, as we saw, humans are created for eternity (see §5.3.2). No human who ever lived will ever cease to be.[4] Not only did God place eternity (Heb. *olam*) in the human heart (Eccl. 3:11), but humans will *live* eternally. Even the wicked will *exist* forever.[5] In the Garden of Eden humans could have lived forever (cf. Gen. 3:22), but post-Fall believers "will *always* be with the Lord" (1 Thess. 4:17; cf. John 17:24; Rev. 22:5). Now, as I asked earlier, let AEH advocates tell us at what moment, or through what development, hominids—which, like all animals, stop existing when they die (like gorillas and chimpanzees today)—became humans destined for

---

[1] Wright (2014, 37).

[2] Smith in Cavanaugh and Smith (2017, 61); quote italicized in the original.

[3] van den Brink (2017, 223).

[4] About the objectionable annihilation doctrine ("one day unbelievers will cease existing"), see Ouweneel (2012a, §14.4.3).

[5] They are described, after their resurrection, as "the dead," and they will be consigned to "the second death" (Rev. 20:12, 14; cf. 2:11; 20:6; 21:8), so to describe them as "living" is inappropriate here; their torment is "forever" (Rev. 14:11; 19:3; 20:10).

eternity. How does a time-bound animal evolve through natural evolutionary processes into an eternity-bound human? God did not simply decide to *grant* humans immortality (cf. 1 Cor. 15:53), as though, if he had so wished, he could have done the same for "lower" hominids. No, humans are *designed* in their very essence for everlasting existence. Modern believers now share this quality that humans possessed before the Fall: "appointed to eternal life" (Acts 13:48).

2. Perhaps nothing is more defining of being human than having been created "in the image and after the likeness of God." These are *not* honorary titles, which God might have assigned to a certain supposed population of evolving hominids, if he had so chosen. On the contrary, being created in the divine image and likeness expresses the very *essence* of being human. In this quality, they differ essentially from all animals. It is not just a matter of gradually evolving toward higher intellect, more abstract thinking, more complex lingual capacities, higher moral and religious consciousness, as suggested by the authors whom we quoted above. It would be perfectly ridiculous to say that humans gradually evolved to become "images of God." The biblical message is the opposite: after the creation of the vegetative and the animal kingdoms, something totally new arose, something that demanded a distinct creation: *humans*, created in the image and after the likeness of God (see §6.4).

Even our languages express this: a human person is not a *what* (Heb. *mah*, Gk. *ti*), as the animals are, but a *who* (Heb. *mi*, Gk. *tis*), just as God is; a human being is not an "it," like the animals, but a "he" or a "she," resembling God; and note especially this: humans are *persons*, just as God is (see §6.1.3).

3. It is incredible that van den Brink ventures to suggest that the human "artistic and religious consciousness" is a product of an evolutionary process.[6] In an almost casual and matter-of-fact way, religion, or at least some human religious capacity, is being viewed here as a product of evolution. In the Bible, religion is not merely one of the specific characteristics of human beings, but is viewed as perhaps their most *essential* characteristic. The human person is really *homo divinus*—an expression coined by John Stott,[7] and highly appreciated by van den Brink.[8]

"Religion that is pure and undefiled before God the Father is this: to visit orphans and widows in their affliction, and to keep oneself unstained

---

6 van den Brink (2017, 224).
7 Stott (1999, 63).
8 van den Brink (2017, 217, 237).

from the world" (James 1:27)—is such behaviour the product of evolution, which is ruled by the law of the jungle? Or, even if van den Brink were speaking merely of the "religious consciousness…through which people for the first time could know themselves to be addressed by God (or the transcendent),"[9] is this capacity a product of evolution? In van den Brink's presentation, religiosity is one among many characteristics of modern humans. I repeat, in the Bible, this is not at all merely one among many characteristics, but belongs to the essence of being human. More than *homo sapiens* (or *homo faber* or *homo socialis* or *homo economicus* or *homo ludens*; see §6.1.2), we are dealing with *homo religiosus*.[10] This is not meant in the immanent-phenomenological sense of the term "religious," the sense of praying and churchgoing people, but in its transcendent sense (see below). The human relationship to the transcendent, and even a person's *own* transcendence, is not a product of evolution but the essential creaturely difference between animals and humans.

German philosopher and sociologist Helmuth Plessner spoke of the "eccentricity" of human beings.[11] I would put it this way: we cannot speak of humanity as such,[12] a person considered independently, distinct from his/her relationships. The essence of human beings lies outside themselves (is "eccentric"), namely, in their relationship to God (or to transcendent substitutes for God: idols). Here we touch upon a matter that, perhaps in its essence, is the very same as what was explained under point 2.

Claim: the AEH *idea of human beings who are the product of hominid evolution is irreconcilable with the biblical notion of human beings as transcendent eternity beings, created in the image of God, whose essence lies in their transcendent relationship to God.* This will all be elaborated in the rest of this chapter.

6.1.2 THE BIBLICAL ADAM

The American-Jewish thinker Abraham Joshua Heschel has argued that the question "What is Man?" is mistaken; we should ask, "*Who* is Man?"[13] As physical, and even psychical objects humans are fundamentally explicable; but as *persons* they are both a mystery and a surprise. As physical-psychical objects human beings are finite, but as persons they are inexhaustible. The question "*Who* is Man?" is a question of dignity and nobility, a question of position and status within the order of beings. *A human being is a "who"*

[9] Ibid., 224.
[10] I prefer this expression to *homo divinus* to avoid the suggestion that humans are "divine."
[11] Plessner (1975).
[12] Cf. Immanuel Kant's phrase *Ding an sich*, "thing in itself."
[13] Heschel (1965, 28).

because God is a "who." I may add that the first time "who" appears in the Bible as an interrogative pronoun is in God's question to Adam (Gen. 3:11), "Who [sing.] told you [sing.] that you [sing.] were naked?" An interesting question! What other "who's" existed other than God, Adam, Eve, and the serpent (i.e., Satan)? No animal could have whispered this assertion into Adam's ear, and God and Satan did not tell him either. Adam himself was the "who"; he enlightened himself, as an intelligent and responsible being.

Apart from Satan (unmistakably also a "who"), God and Man, as the two great "who's," are ontically and eternally connected. They are connected as Creator and creature, as Original and image. In his soteriological narrative about humans, the Flemish-Dutch Catholic theologian Edward Schillebeeckx referred to the beautiful childlike statement that humans are the words with which God tells his story.[14] Anything God has to say he says in terms of human beings. God and human beings belong indissolubly together. If God is an abysmal mystery, then human beings are too—if human beings are an unfathomable mystery, then God is too. In this respect they are counterparts. I add to this that, if human beings are nothing but evolved animals, there is no mystery surrounding them that physics and biology, and possibly psychology, could not solve in principle.

The *humanum*, ie., what is proper to human beings, consists in the reality that a human being is image of God (Lat. *imago Dei*), designed from the outset to live in a relationship of fellowship with God. Two biblical examples come to mind: God "formed" Israel for his own "glory" (Isa. 43:7),[15] "for himself" (v. 21; to be God's servant; 44:21; cf. 49:5), as a vineyard in which he would find good fruit for himself (5:1–2). God also predestined his people for adoption "to himself" (Eph. 1:5, CSB: "for himself"). Similarly, we may say that, in the beginning, God created human beings "for himself," to find pleasure in them. Human beings were *designed* for this purpose of pleasing God, of serving him, of enjoying fellowship with him.

This biblical picture is in glaring contrast to so many other *humana* that have been suggested in past centuries and decades.[16] These alternative portraits view human beings as physico-chemical machines, as incidental products of a fundamentally blind evolution, or as ennobled animals dominated by their drives and instincts; as *homo faber* (technically capable humanity), *homo socialis* (social humanity), *homo economicus* (economic humanity), *homo ludens* (playing humanity), *homo liber* (free humanity,

---

[14] Schillebeeckx (1989, 5).

[15] Notice the three verbs here, which were also used for Adam: *bara* ("to create," Gen. 1:27), *yatsar* ("to form," 2:7–8), and *asah* ("to make," 1:26).

[16] Cf. Verkuyl (1992, 102–103); Erickson (2007, 486–93).

with many, partially false meanings). Viewing humanity even as *homo religiosus* would be mistaken if religiosity were taken only in the immanent-phenomenological sense. This would entail viewing the person who is religiously active (doing things that are pistically qualified, such as praying and preaching; see below), apart from the question concerning the condition of a person's heart (see §6.4).

The point is not whether all the descriptions just mentioned contain elements of truth; as we will see, humans *are*, among many other things, physical, biotic, psychical, cultural, social, economic, and pistical beings. However, none of these expresses what is the actual essence of human beings. That essence consists of humanity's capacity as *homo religiosus* in the *transcendent* sense, beings who, *because of their special creation*, are totally exceptional, beings who are both temporal and eternal, both immanent and transcendent, beings who from the outset were designed to be in a transcendent relationship with God (although, through the Fall, God is often replaced with idols; Rom. 1:23).

John Calvin wrote the famous statement that "man never achieves a clear knowledge of himself unless he has first looked upon God's face, and then descends from contemplating him to scrutinize himself."[17] No true knowledge of the self exists without knowledge of God; no true knowledge of God exists without knowledge of the self. Reformed theologian Gerrit C. Berkouwer argued,

> The most striking thing in the Biblical portrayal of man lies in this, that it never asks attention for man *himself*, but demands our fullest attention for man in his relation to God. We can doubtless characterize this portrayal as a *religious* one.... The Biblical portrayal of man, as a religious portrayal, also emphasizes that this relation to God is not something *added* to his humanness; his humanness depends on this relation.... Man *without* this relation cannot exist, he is a phantom, a creation of abstracting thought, which is no longer conscious of the relationships, the basic actuality, of humanness, which concerns itself with that which can never exist: *man in himself*, in isolation.[18]

In my view, it is inconceivable how such a lofty view of humanity can be reconciled with the portrait supplied by the theory of evolution. In Genesis

---

[17] Calvin (1960, 1.1.2; see also 1.15.1, 2.1.1); and Calvin on Jer. 9:23–24 (2005, ad loc.); on Calvin's anthropology see extensively, Torrance (1957, especially chapter 1).

[18] Berkouwer (1962, 195–96, 197–98; italics original, but omitted in the published English translation); cf. Dooyeweerd (1984, 1:377).

1, and also in the rest of the Bible, the absolute uniqueness of humanity is evident. This comes to light already in the way the matter is presented in Genesis 1. The sixth day of creation is "the day that God created man on the earth" (Deut. 4:32). Animals belong to another realm. We will forever honor the sixth day of creation, not as the day that the land animals appeared (Gen. 1:24–25) but as the "day of Man"[19]—a new and separate creation of God. There is something solemn and serious and exalted in God's announcement, so different from anything he said before: "Let us make *adam* [*Adam*, Man, humans]" (Heb. *na'aseh adam*, v. 26). And as we saw, the language becomes poetically exuberant when the text tells us: "So God created man [Heb. *ha-adam*] in his own image, in the image of God he created him; male and female he created them" (v. 27). The middle line is parallel to both the first and the third lines, in different ways.

6.1.3 IMAGE AND LIKENESS

The Message paraphrase of Genesis 1:27 reads, "God created human beings; he created them godlike, reflecting God's nature." It is one of the many attempts to express the meaning of "image" (Heb. *tselem*, cf. 5:3; 9:6) and "likeness" (Heb. *demut*, cf. 5:1, 3). An enormous variety of views have sought to explain these expressions.[20] Hendrikus Berkhof summarized these attempts by saying, "By studying how systematic theologies have poured meaning into Gen. 1:26, one could write a piece of Europe's cultural history."[21]

The expressions "image" and "likeness" are closely related, and therefore have often been viewed as more or less synonymous.[22] Yet, it is remarkable that different prepositions are used:[23] humans *are* images of God (cf. 1 Cor. 11:7), but *have* (*bear*) the likeness of God. In my view, "in" (our image) has the sense of "as," "in the quality of"; as Rashi understood it, "in our type," that is, as later rabbis put it, "in the mould that We [i.e., God] have prepared

---

[19] Not to be confused with Job 10:5 ("Are your days as the days of man"); God's "days" are not as short and fragile as the "days of man" (cf. 14:1; Gen. 6:4; Ps. 103:15; 144:4).

[20] Hamilton (1990, 137); see also Berkouwer (1962); Gispen (1974, 73–77); Hoekema (1986, chapter 5); Wenham (1987, 28–33); Westermann (1984, 146–60).

[21] Berkhof (1986, 179); see extensively, Kuyper Jr. (1929); Van der Zanden (n.d.).

[22] Thus Luther and Calvin (1960, 1.15.3); Erickson (2007, 523); Bavinck (*RD*, 2:532, 548–50); Berkouwer (1962, 68–69); Wenham (1987, 29–30); Heyns (1988, 125); Hughes (1989, 7); Noordegraaf (1990, 32); Pannenberg (2010, 211, 220); Spykman (1992, 224). However, many church fathers did make a distinction; see Berkhof (1949, 219); examples include Irenaeus (cf. Bratsiotis (1951/52, 297), Clement of Alexandria, and Origen; see Hughes (1989, 8). In modern times, cf. Schlink (1942), as well as Roman Catholic dogmatics.

[23] The LXX (Gk. *kata*) and the Vulgate (Lat. *ad*) use only one preposition, though, and in Gen. 5:3 the prepositions are transposed; cf. Schmidt (1947, 165–69).

for his creation [i.e., to mould him in it]."[24] The preposition "after" (or "according to") suggests similarity, agreement: God made humans in such a way that they resemble God (cf. James 3:9). Hans Walter Wolff suggests that "image" points to representation of what is imaged, while "likeness" points to mere similarity.[25] Let us briefly examine these two notions.

(a) *Representation*. Perhaps the most important meaning of "image" is that humans *represent* God on earth.[26] AEH advocates say, as it were, that when hominids had evolved far enough, God decided it was time to enter into a covenant relationship with them. The Bible says that when the rest of the creation was finished, God introduced an entirely new being who would be placed over all of creation as his (God's) representative. All creation is subjected to God who is enthroned in heaven (cf. Ps. 115:16). However, God has a substitute here on earth, representing him and exercising God's power over creation. Therefore, all creation must have "fear" (respect, awe) for humanity (Gen. 9:2),[27] just as humans in turn "fear" God (cf. 20:11; 22:12; 31:42, 53; 42:18). This is linked to what Genesis 1:26 says about humans having dominion over all the animals (cf. Ps. 8:6–8). This idea of representation is usually associated with the role of an Eastern prince as well.[28]

The position of AEH is that the first humans were basically usurpers, who arose from and rose above the other hominids, and began to rule over them. In the Bible, humans come from elsewhere, from the hand of God himself (Gen. 2:7). They are entitled to kingship because of their exalted lineage: they are not sons of hominids but sons of God (Luke 3:38). The first humans were *regents*, not rising to their position from among their animal subjects, but assigned to their position from above, called to rule over their subjects (Gen. 1:26, 28), "*crowned* with glory and honor" (Ps. 8:5), as befits kings. Whereas among the Egyptians and Assyrians only the king was described as image of God, in the Old Testament the notion involves all of creation: *human beings* are images of God, and as such they are rulers over the entire creation.

(b) *Similarity*. Humans take after God, look like God, and resemble God. Animals are strictly bound to their drives and instincts. They cannot think rationally; humans think, just as God thinks. Animals have no free will (they are guided by their instincts); humans have a free will, just as God

---

[24] Cohen (1983, 6).

[25] Wolff (1973, 236); cf. Westermann (1984, 146–60).

[26] Cf. Jacob (1974, 59); Schilder (1939, 255–56); Berkouwer (1962, 114–15).

[27] Hughes (1989, 6).

[28] Kruyswijk (1962, 192–96); Kamphuis (1985, 20–22); Wenham (1987, 30–31); Noordegraaf (1990, 32–33).

does. Animals cannot make free decisions; humans can, just as God does. Animals have no imagination; humans do, just as God does. Because they wish to narrow the cleft between animals and humans as much as they can, I know that evolutionists have ascribed intellect, volitional decisions, imagination, and so on, to certain "higher" animals (primates, but also, for instance, dolphins). Conversely, they have tried to ascribe as many animal features to humans as they can (cf. §6.3.1). But who could accept this as genuinely plausible? Consider that between the most intelligent animal and a three-year old child, the differences are enormous.[29]

As I said, animals are not persons; humans are, just as God is.[30] In humans we find not only affections and emotions—which are found in (higher) animals as well—but also self-awareness, intellect, volition, creativity, awareness of time and eternity, and all that belongs to a person—just as we find these in God. As part of the empirical, physical cosmos, animals and humans belong together; both are made from the same physical materials, and both stand in a creaturely, ontic relationship of dependence toward their Creator. However, in their resemblance to God, in their being designed for fellowship with God, humans belong to the world of God. In their immanent mode of existence, humans resemble animals to a large extent. In their transcendent Ego, they resemble God to a large extent.

Of course, atheists have used these similarities between humans and God as an argument that God is simply an invention (a projection) of humans.[31] The Bible reveals the opposite: humans are an invention (a design) of God.

6.2 THE IMMANENT STRUCTURES OF HUMANITY

6.2.1 FIVE LAYERS

Elsewhere, I have described extensively what I call the five levels (layers) of the *immanent* structure of the human corporeal mode of existence. I refer the reader to these writings; here, I limit myself to a brief survey, and to pointing out the relevance of this model for philosophical and theological anthropology, and thus for our discussion concerning the historical Adam. The Bible itself does not contain such a philosophical and theological anthropology, which some might suppose needs simply be extracted from the Bible by a philosopher or theologian. Rather, anthropologies, just like any human theory or model, are *designed* by scientists and scholars, but in

29 See extensively, Collins (2011, 93–100), especially regarding the point of language.

30 Cf. Ouweneel (2007b, 92–93).

31 See especially Feuerbach (2008; orig.: 1841).

such a way that they incorporate all the data as comprehensively as possible. The theory of evolution may have been identified by certain evolutionists as a "proven fact," but of course it is nothing more than a theory, a model. Similarly, the anthropology that I present is a theory, a model—but one, I trust, that does justice to the contents and spirit of Scripture. The model is inspired by the radical (root-based, from Gk. *radix*, root) Christian philosophy of Herman Dooyeweerd and Dirk H. Th. Vollenhoven (see §§3.2.2 and 5.1.2).

By way of applying the work of Dooyeweerd and Vollenhoven I distinguish five *structural layers* within human beings, five levels of organization, one above the other, so to speak.[32] The five structural layers or organizational levels that are distinguished in this model are the following.

1. *The physical structural layer*. This organizational level includes the chemical elements, physical processes, and chemical reactions that, on the one hand, guarantee the unity of the material structure of humans and by which, on the other hand, the material components are yet able to undergo constant change. All animate as well as inanimate material things share this structural layer.

2. *The biotic structural layer*. This organizational level includes the cell structure, the tissue structure, the organic structure, and the physiological life processes of a living organism (breathing, digestion, metabolism, reproduction, hormonal processes, and the like). All living organisms share both the physical and the biotic structural layers in such a way that all living organisms presuppose the physical structural layer, but the biotic structural layer, which lies above it, *cannot be derived from it or reduced to it*. This is the same as saying that, from the biblical perspective, the biotic structural layer required a new, distinct creative act of God. In other words, abiogenesis (the *unguided* origin of life from non-life) is impossible.

3. *The perceptive structural layer*.[33] This organizational level includes the sensory phenomena, which are "supported" or "borne" by physico-chemical structures and processes, and also by biotic processes, but these structures and processes themselves are subject to perceptive laws (such as the laws of stimulus and response). All animals share the physical, the biotic, and the perceptive structural layers. In my view, the origin of such creatures that

---

[32] In Ouweneel (2014a) I called these layers *idionomies*; the term was coined by Pieter A. Verburg and first published by Verbrugge (1984).

[33] Most anthropologists who work in the line of Dooyeweerd and Vollenhoven acknowledge only one psychical (perceptive-sensitive) structural layer. See Ouweneel (1986) for my defense of distinguishing the perceptive and the psychical layers. For my present argument, this point makes no difference.

possess a certain capacity of awareness (which plants do not have) required another distinct creative act of God because the perceptive structural layer cannot be reduced to the lower structural layers. That is to say, awareness cannot develop from unawareness.

4. *The sensitive structural layer*. This organizational level includes feelings, such as affections, emotions, and urges. There are no feelings without physiological and perceptive processes, but feelings cannot be reduced to these processes. Sensitive life is a level of functioning that presupposes (is "borne" by) the previous structural layers, which form the foundation for it, while at the same time, the sensitive structural layer is entirely new. All higher animals share the physical, the biotic, the perceptive, and the sensitive structural layers. Again, in my view, the origin of this sensitive structural layer required a distinct creative act of God. That is to say, feelings cannot develop from non-feelings.

6.2.2 THE MENTAL STRUCTURAL LAYER

5. *The mental structural layer*.[34] In addition to the structural layers already mentioned, humans also have a mental structural layer, which includes thoughts, deliberations, decisions, imagination, and so on. We must note carefully what this means. We cannot construe here some evolutionary process in which higher forms of life gradually acquire a new structural layer, *which supposedly sprouts from the previous structural layers*. There is no sprouting here; each structural layer represents an essentially new and different mode of existence, a new level of organization, *irreducible* to previous structural layers. *There is no developmental process involved*; I have stressed that this is the same as saying that each new structural level required a new, special creation of God.

One more point is vital for our discussion. Plants, animals, and humans share, for example, the physical structural layer, but this does not at all mean that this physical structural layer is the same in each of these categories of life forms. What I mean is that the physico-chemical matter in humans must be capable of "carrying"[35] not only the physiological ("life") processes, as well as the perceptive and the sensitive processes, but also the mental life of humans (thoughts, deliberations, decisions, imagination,

[34] The words "mental" and "mind" as we use them here (a) are meant in an immanent sense, which means that "mind" must be distinguished from the transcendent Ego (see §6.4); and (b) must not be viewed as a substance that exists independently of the body.

[35] In what follows, the verb "carry" and its various synonyms, placed in quotation marks, are intending to communicate the notion of supporting, upholding, bearing along, bearing up, enabling, conditioning, and rendering possible.

etc.). This makes the physical structural layer in humans absolutely unique: it is capable of doing things that the physical structural layer in gorillas and chimpanzees cannot do.

Human mental functioning qualifies the essence of this new, fifth organizational level, which I have called the mental structural layer. This structural layer entails that humans can think, deliberate, make volitional decisions, speak, invent, create, trade, make art, love, believe, and so on. However, I repeat: on the one hand, all these activities cannot be reduced to the physical, the biotic, the perceptive, or the sensitive, but on the other hand, these mental processes would not be possible without the lower structural layers, which "carry" (form the indispensable substratum for) them.

To state the point bluntly: no thinking, no loving another person, and no believing in God (or the gods) occurs without an exchange of potassium and sodium ions on both sides of cell membranes.[36] Two kinds of errors can occur at this point. Those materialists who hold that thinking, loving, and believing can be *reduced* to such physico-chemical processes ignore the true nature of such mental acts.[37] Thinking, loving, and believing are "borne" by these processes, but they are much more than these processes. However, those spiritualists (including many Christians) who suppose that thinking, loving, and believing are activities of some distinct substance (soul, spirit, mind, whatever you call it), which supposedly operates independently of the body, are equally mistaken. No praying and preaching occur without a lot of physics.

As I have explained elsewhere,[38] the physical, the biotic, the perceptive, and the sensitive structural layers are governed by natural laws. But the mental functioning of a human being operates according to its own specific laws. These laws are not to be considered natural laws but *norms*, which possess an entirely different quality. The processes that "carry" mental acts *are* ruled by natural laws, whereas the mental acts themselves are not. Natural laws tell us how things are ("if you heat iron, it expands"). But norms tell us what humans *ought* to do ("if you wish to reason, write, construct, trade, paint, in a proper way, you must follow these and these rules"). Norms can be ignored, but natural laws cannot.

Especially moral and religious norms are characteristic of human beings. Morality and religion cannot be reduced to animal features, although people have tried to do so; to put it bluntly, the "primeval soup" did not contain any morals. Morals come from the Creator, who laid them in the human heart (see §6.3).

---

[36] Thus Wilder-Smith (1981).

[37] A recent example of such a reductionist attempt is Swaab (2014).

[38] Ouweneel (2014a).

6.2.3 FIVE HUMAN STRUCTURES

The five structural layers that together enable immanent-human existence may be called "human structures" (German: *Humanstrukturen*; Dutch: *humaanstructuren*), in order to make clear that humans differ from plants and animals not only in possessing a mental structure, but also in the fact that the four lower human structural layers differ fundamentally from the corresponding structural layers found in plants and animals. For instance, the physical structural layer in plants and animals can "carry" biotic processes, but no mental processes. The important conclusion from this is that the creation of humans *required not only the special creation of the mental structural layer but also the special creation of the previous (physical, biotic, perceptive, and sensitive) structural layers*. In other words, human beings differ from plants and animals not only because people have a mental structural layer while animals do not, but also because people have different physical, biotic, perceptive, and sensitive structural layers than animals.

Of course, I am describing all these things in terms of an anthropological model that the reader might never have encountered before, or does not particularly like. I am using this model only because it helps me explain complicated matters in a simpler way; that's what models are for. But my point here is not to present my model as the final truth about humans. The model is only an aid; what matters are the important insights that I am trying to express through it. And these boil down to one conclusion (among others): human beings cannot have evolved, just as the biotic cannot have evolved from the physical (this would be abiogenesis), nor the psychical from the biotic, nor the mental from the psychical.

Once again, the mental structural layer includes the so-called mental functioning, or, if you like, the functioning of the human mind (as long as the mind is not viewed as some kind of "thing" [substance] independent of the body). Within this structural layer human deliberations, imaginings, and decisions occur. Such operations of the human mind are sometimes called *acts*.[39] Acts are inner operations of the mind that are directed toward resulting actions, toward external behaviours. An animal also performs certain *actions*, but these actions do not arise out of acts, for animals do not have a mental life or act-life. The actions of an animal proceed instead from its perceptively determined instincts and/or sensitively determined urges or drives, not from mental deliberations and free decisions of the will.

Such instincts and drives are not absent in human beings but, if humans truly behave like humans, their actions are determined primarily by human

---

[39] The term "act" in this sense goes back to the German philosopher Franz Brentano (1995; orig.: 1874).

act-life or mental life, that is, by the free, deliberate functioning of one's mind. This is why a human being must take responsibility for his or her actions, whereas an animal does not. Responsibility is "response-ability": a human being has the ability, and the duty, to give a response to the question as to why he or she did this or that, in other words, from which (mental) *acts* his/her *actions* arose. In our mode of existence, we cannot account for those phenomena that are governed entirely by natural laws. But we *can* and *must* account for our actions insofar as they are governed by *norms*.

A human being can, and should, "answer" for his or her deeds. This is why some Christian philosophers like to refer to a human being as *homo respondens*.[40] The first question that God asked Adam was this: "Where are you?" (Gen. 3:9). This means not only, Where did you hide? but also, Into what condition did you get yourself? What normative actions brought you here? Similarly, the first question that God asked Eve was this: "What is this that you have done?" (v. 13). That is, How do you account for the normative actions you performed? The two of you are not responsible for what happened to the fruit of the forbidden tree once you had eaten it. But you are definitely responsible for putting this fruit into your mouth, for chewing and swallowing it.

God's third question to a human being was addressed to Cain: "Why are you angry, and why has your face fallen?" (Gen. 4:6); his fourth ("Where is Abel your brother?" v. 9) and fifth ("What have you done?" v. 10) questions were addressed to him as well. These are the questions: "Where?," "What?," "Why?" To each of these questions human beings must be able to give a sensible answer, one that is deliberate and thought through. This is because they are *human beings*. Animals cannot do this, not only because they cannot talk, but especially because they have nothing to talk about.

6.3 MORE ABOUT THE HUMAN STRUCTURES

6.3.1 MENTAL ACTS

Humans and chimpanzees may have many genes in common,[41] which evolutionists explain as evidence for common ancestry. An *historical* explanation is being given of a *biological* fact (similarity in the genome) without pondering all the consequences of this type of explanation. Other possible

---

[40] Buijs et al. (2005).

[41] Although apparently many fewer than has often been asserted; see Ebersberger et al. (2007).

explanations, such as common design, are discarded. Yet, everybody may notice that humans and chimpanzees may differ little with respect to their genes but very much in terms of their behaviour. In fact, the differences are so large that, apparently, human features cannot all be, or cannot be fully, explained by the genes. And they are not simply differences of *degree* (as if we were simply more developed than the chimpanzees) but of *kind* (for instance, human language is discontinuous with animal communication).[42]

People have tried repeatedly to "animalize" or "zoomorphize" (a neologism that parallels "anthropomorphize") these features. Denis O. Lamoureux, who presents himself as an orthodox Christian and should be aware of the uniqueness of human beings, minimalizes the differences between humans and animals.

> [John] Collins argues that language, art and craving for a just community[43] are evidence that "at the creation of man, the result was discontinuous in some way from what had preceded."[44] However, any introductory textbook on evolutionary psychology offers explanations for Collins's purported discontinuities, which reflect gaps in his knowledge, not gaps in nature.[45]

In other words, any evolutionist textbook that downplays the differences between animals and humans should show Christians that humans are not that unique after all.

Without elaborating this here, I claim that each aspect of mental life exhibits an essential difference between animals and humans.[46]

1. *The logical aspect*: Carl Safina tells us that animals can "think" but his juxtaposition of "think" and "feel" makes me suspicious.[47] His book gives "stories of animal joy, grief, jealousy, anger, and love" (says the publisher), but these are all examples of feeling. Earlier I argued that higher animals have feelings. But logical thinking—thinking that is ruled by the laws of

---

[42] Cf. C. J. Collins in Barrett and Caneday (2013, 165).

[43] Ibid., 165–66.

[44] Ibid., 168.

[45] Denis O. Lamoureux in Barrett and Caneday (2013, 179–80).

[46] For many fascinating examples, see extensively, Le Fanu (2010). A Jewish tradition says, Man "unites both heavenly and earthly qualities within himself. In four he resembles the angels, in four the beasts. His power of speech, his discriminating intellect, his upright walk, the glance of his eye—they all make an angel of him. But, on the other hand, he eats and drinks, secretes the waste matter in his body, propagates his kind, and dies, like the beast of the field"; see Ginzberg (1969; https://philologos.org/__eb-lotj/vol1/two.htm#1).

[47] Safina (2015, subtitle: *What Animals Think and Feel*).

logic, and that can be evaluated according to these laws—is a very different matter. Something that resembles this type of thinking is found among animals only at the most elementary level, namely, that of conceptualizing (e.g., recognition of pictures).

2. *The formative aspect*: No one wishes to deny that certain animal species know elementary forms of creativity; we are impressed that beavers have been constructing the same type of dams throughout the centuries, and birds the same type of nests (sometimes changing the building materials when newer materials become available). But this is significantly different from the formative potential of the three-year old child that I mentioned earlier.

3. *The lingual aspect*: Today people generally acknowledge the poor results of numerous attempts to teach certain animals conceptual languages that go beyond the instinctive communication methods they already employ.[48] No animal can be taught to talk—but a child born into a community is *ready* to talk: the child will begin to master the language of his or her surroundings in one or two years. Human children have a "Language Acquisition Device"[49] or LAD, which even the "highest" animals evidently do not.

4. *The social aspect*: Of course, animals practice certain forms of social life; some are very social animals (think of beehives, anthills, and herds). Yet, such anthropomorphisms can easily obscure the fact that such social animal behaviour is strictly limited by the animals' innate instincts. Humans, however, have built families, marriages, villages and cities, nation states, schools, companies, associations, clubs, you name it—and their creativity in doing so is inexhaustible.

5. *The economic aspect*: Many animals know about saving and exchanging but what are these instinctive behavioural patterns in comparison to the production and consumption of goods and services, buying and selling, the markets and stock exchanges, as we practice and develop these in the human world?

6. *The aesthetic aspect*: What animal species has any demonstrable sense of beauty and ugliness, of harmony and disharmony, that is, of music, of literature, of the visual arts, or of something remotely similar? Collins insists that "we can find artifacts such that, when we see them, we have no doubt that the divine image is there."[50]

7. *The juridical aspect*: Collins again: "Think as well of the craving for a safe and just community—something we see all over the world, from

---

[48] VanDoodewaard (2016, chapter 4).

[49] The phrase was coined by the American linguist Noam Chomsky (1965).

[50] C. J. Collins in Barrett and Caneday (2013, 166).

ancient and modern culture, whether or not they believe in the true God."[51] Animals may have a sense of what is advantageous and disadvantageous to them—but even young children have a clear idea of what is fair and unfair, which is a very different notion.

8. *The moral aspect*: "Only Humans Have Morality, Not Animals," says Helene Guldberg in *Psychology Today* (2011),[52] protesting against a book by Dale Peterson, *The Moral Lives of Animals*.[53] This book downplays what is unique about human morality; Peterson is a striking example of ongoing attempts to minimize the uniqueness of humans. Of course, he and others register the opposite accusation against people like Guldberg, namely, "false anthropo-exemptionalism." The atheist philosopher Thomas Nagel follows a wiser approach: How can a natural world produce features that are characteristic of what we call a "personality": intellect, moral awareness, and the like?[54]

9. *The pistical aspect*: This is the structural aspect of faith, trust, confidence, not as a fundamentally sensitive notion, but especially as an awareness of the transcendent, of the Ultimate Ground of being, of life, and of the world, and the inner desire to surrender to this. This is what we call "religion" in its broadest sense. Who would wish to suggest that even the highest animals have any such awareness of the transcendent? The pistical aspect is a boundary aspect; that is, it marks the boundary between what is immanent in humans and what is transcendent (see further in §6.4).

6.3.2 DIMENSIONS

We may distinguish three *dimensions* in the mental structural layer, which may provide deeper insight in this particular structural layer.

1. *The cognitive dimension*: This involves *knowing* and *coming to know* by means of thinking. Those acts—internal processes within the mental structural layer—that deal especially with such thinking and coming to know, with deliberation and reasoning, we call cognitive acts. The highest animals do not perform such acts.

2. *The creative dimension*: This involves *imagining*, fancying, or picturing (visualizing) something, as well as devising and inventing something. Acts

---

[51] Ibid.; he referred to Aristotle, who observed that "the human being is by nature a political animal" (see extensively, http://www.perseus.tufts.edu/hopper/text?doc=Perseus%3Atext%3A1999.01.0058% 3Abook%3D1).

[52] https://www.psychologytoday.com/blog/reclaiming-childhood/201106/only-humans-have-morality-not-animals; cf. extensively, Guldberg (2010).

[53] Peterson (2012).

[54] Nagel (2012).

that consist primarily of imagining (picturing, visualizing) something as existing (Dutch: *zich indenken*), or of inventing something new (Dutch: *uitdenken*), we call *creative* acts. The highest animals do not perform such acts.

3. *The conative (or volitional) dimension*: This involves willing or conscious striving after something, that is, with desiring, choosing, and deciding on the mental (not the animal sensitive) level. Acts that involve primarily an inner choice or a decision we call *conative* (or volitional) acts. The highest animals do not perform such acts.

There are several other ways to look at acts. In the previous section, we have considered nine aspects of mental life. We can now further explain that every act is always specifically characterized or qualified by one of these aspects. Here are a couple of examples to make the point clear.

The acts of a scientist are *analytical* in nature; that is to say, they are characterized by the logical aspect, that is, by *differentiating thinking*, which is thinking that takes apart or dissects (analyzes). The scientist's investigating involves the cognitive dimension, the scientist's designing theories involves the creative dimension, and the scientist's choosing between competing theories involves the conative dimension. *Animals do not perform such acts.*

The acts of the person who likes to work with material, as well as the acts of the political leader, are characterized by the *formative* aspect. That is to say, they are characterized by the exercise of power, domination, or control of something, whether control of a material, like wood, or of people, like citizens of a nation state. The artisan's investigation of materials involve the cognitive dimension, their designs involve the creative dimension, and their choosing between competing designs involves the conative dimension. *Animals do not perform such acts.*

The acts of a person directed at formulating thoughts into words and sentences are characterized by the *lingual* aspect; that is to say, this person considers which sound symbols (the things we call "words") one must choose in order to bring one's thoughts to expression in the most proper way. A person's knowledge of the language involves the cognitive dimension; he or she discovers a (lofty or plain) way of formulating, which involves the creative dimension; and engages in choosing words, which involves the conative dimension. *Animals do not perform such acts.*

Similarly, the acts of someone whose aim is association with others are *social* in nature; that is to say, these acts are characterized by interpersonal contact, by social living, and by community. The acts of a businessman who takes inventory and formulates a budget are *economically* qualified; that is to say, they are governed by principles of management, efficiency, and pricing. The acts of an avenger or of a judge, which are aimed at retribution, are *juridical* in character; that is to say, they are qualified by principles of justice and righteousness (principles that are obeyed or disobeyed). The acts of a person in love or of a philanthropist are *moral* in character; that is to say, they are characterized by benevolence, by love. Finally, the acts of someone who prays or worships or participates in the Lord's Supper or reads the Bible believingly—or worships in a mosque or a synagogue, and studies the Quran or the Talmud believingly—are *pistical* (the Gk. word *pistis* means "faith") in character; that is to say, they are characterized by the certitudes of a certain religious or ideological faith. *Animals do not perform any of these acts.*

In each of these cases, we may distinguish between the cognitive, the creative, and the conative dimensions.

6.3.3 TWOFOLD PURPOSE

Please notice again the twofold purpose for describing this anthropological model in some detail.

(a) Our purpose is to make more obvious that no structural layer can be reduced to a lower structural layer (*contra* evolutionism). In other words, no higher structural layer could have ever arisen from a lower structural layer through an evolutionary process. This is what Western scholars in earlier centuries have always believed, even if they used very different terms to explain the matter. The burden of proof is with the evolutionists, who do believe that morals, faith, and love evolved from some primeval soup.

(b) The lower structural layers in humans are different from those in plants and animals because the physical and physiological processes, the sensory phenomena, and the affections and emotions in humans *are such that they can "carry" mental processes.* A simple example: humans can blush, but chimpanzees cannot. Blushing itself is a purely physiological process, but here it is subservient to human mental life (specifically: to shame).[55] Many more examples could be mentioned, of course; think of the activation of the autonomic nervous system:

- This is a characteristically biotic activity, involving both physical and physiological processes.
- It is manifested in various physical and physiological symptoms: variation in facial color, increased heart rate, altered breathing, dilated pupils, sweating, and so on.

---

[55] I found the example in Vollenhoven (2005).

- This biotic activity is subservient to the higher structural layers, especially the sensitive structural layer, in "carrying" emotions such exhilaration, deep grief, fury, and fear. This is what humans have in common with higher animals.
- It is subservient to the mental structural layer in "carrying" cognitive, creative, and conative acts and actions, as, for instance, when coming close to an imminent discovery, or at the birth of a revolutionary idea, or when proposing marriage to a lady, during ecstatic religious experiences, and so on. This, too, differentiates human beings from higher animals.

Conclusion: Humans differ essentially from animals, not only in mental, but also in sensitive, perceptive, biotic, and physical respects. The human condition cannot be explained from the animal world. This would be true even if we were to consider only the five structural layers that we have distinguished in the human immanent mode of existence. Here the differences are unbridgeable. However, this is far more strongly the case when we now come to the *transcendent* dimension of human existence. This is the subject of the following sections. In principle, it is possible to downplay all the unique human qualities, as described especially in §6.3.1. Indeed, this has been done many times by—I repeat—"anthropomorphizing" animals and "zoomorphizing" (animalizing) human beings. People may well suppose that some authors have succeeded quite well in this respect. However, we now turn our attention something that surpasses all that is immanent-empirical.

6.4 THE HUMAN HEART

6.4.1 BIBLICAL DATA

In the Bible, humans are always viewed in a *holistic* way, that is, as a unity.[56] Thus, for instance, what we call the psychical and the mental is in the Bible sometimes associated with physical body parts, such as the heart (see below), the kidneys (e.g., Ps. 16:7; 73:21 KJV), or the bowels (Job 30:27; Ps. 40:8 JUB; Isa. 63:15; cf. John 7:38 KJV). We hear about the "soul" (Heb. *nephesh*) of a dead person and, under the influence of a long Hellenistic-Christian tradition, we inadvertently think of a substance that has left the body and continues to exist independently. But the context shows that this is nothing but the last breath or the shed blood (e.g., 1 Kings 17:21–22). Astonishingly,

[56] Again, for details and references, see Ouweneel (1986; 2008a).

even the remaining body is sometimes called *nephesh* (Lev. 24:18; Num. 6:6). Thus, the Bible corrects all thinking that divides a person into "parts," "substances," and "compartments" (this includes 1 Thess. 5:23; see §6.5.3).[57] Not just the spirit but the entire person is directed toward God: the body is devoted to God (Rom. 12:1), the flesh faints for God (Ps. 63:1), soul and spirit magnify the Lord (Luke 1:46–47). Precisely this is what constitutes the human person: not a composition of various parts (spirit, body, mind, soul, etc.), but a unity, which is referred to as "image of God," with the *heart* as the centre and point of concentration of the human being.[58]

The heart (Heb. *leb*, *lebab*) is the core of something, the centre, in which the whole is represented. Thus for instance "the heart of the sea" (Exod. 15:8; Jonah 2:3), the "heart of heaven" (Deut. 4:11), the "heart of the oak" (2 Sam. 8:14)—phrases pointing to where the sea is everything (where the sea is at its "sea-est"), where heaven is everything, where the oak tree is everything. In the animal kingdom, the term "heart" may indicate what is characteristic of the animal in view (2 Sam. 17:10; Job 41:15; Dan. 4:16). Also in humans, the heart is what is most characteristic of them, what represents them most clearly, what best expresses their essence. The heart is that where a human being is most "humanish," to express the matter a bit strangely. The heart is not some physical or mental part of a person, but is what represents the person as a whole. It is what is most genuine, authentic, and essential in a person.

Of course, this most profound meaning of "heart" does not come out equally clearly in every Bible passage. The Bible does not employ the unambiguous idiom that we expect in science and scholarship. Sometimes, the word "heart" simply refers to the physical heart in the body (e.g., 1 Sam. 24:5). But even then, it is the representative organ, which takes the honorary position among the organs (cf. Song 8:6). Sometimes, the "heart" is the bearer of human vital power (Gen. 18:5; Ps. 104:15; *biotic*). Often, the term has a psychical meaning, that is, the *psychical* functions of the "heart" are stressed such that the heart is the node of human affections, desires, and tendencies (e.g., sexual desire), and human emotions (joy, sadness, fear, despondency, consternation, courage, cowardice, weakness, passion, fury; dozens of verses could be quoted here).[59]

---

[57] König (2006, 314–15); a shocking example of such scholastic compartmentalizing is Wommack (2010) (notice the cynical subtitle: *What You Didn't Learn in Church*—but Wommack's outdated trichotomism is the very thing we have been learning in church for centuries, alternating with dichotomism). Watchman Nee wrote wonderful books, but in my view, the book on human trichotomy (1968) is one of his weakest.

[58] See extensively, Von Meyenfeldt (1950).

[59] See again Ouweneel (1984a; 1986; 2014b).

On the mental level, the heart sometimes has a *cognitive* meaning (Prov. 4:4, 21; 15:7, 14; in Job 34:10, 34; Prov. 15:32, "intelligence" is *lebab*), or an *imaginative-creative* meaning (1 Kings 12:33; Neh. 6:8), or a *conative* meaning; in many Bible passages, the heart is the origin of all motives and decisions, the place where actions are devised.

In the moral-religious sense, the heart can stand for the *conscience*, for which the Hebrew language has no special term (Gen. 20:5–6; 1 Sam. 24:6; 25:31; 2 Sam. 24:10; 1 Kings 2:44; 8:38; Job 27:6; Eccl. 7:22).[60] Proverbs 15:21 uses the Hebrew phrase *chasar-leb*, referring to the person who lacks "heart," here, moral determination (cf. v. 32; 2 Chron. 13:7; Eccl. 7:7). Conversely, the heart is the storehouse of God's ethical commandments (Deut. 4:9; Prov. 3:1; 6:21; 7:3). It is the place where the person is social-ethically turned toward, or turned away from, one's fellow human (Exod. 14:5; Deut. 15:7,10; Judg. 5:9; 9:3; 2 Sam. 15:13; 19:14; Ezra 6:22; Prov. 23:7). Most of all, it is the seat of love (1 Sam. 10:26; 2 Sam. 24:10; 2 Chron. 7:16; Prov. 31:11), but, since the Fall, also of hatred (Lev. 19:17; Prov. 26:25). And, of course, it is the organ of faith (negative: Gen. 45:25; positive: Mark 11:23; Luke 8:12; 24:25; Acts 15:9; Rom. 10:9–10; Eph. 3:17; Heb. 10:22).

6.4.2 THE PERSON'S RELIGIOUS CENTRE

In the sense just described, the heart is the deepest node in the inner person, the point of one's most genuine and essential concentration and integration. Often, the heart is the authentic inner person, standing in opposition to the, often insincere, outer person (Deut. 8:2; 1 Sam. 16:7; Ps. 28:3; 55:21; Prov. 23:7; 26:23, 25; Eccl. 7:3; Isa. 29:13; Jer. 3:10; 17:9; Hos. 10:2). In the heart, all aspects and all structural layers of the human mode of existence converge as in a node or in a focal point. At the same time, this heart elevates the person within one's deepest being above the entire temporal-immanent reality. This brings us to the most important point: *the heart in its transcendent-religious meaning refers to the deepest creaturely orientation of the human being toward one's personal Creator.* In fact, many of the passages already mentioned have a clearly transcendent-religious meaning. In the heart, it becomes evident whether a person is directed toward God, or is turned away from him (toward idols[61]).

---

[60] See more extensively, Ouweneel (1984a, §5.2.1).

[61] We are referring to idols in the broadest sense of the term; before the Flood we do not hear of idols, but in Lamech and his sons (Gen. 4:18–24) we do meet the idolatrous attitude, which consists of making a world without God as pleasurable and self-gratifying as possible. The Self was humanity's first idol.

Many times, the heart is referred to as the place where sin is planned, as the source of false prophecy (where it is not God revealing but the person making up things), of hubris (Gk. *hybris*, anti-divine over-confidence), of self-deification, of wicked arrogance, and so on (Isa. 14:13; 47:10; Ezek. 28:2, 6; Obad. 3; Zeph. 2:15). However, it is also the focal point of the awareness of guilt (2 Chron. 6:37; Lam. 1:20, 22), of internal divine activity and conversion (e.g., Jer. 24:7; 31:33), the "organ" for seeking God (e.g., 2 Chron. 11:16; 12:14) and for enjoying fellowship with God (1 Kings 8:61; 14:8; 15:14), and more: for fearing the Lord, serving God, walking with God, experiencing grief toward God, rejoicing in God, worshiping God, studying and keeping God's commandments, praying to God, and trusting in God.

In the New Testament, the use of the Greek word *kardia* ("heart") is different from that in the Hellenist world, but comes close to the meaning of "heart" in the Old Testament.[62] In addition to expressions such as the "heart" (most inner part) of the earth (Matt. 10:40), and the biotic meaning (Acts 14:17; James 5:5), the heart is the node of the psychical life (affections, impulses, emotions) and the mental life, with its thinking and knowing, its imagination and creativity, and its volitional life. Here again, the heart stands in contrast to the "outer" person (2 Cor. 5:12; 1 Thess. 2:17), to the mouth and lips (Matt. 15:18; Mark 7:6; Rom. 10:8–10), and to the outer flesh (Rom. 2:28–29; 1 Pet. 3:3–4). The heart is a person's most inner life, what represents the Ego, the personality (Col. 2:2; 1 John 3:19–20). It is the source of sin (Matt. 15:19), but also the point where God can work (Acts 15:9; Rom. 2:15; Eph. 3:17).

Here, in "the hidden person of the heart" (1 Pet. 3:4), we find the root of the human transcendent-moral-religious life, of the human relationship with God or, since the Fall, with idols. Here, in the heart, the person finds his or her indissoluble unity; here, all the dimensions of his or her existence converge. Here lies what is most proper to the person; it is fundamentally inaccessible to the sciences, fundamentally known only through the divine revelation concerning nature and essence of human beings. It is that which is spiritually grasped, not by feeling or intellect, but only by the heart itself. As Blaise Pascal put it: "The heart has its reasons, which reason does not know" (Fr. *Le coeur a ses raisons que la raison ne connaît point*).[63]

In this context, it is remarkable that the Bible sometimes speaks even of the "heart" of God (e.g., 1 Sam. 2:35; 13:14; 2 Sam. 7:21; 1 Kings 9:3;

---

[62] In addition to standard lexical entries, see the article on the Gk. word *kardia* by Behm (1964, 3:605–13.

[63] Pascal (1995, 154).

2 Kings 10:30; Ps. 33:11; Jer. 3:15; Acts 13:22.); this is God in his most inner being. No matter how anthropomorphic or metaphorical one wishes to view such language, this speaking of God's "heart" is of utmost importance for grasping the relationship of humans ("with all their hearts"[64]) and God. This relationship is one between God's very heart and a human being's very heart.

6.4.3 TRANSCENDENT CONCENTRATION POINT

We call "transcendent" everything that in some way surpasses our immanent, cosmic, empirical, created reality. One might think that transcendence is a unique characteristic of God, but this would be mistaken. In my view, the very fact that humans have been created in the image and after the likeness of God entails that they are not only immanent beings but also transcendent beings. With respect to their corporeal existence, humans belong to the cosmic, empirical reality. However, in their Ego, their heart, their personality centre, or whatever one wishes to call it, humans transcend this corporeal mode of existence, and all immanent, cosmic, empirical, created reality.[65]

This polar duality of the transcendent and the immanent is of utmost significance. Thus, the pole of the one, transcendent, imperishable *heart* stands over against the multiplicity and diversity of immanent, perishable (biotic, psychical, and mental) *functions*. Consider in this context Proverbs 4:23, "Keep your heart with all vigilance, for from it flow the springs [issues, outflowings] of life."[66]

Please note that the expression "over against" here has nothing to do with a dual*ism*;[67] what we have here is the dual*ity* of the heart and its functions. This is not a dual*ism* (in the ancient, substantialist sense of the term) because the heart is nothing but the transcendent focal point *of the functions themselves*, and because the immanent functions are nothing but the diversity proceeding *from the heart itself*. The heart *is* the functions in their transcendent unity, the functions *are* the heart in its immanent diversity. We are dealing here with the one self-awareness of the human person, which on the one hand, in its religious orientation toward God (*or*

---

[64] Cf. the expression "with all your heart" in Deut. 4:29; 6:5; 10:12; 11:13; 13:3; 26:16; 30:2, 6, 10 (and other passages).

[65] See extensively Ouweneel (1986, chapter 5), where I work with the philosophical anthropology of Herman Dooyeweerd. Here in this chapter, I am further developing the material offered there. For the numerous quotations from Dooyeweerd's writings, see that chapter.

[66] To reduce "heart" here to thinking (as do the ERV and GNT) is to reduce this transcendent node to just one of its immanent functions, ie., the logical one.

[67] See extensively, Ouweneel (1986, chapter 6); cf. Berkouwer (1962, 211–12), who rightly makes a sharp distinction between duality and dualism.

the idols) transcends immanent reality, and on the other hand, functions within the entire diversity of immanent human existence.

This emphasis on the heart's transcendence is enormously significant for opposing those who have frequently wanted to restrict the human Ego to one or some of its *immanent* functions (cf. note 66). For instance, Immanuel Kant spoke of a "transcendental-logical," or an "empirical-psychological" Ego. Today, AEH advocates do something similar; van den Brink mentions "symbolic thought," communication, the cognitive (the logical and lingual functions), creativity, the artistic (the aesthetic function), and the religious (which here is obviously what I call the immanent-pistical function).[68] Such a summary shows not the slightest awareness of the *transcendent* nature of human beings—recognition of which hardly fits with a theory of human evolution. However, the psychical, the logical, the lingual, the aesthetic, the pistical, and so on, involve nothing but immanent functions of the Ego. This Ego or the heart itself is the necessarily *transcendent concentration point* of *all* these immanent functions. It is in this heart, not in any of the immanent human functions, that the image and likeness of God find expression.

In my view, the human heart *must* be transcendent because, on the one hand, within immanent cosmic reality we find only a diversity of functions, of structural layers, and of acts, but no deeper unity and fullness. On the other hand, it must be transcendent because it is ontically and essentially oriented toward God, and designed for religious fellowship with him. Throughout the entire Bible, humanity is viewed "under the sight of God" (Lat. *sub specie Dei*) and "in the presence of God" (Lat. *coram Deo*).[69] As we saw before, there is no Man-as-such, no Man-in-himself; there is only the "eccentric" Man-in-relationship-with-the-transcendent, being transcendent himself (cf. §6.1.1).

6.5 THEOLOGICAL CONSEQUENCES

6.5.1 THE ROOT OF THE COSMOS

The notion of the transcendent human heart has not only profound philosophical and psychological significance, but also deep theological importance. This is especially because creation and Fall, as well as redemption and restoration, involve the transcendent "root" (the transcendent heart) of human existence. Only from that starting point do they affect the many

---

[68] van den Brink (2017, 223–24).

[69] Verkuyl (1992, 101).

immanent "ramifications" thereof (the immanent functions). I will arrange my analysis in terms of the well-known quartet of creation, Fall, redemption, and restoration.

(a) *Creation*. The human heart—or the human soul or spirit in the most pregnant biblical meaning of these terms (e.g., Matt. 10:28; 1 Cor. 14:32)—is not only the transcendent focal point of the individual's human existence, but at a more profound level we are dealing here with the transcendent, imperishable root of the *entire* human race. This is one of the many reasons why it is so important that, at the beginning of human history, God made one pair of humans, not born of hominids but directly created, who became the ancestors of all humanity.

(b) *Fall*. Sin in its transcendent meaning is not just an immanent trespassing of some commandment but touches the *heart*, the root of the human race, and thus ultimately the entire cosmos. Thus, the immanent transgression of Eve and Adam had transcendent consequences for their innermost being, and consequently for all creation (humanity as well as the rest of the cosmos). Since then, sin is manifested in immanent rebellious actions of humanity, in continual violation of God's commandments. But this immanent sinful behaviour proceeds from the human corrupted transcendent heart (cf. Matt. 15:19; see further in chapters 8–10).

(c) *Redemption and Restoration*. Not only the Fall but also rebirth (regeneration) involves the transcendent human heart, and from there it permeates all immanent human existence. However, not only does rebirth occur "in Christ," but Christ is the covenant head of the redeemed part of humanity. Rebirth involves the individual person, while restoration involves the entire cosmos. Jesus will come again for the "restoration of all things" (Acts 3:21 NKJV). Just as through the fall of the first head of creation, all creation was corrupted, so through the redemption brought about by the second and last head of creation, all creation is brought to its fulfillment and completion (restoration). Not just all (renewed) humanity but all creation: one day, sin (as the negative power of evil) will be abolished from the *cosmos* (John 1:29). It is not only the new humanity but also the new heavens and the new earth that will be the domain where righteousness will dwell (2 Pet. 3:13).

6.5.2 AEH RESPONSE

Look again at the three points mentioned in the previous section. What must be the necessary response of AEH advocates to these three points?

(a) *Creation*. AEH advocates can hardly posit some historical Adam and Eve who were going to be the ancestors of all living humans who were to be born until the end of time. Therefore, it cannot posit the unity of the human race, as we find that unity manifested in the human heart. This unity is first in Adam, and then in every individual, all human hearts being linked together through common ancestry.

(b) *Fall.* Here again, it is essential that sin did not, as AEH advocates suggest, involve a couple of hominids. We can imagine how the sin of a few affected the other hominids, just as, for instance, David's sinful census affected the entire nation (2 Sam. 24). But we cannot imagine how sin corrupted the entire human race *in its root* if the so-called Adam just was one individual, or a limited group, from a mass of hominids. David's sin had severe consequences for all his people; but they themselves did not become sinful through David's sin. Adam's sin, however, did make every human being sinful because they all inherited his sinful nature. This essential point will be developed further in chapters 8–10.

(c) *Redemption and Restoration.* AEH advocates have no room at all in their ideology for this notion of the sin of one human pair affecting all creation *in its root.* They cannot explain how the fall of a few could affect every other human *in the root of humanity.* The historical second head of creation is inconceivable without the historical first head of creation; the redemption by the former is inconceivable without the fall of the latter (see chapter 8–10). In one's Ego each individual person transcends one's immanent mode of existence, as well as all societal relationships in which one is involved. In this individual, transcendent Ego, a person is identified by the transcendent root of all humanity, first (before regeneration) in the first Adam, and then—through regeneration, that is, the inner, transcendent renewal by the Holy Spirit—in the last Adam (cf. Col. 1:16–17).

AEH advocates have little room in their ideology for the transcendent human heart; if they did, they would have to show how the transcendent evolved from the immanent. Their ideology knows precious little about Adam as the root of humanity (how could he be this if he were merely one, or a few, among the human race, and not its father and head?). But if AEH advocates do not accept the notion of Adam as "beginning" and "origin" and "head" (Gk. *archē*), both in the creation and in the Fall, how could it acknowledge a redemption that creates the new humanity in its head, Christ, who is the *archē* of the new creation (Col. 1:18; Rev. 3:14)?

In chapters 8–10 below, this will be developed further. But at this point I must say: if there is no historical Adam (in the plain biblical sense, i.e., the sense in which Jews and Christians have understood this for many centuries), then there is no historical Fall (again, in the sense in which Jews and Christians have understood this for many centuries). And therefore, there is no "last Adam" as the counter-part of the first Adam, and thus no redemption (again, in the sense in which Jews and Christians have understood this for many centuries). In this way, AEH ideology necessarily refashions not only ktiseology (the doctrine of creation) and hamartiology (the doctrine of sin), but also soteriology (the doctrine of salvation) and eschatology (the doctrine of the last things), and even Christology (the doctrine of Christ).

6.5.3 AGAIN, ETERNITY BEINGS

Back in §5.3.2 we considered human beings as "eternity beings." AEH ideology has no proper answer about where, when, and how during the supposed human evolution[70] certain hominids (which, as animals, cease to exist when they die) gave birth to human beings, who had now become *persons*, each of them a "who" (no longer a "what") divinely designed to live forever. Until now, I have not seen an AEH study in which this question was addressed. I cannot imagine any sensible answer with which AEH advocates could respond. Special divine intervention?[71] But in this case, we would not be dealing with evolution at all, but rather with creation. And if this is conceivable, why would it be inconceivable that God created the first human being the way Genesis 2:7 describes him doing?

This is our dilemma time and again: either AEH advocates accept the notion of human evolution, but then they cannot smuggle God in whenever the matter becomes complicated—or AEH advocates accept the biblical testimony, but then we do not need to read human evolution into it: let Genesis 2:7 speak for itself. If God *could* have implanted some kind of a soul into some hominid, but he *said* he breathed into a heap of dust the breath of life, then why not believe the latter? This has nothing to do with accommodation for the sake of the reader; I do not see why, had it been the case, Genesis could not have said that God created many primeval humans, and from them he selected Adam and Eve in order to turn them into eternity beings, images of himself.

---

70 I am not engaging the arguments for this supposed human evolution itself; I would merely point the reader to some wise words by staunch evolutionist Stephen J. Gould about human evolution: "Needless to say, no true consensus exists in this most contentious of all scientific professions—an almost inevitable situation, given the high stakes of scientific importance and several well known propensities of human nature, in a field that features more minds at work than bones to study" (2002, 910; cf. Fishman [2011]).

71 The Roman Catholic reply would be, "Yes": if God creates a separate soul for every human in his/her mother's womb (psycho-creationism), at a certain time during human evolution he would also have created the first human soul, to be implanted in the hominid that would now be called "Adam" (cf. §5.1.2).

The reality that humans are eternity beings can now be expressed in this way: humans have a "heart," or a "soul," or a "spirit" in the pregnant biblical meaning of these terms. We have seen (§§5.3.2, 6.1.1) that God has embedded eternity (Heb. *olam*) in this heart (Eccl. 3:11). It is this concentration point (heart/soul/spirit) in which humans transcend time and space, as images of God designed for eternity. It is this heart/soul/spirit that, in some way or another, survives physical death. (I do not know a better way to express it; this is what Gen. 2:7 means by the "breath of life" that God breathed into man, whence he became a living being). Elsewhere I have argued in the strongest terms that this does *not* imply some Hellenistic-scholastic soul–body dualism (or spirit–soul–body trichotomy), as has been often objected.[72] Speaking of the human immanent-empirical corporeal existence as well as the human transcendent point of concentration of this human corporeal existence constitutes a dual*ity*, but not a dual*ism*, that is, a juxtaposition of two different substances that are essentially foreign to each other (Hellenist-scholastic substantialism).

A good biblical example is Jesus' saying in Matthew 10:28, "Do not fear those who kill the body but cannot kill the soul. Rather fear him who can destroy both soul and body in hell." This is not dualism, for, as Jesus explains in Luke 16, it is the *entire* person (summarized as "body") who dies, and the *entire* person (summarized as "soul") who in some way survives personal physical death.[73] This is the person who, beginning at the resurrection, will exist forever, either with the Lord ("always with the Lord," 1 Thess. 4:17; cf. 2 Cor. 5:8) or in hell ("the lake of fire and sulfur…and they will be tormented day and night forever and ever," Rev. 20:10; cf. Matt. 25:41, 46).

Another example is this statement by the apostle Paul, "May your whole spirit and soul and body be kept blameless at the coming of our Lord Jesus Christ" (1 Thess. 5:23). Numerous theologians, including some modern theologians, have "discovered" a trichotomy here ("a person consists of three *parts*").[74] But Paul was no more speaking here of three "parts" than Jesus was speaking of four "parts" in Luke 10:27, "You shall love the Lord your God with all your *heart* and with all your *soul* and with all your *strength* and with all your *mind*." There are no "parts" here, but merely different ways of expressing the human essence. Andrew R. Fausset came

---

72 Ouweneel (1984a; 1986; 2008, chapters 5–8; 2012a, chapters 2–3).

73 Luke 16:22–23, "The poor man died [not just his body] and [*he*, not just his soul] was carried by the angels to Abraham's side. The rich man also died [not just his body] and was buried [*he*, not just his body], and in Hades, being in torment, *he* [not just his soul] lifted up his eyes."

74 Ouweneel (2008, §8.1.1).

close to making this claim when he said about this verse (1 Thess. 5:23), "All three, spirit, soul, and body, each in its due place, constitute *man 'entire'*." But he then introduced the word "part": "The 'spirit' links man with the higher intelligences of heaven, and is that highest part of man which is receptive of the quickening Holy Spirit (1 Cor. 15:47)."[75]

Aside from the issue of dichotomy and trichotomy, do not overlook the important point that the Bible says about *every* believer that at Christ's second coming the believer's *entire* personality will be "kept" (1 Thess. 5:23); no power can destroy any aspect of a person's personality. Believers are designed to live *forever*. And so are unbelievers (see note 5).

6.6 OTHER AEH STUMBLING BLOCKS

6.6.1 DOMINION

AEH advocates do not know how to handle the biblical given that the first humans received from God dominion over all creation because, in their view, this would necessarily imply the dominion of "Adam" (whoever he might be) over the rest of the hominid/human world. In the divine plan involving humans ruling over the rest of creation the image of God comes to manifestation in a remarkable way. There has been much discussion about the question whether the image of God *involves* dominion, or whether dominion is only an *effect* of the image of God.[76] I prefer the former because the term "image" points to representation (§6.1.3), and thus inherently to human dominion over the world on behalf of God.[77] There are several reasons why humans were called to this dominion, and each of these reasons constitutes an enormous challenge to the AEH ideology.

(a) Humans are capable of such dominion, not because they are most advanced in terms of supposed evolution, but because God breathed the breath of life into them, and thereby embedded something of his own being within them. Thus, we read that, when a person dies, "the dust [cf. Gen. 2:7!] returns to the earth as it was, and the spirit returns to God *who gave it*" (Eccl. 12:7). In a purely physical-biotic respect, humans are inferior to animals in many ways. Their sight is far weaker than that of birds of prey (cf. Job 28:7; 39:32), and their speed is much less than that of the ostrich (30:21). However, all these shortcomings in them are richly compensated by

---

[75] Fausset in Jamieson et al. (17871, 3:469; http://biblehub.com/1_thessalonians/5-23.htm); italics added.

[76] Cf. Berkouwer (1962, 70–72).

[77] Cf. Hoekema (1986, 14).

human intellect, which makes them superior to all animals. This superiority is evident from Adam giving names to the animals (Gen. 2:19–20; see §5.2.3). This was a historical event demonstrating both his insight into the essence of each animal, as well as his dominion over the animals; but most especially, this naming proceeded from Adam's transcendent-religious relationship toward God, who was the first to give names (1:5, 8, 10).

Please notice that AEH advocates are able to reduce all this to figurative language, *not* because the text itself compels them to do so—as proper exegesis would require—but only because of their evolutionary bias.

(b) Humans are capable of dominion because God blessed them to this end (1:28). Of course, all creation stands under God's blessing (see v. 22 with respect to the water animals and the flying animals). But the blessing that God pronounced over the first pair of humans had dominion in view: "And God blessed them [granting them certain authority] and said to them, 'Be fruitful, multiply, and fill the earth, and subjugate it [putting it under your power]; and rule over (dominate) the fish of the sea, the birds of the air, and every living thing that moves upon the earth'" (v. 28 AMP). Such a blessing is not just wishing someone well but—much more strongly—it *brings about* a state of being blessed.[78]

Again, AEH advocates ought to wrestle here with the notion of the dominion of the first humans over all creation because AEH advocates usually admit that, when the persons whom we might call Adam and Eve were alive, there were many more humans on earth. Whoever these Adam and Eve might have been, did they also receive dominion over these other humans? Or does the notion of dominion refer to *all* humans? Did all evolved humans receive divine rule over all less evolved hominids? Who believes this? Or must Genesis 1 and 2 be taken figuratively again? This is the problem facing especially those AEH advocates who wish to convince us that, despite their defense of hominid to human evolution, they are still maintaining the notions of a historical Adam and a historical Fall (as does van den Brink). They constantly switch between, on the one hand, the figurative exegesis whenever it fits their evolutionist bias, and on the other hand, literal exegesis whenever they wish to strengthen the impression that their approach to Genesis 1–3 is basically historical, and therefore orthodox. In this way, they think to elude the objections of their critics.

(c) Humans are *entitled* to dominion, not because they supposedly are most advanced in evolution (a gradualist argument), but because they were

---

78 See Ouweneel (2004, 224–26); cf. the Gk. verb *eulogeō* and the Lat. verb *benedicere*, "to bless," literally "to speak well"; these terms mean not to *wish* well, but to *pronounce* God's goodness over someone.

created in the image of God (a distinctive gap argument). On earth, humans are the visible expression and representation of God's invisible power over his creation. We see the image of God in them not because they evolved to such a high level but because God created them in such way that they are in the image and after the likeness of God, and as such are essentially distinct from the animals, including hominid animals.

AEH ideology has hardly any room for this. Because it starts from the notion of supposed human evolution, it will always be tempted to reduce the characteristic features of humans to those of animals ("zoomorphizing"), unless they assume some special divine intervention here. But again, what will it be: natural human evolution purely according to God's own natural laws, or some special act of divine intervention after all? What will the genuine advocates of evolutionary thought think of this emergency appeal to a Creator-God? With a variation upon 1 Kings 18:21 we may ask, how long will AEH advocates go limping between two different opinions? If the God of creation is God, follow him; but if the God of evolution is God, then follow him. Notice what comes next in this verse: "And the people did not answer him a word." For what could they have said? (Cf. Lev. 10:3; Matt. 22:12.)

6.6.2 FAILURE

There has been much discussion about whether, and to what extent, humans remained images of God after the Fall, continuing to exhibit his likeness.[79] At least we can say that, through the Fall, this image was marred or distorted. On the one hand, there seems to be a remarkable contrast in Genesis 5, mentioned earlier: "When God created man, he made him in the *likeness of God*.... When Adam had lived 130 years, he fathered a son *in his own likeness*" (vv. 1, 3). On the other hand, James 3:9 seems to suggest that the likeness continues: "With it [i.e., the tongue] we bless our Lord and Father, and with it we curse people who are made in the likeness of God." Even the most wicked people show vestiges of the splendour that God had bestowed upon the "first man" (the *grandeur* to which Blaise Pascal referred[80]).

Several times the image and likeness of God in wicked humans became so badly distorted that God intervened when humans abused their power over creation. Since the Fall, humans have become similar to the servant in Jesus' parable: "that wicked servant says to himself, 'My master is delayed,'

[79] See Berkouwer (1962, chapter 4); cf. Bratsiotis (1951/52); Zenkowsky (1951); Zenkowsky and Petzold (1969); Noordegraaf (1990, 74–76).

[80] See chapter 5, note 66.

and begins to beat his fellow servants and eats and drinks with drunkards" (Matt. 24:48–49). Animals are not held responsible for the way they fulfill their creaturely task—but humans are. As God said (in a different context), "Among those who are near me I will be sanctified" (Lev. 10:3). The nearer one is to God in terms of status, the higher one's responsibility. No animal will ever appear before the judgment seat of God because there is nothing for which animals will have to give account.[81] Humans without exception will give an account: "We must *all* appear before the judgment seat of Christ, so that *each* one may receive what is due for what he has done in the body, whether good or evil" (2 Cor. 5:10). "I am coming soon, bringing my recompense with me, to repay *each* one for what he has done" (Rev. 22:12).

Originally, humans exhibited a clear resemblance ("likeness") to God, at least in the sense of purity and innocence. The destiny of the domain over which the first humans received dominion depended on their maintaining this divine resemblance. After the Fall, not only do humans no longer resemble God in this beautiful way but they also brought the entire earth under the curse (Gen. 3:17; 5:29), and under the yoke of perishability (cf. 1 Cor. 15:42, 50–54).

Little remains of this in the portrait that AEH advocates present to us. Throughout the time of supposed human evolution, there never was a phase of genuine purity, innocence, and imperishability. There never was a period during this supposed human evolution when males did *not* exert dominion over females, that the ground was *not* cursed as it produced thorns and thistles, and that people did *not* have to eat their bread "by the sweat of their faces" (cf. Gen. 3:17–19). Imagine what this means: according to AEH advocates, there never was a time when the creation was *not* "subjected to futility, not willingly, but because of him who subjected it"; there never was a time when the creation was *not* in "bondage to corruption," and when it was *not* "groaning together in the pains of childbirth" (cf. Rom. 8:20–22). No matter what impression of being orthodox that AEH advocates would like to give their readers—and I wish to take seriously their motives in doing so—they can never do justice to the Bible's true picture, and to so many other passages in both the Old and the New Testaments.

---

[81] The term "dogs" in Rev. 22:15 is a figurative reference to particular wicked people (cf. Deut. 23:18).

THE TREES AND THE SERPENT

> For I feel a divine jealousy for you, since I betrothed you to one husband, to present you as a pure virgin to Christ. But I am afraid that as the serpent deceived Eve by his cunning, your thoughts will be led astray from a sincere and pure devotion to Christ (2 Corinthians 11:2–3).

THESIS

No matter how many metaphors and anthropomorphisms and other figurative elements one discerns in Genesis 2 and 3, the tree of life, and the tree of the knowledge of good and evil, and the serpent are thoroughly historical. Understanding their true nature and significance is essential for understanding the story of humanity's creation, fall, redemption, and restoration, and thus for understanding Christianity itself.

7.1 THE TWO TREES IN THE GARDEN

7.1.1 INTRODUCTION

Up to this point in the book, I have explained why I believe that Genesis 2–3 must be interpreted as giving an account of historical events. These chapters tell us five features about the first humans (1) who ever existed on earth, (2) from whom all present humans descended, (3) who had been created in a "very good" state, and (4) at a given moment in time fell into sin, and (5) thereby entered into an evil state. They fell from a very good state to a very bad state. In §4.5.2 above, I referred to Ecclesiastes 11:3, "In the place where the tree falls, there it will lie." This is what Genesis 3 tells us: one moment the "tree" was still standing, the next moment it lay on the ground.

Speaking of trees: trees will help us to establish what Genesis 2–3 is *not* telling us. I mentioned earlier that we should not inquire from the book of Jonah what species of fish swallowed the prophet—it was a sea monster—so too we should not ask what species of trees was present in the garden. Were they of the rose family, of the beech family, of the cashew family, of the papaya family, of the mulberry family, of the palm family? What genus were they? These are foolish questions, just like the question about the species of the serpent in the garden (see §7.4).

Naturally, people have speculated about the species of trees involved. Louis Berkhof said of the tree of knowledge, "We do not know what kind of tree this was. It may have been a date or a fig tree, or any other kind of fruit tree."[1] In the Talmud we read, "Rabbi Meir holds that the tree of which Adam ate was the vine, since the thing that most causes wailing to a man is wine.... Rabbi Nehemiah says it was the fig tree, so that they [i.e., Adam and Eve] repaired their misdeed with the instrument of it [Gen. 3:7]."[2] Christian tradition has often spoken of an apple tree, presumably because the Latin word *malus* means both "apple" and "evil, mischief, crime." The Swedish natural scientist Carl Linnaeus, inventor of botanic nomenclature, gave the banana the Latin name *Musa paradisiaca*, "banana of Paradise," apparently because he believed the forbidden tree to have been a banana tree.[3]

Genesis 3 is thoroughly historical in the sense that Moses and Paul were historical, *not*—on the one hand—in the necessarily figurative sense of AEH, *nor*—on the other hand—according to the criteria of a journalistic documentary. At the same time, we should beware of claiming that these trees were merely figurative trees (as AEH advocates are compelled to believe). Genesis 2:9 says, "And out of the ground the LORD God made to spring up every tree that is pleasant to the sight and good for food. The tree of life was in the midst of the garden, and the tree of the knowledge of good and evil." Here, the latter two trees belong among *all* the trees that the ground brought forth; it is hard to maintain that the former trees were literal ("biological") trees, whereas the latter were merely figurative. The same is true in verses 16–17: "You may surely eat of every tree of the garden, but of the tree of the knowledge of good and evil you shall not eat, for in the day that you eat of it you shall surely die." Here again, the two trees may have a special function, but they are trees like all the other trees in the garden. The same is true again in 3:1–3; the serpent

---

[1] Berkhof (1949, 242).

[2] Tract Berakoth 40a.

[3] Was this perhaps because of its phallic significance? Cf. http://www.promusa.org/Musa+paradisiaca; The name Linnaeus gave to the banana that is eaten as a dessert is *Musa sapientum*, "wisdom banana," possibly a reference to the tree of the knowledge of good and evil.

"said to the woman, 'Did God actually say, "You shall not eat of any tree in the garden"?' And the woman said to the serpent, 'We may eat of the fruit of the trees in the garden, but God said, "You shall not eat of the fruit of the tree that is in the midst of the garden."'"

Together the three passages implicitly tell us something about the significance of these trees: humans cannot eat of both trees. That is, refrain from the one tree, and you will retain access to the other tree, and thus live forever. Conversely, eat from the one tree, and you will no longer have access to the other tree, and thus, one day you will die. The first humans were confronted with a *choice*, which had rational as well as moral and pistical features. That is, in almost the first action of human beings that we find recorded in the Bible—which tree do you choose?—their unique humanness comes to light. Animals make choices, too (what food to eat, where to build a nest), but such choices are always strictly bound to their instincts; they are never *responsible*, neither for their instincts, nor for their choices. In the very first choice that human beings had to make, they morally failed and had to account for this. They were fallen humans—but they remained *humans*.

7.1.2 THE TREE OF LIFE

Before we consider what AEH advocates have to say about the Fall (see chapters 8–10), let us look first at the biblical text (Gen. 2–3). We begin by considering the two trees in the Garden of Eden. Both trees are significant for hamartiology (the theological doctrine of sin) because by eating of the tree of the knowledge of good and evil, the first humans fell into sin (Gen. 3:6, 11, 16–19), and subsequently their access to the tree of life was blocked (v. 22). The tree of life is also significant for soteriology because this tree reappears later in the Bible, in an eschatological context: it supplies the (spiritual) food for the "conquerors" (those who have conquered the dragon by the blood of the Lamb; Rev. 12:11) in the New Jerusalem (2:7; 22:2, 14, 19).

Interestingly, in the portrait of the New Jerusalem as the bride of the Lamb (Rev. 21:9–10), notice how the Bridegroom describes his bride: "A garden locked is my sister, my bride.... Your shoots are an orchard [Heb. *pardes*, from which the word "paradise" developed!] of pomegranates with all choicest fruits...with all trees of frankincense, myrrh and aloes, with all choice spices—a garden fountain, a well of living water, and flowing streams from Lebanon" (Song 4:12–15). Human history begins with a paradise,[4] and

[4] We call the Garden of Eden the original "Paradise" because the Septuagint renders "Garden" as *paradeisos*.

it ends with a paradise; in both cases, mention is made of the tree of life.[5]

The trees in the Garden of Eden are first mentioned in Genesis 2:9, "Out of the ground the LORD God made to spring up every tree that is pleasant to the sight and good for food. The tree of life was in the midst of the garden, and the tree of the knowledge of good and evil." Subsequently, we hear God's commandment: "And the LORD God commanded the man, saying, 'You may surely eat of every tree of the garden, but of the tree of the knowledge of good and evil you shall not eat, for in the day that you eat of it you shall surely die'" (vv. 16–17). This "dying" contains an implicit reference to the tree of life: eating of the one tree involves losing access to the other tree and to the life it supplies. The one tree epitomized death, the other tree epitomized life.

For centuries, Jewish and Christian expositors have wondered about the precise significance of these trees. To what kind of "life" does the one tree refer? And what kind of "knowledge" (knowledge of good and evil) is intended with the other one? And what exactly do "good" and "evil" mean here?

Let us begin with the tree of life (Heb. *ets hachayyim*). Its significance comes to light for the first time in Genesis 3:22, where God says, "Behold, the man has become like one of us in knowing good and evil. Now, lest he reach out his hand and take also of the tree of life and eat, and live forever." God had designed humans for life, and apparently, eating from the tree of life was an essential aid for this. However, when the first humans fell into disobedience by eating from the tree of knowledge, any further access to the tree of life was denied to them. But what "life" is intended here?

Usually, expositors think of natural human life.[6] In this view, it is important to notice that prelapsarian humans apparently did not possess any immortality in themselves (cf. 1 Tim. 6:15–16, immortality belongs to God alone). Genesis 3:22 is usually understood to imply that the first humans were perpetually dependent on the tree of life in order to continue their earthly physical lives. They were mortal in the sense that they *could* die but did not *have* to die. Like Christ on earth, before sin they

---

[5] Cf. 2 Esdras 2:12 (NRSV): "The *tree of life* shall give them [i.e., restored Israel] fragrant perfume, and they shall neither toil nor become weary," and 8:51–52, "But think of your own case, and inquire concerning the glory of those who are like yourself, because it is for you that *paradise* is opened, *the tree of life* is planted, the *age to come* is prepared, plenty is provided, a city is built, rest is appointed, goodness is established and wisdom perfected beforehand." Notice how various relevant ideas (printed in italics) are combined here.

[6] See Gunkel (1997, 8); Böhl (1923, 66, 71); Leupold (1942, 119); Morant (1960, 112–13); Rice (1975, 105); see Oosterhoff (1972, 127–28) for references to other writers (to H. Th. Obbink, M. A. van den Oudenrijn, J. de Fraine, A. van Selms, A. Kuyper, Th. C. Vriezen).

were in the position of "being capable of dying" (Lat. *posse mori*) and of "being capable of not dying" (Lat. *posse non mori*). Since the Fall, humans are in the position of "not being capable of not dying" (Lat. *non posse non mori*).[7] In eternity, believers will be in the position of "not being capable of dying" (Lat. *non posse mori*), while unbelievers will be in a state of eternal death, called the "second death" (Rev. 2:11; 20:6, 14; 21:8). We can compare all this with the analogous positions relating to "(not) being capable of (not) sinning (Lat. *[non] posse [non] peccare*), which the reader can readily discern.[8]

Because access to this tree was denied to the first humans after the Fall, their natural, physical death ultimately became inevitable, in accordance with God's warning: "Of the tree of the knowledge of good and evil you shall not eat, for in the day that [or, when[9]] you eat of it you shall surely die" (Gen. 2:17). Henry M. Morris speculated—quite uselessly, in my view—about what substances in the fruit of the tree of life might have prevented the aging process.[10]

South African dogmatician Johan A. Heyns looked at the eating of the tree on a more spiritual level: the choice between the two trees

> does not mean, of course, that, through the physical working of their fruits, these two special trees could supply, on the one hand, immortality or everlasting life, and, on the other hand, knowledge of good and evil.... The tree of life was a symbol of what humanity would receive in the pathway of obedience to God; the tree of the knowledge of good and evil was a symbol of what humans would encounter in the pathway of disobedience.... [Humanity] received knowledge of sin, not because they ate of the *tree* but because they trespassed God's *Word*.[11]

(The last statement is true, of course, as long as we understand that humans trespassed God's Word by eating of the tree.)

---

7 Berkhof (1949, 247).

8 Ouweneel (2008, §12.2.2).

9 Both grammatically and literarily, the word "day" can be understood here either literally ("in the very day that") or figuratively ("at the time when").

10 Morris (1976, 87).

11 Heyns (1988, 170); cf. Louis Berkhof (1949, 242): "There was nothing injurious in the fruit of the tree as such. Eating of it was not *per se* sinful, for it was not a transgression of the moral law. This means that it would not have been sinful, if God had not said, 'Of the tree of knowledge of good and evil thou shalt not eat' [Gen. 2:17]." Thus also Duffield and Van Cleave (1996, 166).

7.1.3 MORE THAN PHYSICAL LIFE?

The preceding remarks lead us to ask whether more was involved here than just physical life.[12] I ask this especially because of the role that the tree of life plays in the rest of Scripture. In addition to its physical description in Genesis 2:9 and 3:22–24, we find the phrase used figuratively in Proverbs 3:18, where wisdom "is a tree of life to those who lay hold of her";[13] in 11:30, "The fruit of the righteous is a tree of life, and whoever captures souls is wise"; in 13:12, "Hope deferred makes the heart sick, but a desire fulfilled is a tree of life"; and in 15:14, "A gentle tongue is a tree of life." See further Revelation 2:7, "To the one who conquers I will grant to eat of the tree of life, which is in the paradise of God"; and 22:2, "on either side of the river, the tree of life with its twelve kinds of fruit, yielding its fruit each month."[14]

In Proverbs, the tree of life is a figure related to wisdom, to righteous acting, to desires that are fulfilled, and to gentleness. In short: true wisdom leads to the blessed life that God intended for humans (cf. 1 Tim. 6:19); leading such a life of godliness, as a truly righteous person (Heb. *tsaddiq*; cf. the godly man mentioned in footnote 13), is equivalent to eating of the tree of life. In the book of Revelation, too, the tree of life is symbolic, representing not physical life but true spiritual life, as it will be enjoyed by God's people in the "paradise of God," or in the New Jerusalem (2:7; 22:2, 14, 19; notice the different metaphors: a garden and a city, referring to the same eschatological spiritual reality).

This does not mean, of course, that in Genesis 2–3, the tree of life must be interpreted as being purely symbolic. It does suggest, however, what some have called the "sacramental" meaning of this tree. Although John Calvin understood the tree to be literal, he interestingly underscores the sacramental significance thereof:

> The term "sacrament"...embraces generally all those signs which God has ever enjoined upon men to render them more certain and confident of the truth of his promises. He sometimes willed to present these in natural things, at other times set them forth in miracles.

---

[12] See Sikkel (1923, 150–54); Bavinck (*RD* 2:575); Aalders (1932, 480–81); Kidner (1967, 62); Oosterhoff (1972, 128–37); Gispen (1974, 107–108). Ridderbos (1956, 706–707) wished to see both meanings in the tree of life, and I agree with his view.

[13] Cf. 4 Macc. 18:16 (NRSV), where it is said of a godly man, "He recounted to you [i.e., the man's sons] Solomon's proverb, 'There is a tree of life for those who do his will.'"

[14] In addition to these canonical references, we find the tree of life mentioned in 1 Enoch 24:4; 2 Enoch 8:3, 5, 8; 9:1; and 2 Esdras 8:52.

> Here are some examples of the first kind. One is when he gave Adam and Eve the tree of life as a guarantee of immortality, that they might assure themselves of it as long as they should eat of its fruit [Gen. 2:9; 3:22].[15]

In this way, God himself is the actual giver of life; the tree is merely the sacramental symbol of that life,[16] just as—we may add—*literal physical* bread and wine can be sacramental symbols of the body and blood of Jesus.

If we stress the spiritual meanings of the tree of life, AEH advocates could be tempted to use this to argue that the tree *itself* was only spiritual, figurative, and symbolic. This is why the sacramental aspect is so important, because sacraments, in my understanding of them, always involve certain substances: bread, wine, water, oil.[17] In the sacraments, these substances acquire a symbolic meaning but they themselves are genuinely physical and material. Similarly, the rainbow as well as circumcision—other examples mentioned by Calvin—may have deep spiritual meanings, but they themselves are purely physical phenomena (see Appendix 5).[18]

The same is true of the two trees in Eden's Garden. No matter how spiritually rich was the language that was spoken, as it were, by the two trees, this does not mean that the trees themselves were figurative. This same claim is illustrated by the many types of Christ in the Bible. Some of the greatest types, such as Joseph and David, speak such an eloquent language that the typological meaning shines forth through almost every episode of their life. But nobody would suggest, I suppose, that therefore Joseph and David themselves must be understood in a figurative sense. The apostle Paul says that, in the wilderness, the Israelites "drank from the spiritual Rock that followed them,[19] and the Rock was Christ" (1 Cor. 10:4). This is the rock of Exodus 17:5–6. Despite Paul's strong language—the "spiritual" Rock, the Rock "was Christ"—nobody would be tempted to understand the rock in Exodus 17 figuratively. Thus, the eloquent symbolic language does not contradict the essentially literal meaning of the rock. Exactly the same is true about the trees.

---

[15] Calvin (1960, 4.14.18); cf. Leupold (1942, 120–21); Van Genderen (2008, 390).

[16] Hamilton (1990, 163).

[17] See extensively, Ouweneel (2010, chapter 5).

[18] The rainbow is the sign of the Noahic covenant (Gen. 9:12–13); circumcision is the sign of the Abrahamic covenant (17:11); and if we can speak of an Adamic covenant (cf. Hos. 6:7), we could call the tree of life the sign of this covenant; see Ouweneel (2018c, chapter 2). The sign of the Sinaitic covenant is the Sabbath (Exod. 31:13–17), which is not material, but neither is it essentially figurative.

[19] This allusion was probably derived from an ancient Jewish legend; cf. Fee (1987, 448n34).

7.1.4 THE MEANING OF THE TREE

In the Bible, especially in John's Gospel and John's first letter, eternal life is not only a quantitative notion (everlasting existence) but, more importantly, also a qualitative notion: it is the "true life" of 1 Timothy 6:19, which is the life with God. This life is even God himself, more specifically, God the Son (1 John 5:20).[20] Thus, in the book of Revelation, too, "life" is not just a matter of everlasting existence but also a matter of the *bliss* of the living with God. As Dutch theologian Berend J. Oosterhoff stated the matter, "Through the Fall, humanity lost not only immortality. They lost Paradise as well, they lost peace and happiness, they lost fellowship with God. In a word, they lost life in the full sense of the word. The life that God, through the tree of life, offered to humanity, was the true, spiritual life in his fellowship."[21] And German theologian Wolfhart Pannenberg said, "Since sin is turning from God, sinners separate themselves not only from the commanding will of God but also from the source of their own lives."[22]

David says, "With you [i.e., the Lord] is the *fountain of life*" (Ps. 36:9; cf. Prov. 14:27, "The fear of the LORD is a *fountain of life*"). And the LORD himself says in the book of Jeremiah, "My people have committed two evils: they have forsaken me, the *fountain of living waters*, and hewed out cisterns for themselves, broken cisterns that can hold no water" (2:13). And the prophet says, "O LORD, the hope of Israel, all who forsake you shall be put to shame; those who turn away from you shall be written in the earth, for they have forsaken the LORD, the *fountain of living water*" (17:13). In Revelation 22:1–2 the image of freshly flowing water is connected with that of the tree of life.

Of course, the difficulty with any such exegesis concerning the tree of life is that one cannot possibly receive *spiritual* life through eating of a *physical* tree. Here we benefit from the emphasis on the tree's sacramental significance. Physical water itself cannot regenerate; but God can grant the life of rebirth through what the water figuratively represents (John 3:5; cf. 15:3; Eph. 5:26). Physical oil itself cannot heal; but God can grant healing through what the oil figuratively represents (cf. Mark 6:13; James 5:14–15). In other words, the tree of life was just as physical and literal as the water of baptism, the bread and wine of the Lord's Supper, and the oil of the ministry of healing[23]—but God is the actual Giver of life (in all the

[20] Cf. Visscher (1928, 153–54) in relation to the tree of life.
[21] Oosterhoff (1972, 130).
[22] Pannenberg (2010, 266; cf. 267–69).
[23] Notice how James 5:14–15 refers not to the oil that is said to heal, but to the elders' prayer, which appeals to God's power.

meanings of this word), of strength and comfort, and of healing.[24] Thus, the significance of the tree of life was that life in fellowship with God was reserved for humans only if they continued walking in the pathway of obedience. If they were to fall into sin, access to the true life of God would be cut off for them. In Adam's case, this loss of access was literally experienced when he was driven out of the Garden of Eden. This brings us to the heart of soteriology: only through a work of redemption can the door to true life be opened again: whoever believes in the Son of God, has (possesses) "eternal life" (John 3:15–16; cf. v. 36; 5:24; 6:40, 47).

We find this figurative, soteriological sense of the tree of life in Revelation 2:7 and 22:2. The Orthodox tradition understands these passages to be referring to the cross,[25] as do some Western traditions.[26] However, it seems preferable to think here of Christ himself, for he is our life (Col. 3:4); in him is life, and he grants life (John 1:4; 5:26; 1 John 1:1–2; 5:11–13 etc.).[27] The tree of life in the middle of the Garden was a pointer to, and promise of, something (or Someone) better and greater than all the good with which God had surrounded Adam. This Someone held the "promise of eternal life" (1 John 2:25) before sin entered, and even "before the ages began" (Titus 1:2; cf. 2 Cor. 1:20).[28]

7.2 THE TREE OF KNOWLEDGE

7.2.1 WHAT KIND OF KNOWLEDGE?

One key to understanding the remarkable tree of the knowledge of good and evil (Heb. *ets hadda'at tob wara*) is Genesis 3:6, where we hear "that the tree was to be desired to make one wise [or give insight; Heb. *lehaskil*[29]]." This can be related directly to the ideas of "knowing" (it is the tree of the true knowing; Heb. *da'at*) and of being "like God" (vv. 5, 22). These various

---

[24] Pannenberg (2010, 102).

[25] See https://web.archive.org/web/20070227172919/http://www.unicorne.org/orthodoxy/automne2004/ treeoflife.htm. This identification is understandable because the cross is sometimes called a tree (Acts 5:30; 10:39; 13:29; Gal. 3:13; 1 Pet. 2:24).

[26] The Dutch poet Hieronymus van Alphen said in a well-known Dutch hymn: "We, guilty, driven out by God, we remained far from Eden. But the cross became to us the tree of life, pointed out by the Father himself." Cf. Falk (2004, 53): "What a strange tree of life is this trunk on which God had to suffer and die.... The tree of life, the cross of Christ...."

[27] Grant (1890, 32).

[28] Coates (1920, 23–24).

[29] This Heb. word is related to the *maskilim* ("those who are wise, have understanding" in Israel) in Dan. 11:33, 35; 12:3, 10.

aspects must be closely related: the point is a knowing like God himself knows.[30] This is not intellectual knowledge but existential or experiential knowledge; not an objective knowledge of what good and evil ethically involve, but "knowing" is here also "knowing how to": it is the moral insight into what is good/beneficent as well as the moral capacity to do what is good/beneficent and to abstain from what is evil/harmful.[31] Basically, only God knows how to do this—as well as redeemed and renewed human beings, in whom the image of Christ is being formed (cf. Rom. 8:29; Gal. 4:19; Col. 3:10 [1:15]), and who are being led by the Holy Spirit (Rom. 8:14–15; 2 Cor. 3:18 [4:4]).

It is important to note that in their state of integrity (Lat. *status integritatis*; that is, the state-of-not-having-yet-sinned) humans did not possess this "knowing about" and this "knowing how to." Therefore, it is quite audacious to claim that prelapsarian humans were holy and righteous.[32] They could not have been holy and righteous, because true holiness and righteousness presuppose the knowledge of good and evil, and this is the very thing that Adam and Eve did not yet have before they fell into sin.[33] If we make prelapsarian humans "too holy," we encounter problems like those signaled by Louis Berkhof: explaining "how a holy being like Adam could fall in sin...how temptation could find a point of contact in a holy person."[34] But this misses the point. Adam did not know good and evil; terms like "holy" and "unholy" would have made no sense to him. He would learn what holiness and unholiness were only by falling into *unholiness*, and thereby learning the distinction between holiness and unholiness. After the Fall, humans would attain the state of holiness only through redemption, regeneration,[35] and the power of the Holy Spirit (cf. 1 Cor. 6:11).

As I see it, this was the order of stages: Adam had been created innocent, then (through his fall) became unholy and unrighteous, then (through regeneration) became holy (a saint) and righteous. Compare this with

---

[30] Cf. Vriezen (1937, 146).

[31] Cf. Von Rad (1972, 86–87).

[32] See Heidelberg Catechism Q/A 6; see Ouweneel (2016, ad loc.); see further the Belgic Confession, Art. 14; the Canons of Dordt 3/4.1; also Luther's Formula of Concord. van den Brink (2017, 184) seems to sympathize with this view.

[33] Cf. Bonhoeffer (1997, 87); Falk (2004, 45).

[34] Berkhof (1949, 245).

[35] Of course, the text does not speak of "regeneration"; it is only in the light of New Testament testimony that we must assume such an event to have occurred in the lives of Adam and Eve (cf. John 3:3–5). This is how I understand 1 Pet. 4:6 (NKJV), "For this reason the gospel was preached also to those who are [now] dead [i.e., Old Testament believers-to-be], that they might be judged according to men in the flesh, but live according to God in the spirit [or, Spirit]."

Cain and Abel: Cain was born unholy, and remained unholy. Abel was born unholy, and (through regeneration) became holy (a saint) and righteous (cf. Matt. 23:35, "righteous Abel").

Opinions may differ on the precise nature of the state of integrity: holy and righteous, or "only" pure and innocent (cf. Eccl. 7:29, "God made man upright [Heb. *yashar*]"). Regardless of those differences, however, orthodox Christians, such as Eastern Orthodox, Roman Catholic, and Protestant, have always agreed on the existence of such a state of integrity. However, if we take AEH advocates seriously, it is impossible to maintain such a notion of a state of integrity. No version of AEH, as far as I can assess, posits a group of humans (in my view: just two, Adam and Eve) who were initially pure and innocent. Whatever kind of people they may have been, they had evolved from stealing, murdering, and whoring hominids. Indeed, some modern theologians, living on what I view as the borderline between orthodox and liberal Christianity, simply claim that there never was any state of integrity.[36] But this seems to be inspired by theistic evolutionism, and is a flagrant contradiction of both Scripture and the Christian theological tradition.

Bram van de Beek may be right in saying that there never was some idyllic dwelling of Adam and Eve in the bliss of Paradise. Earlier I quoted a Jewish tradition holding that Adam and Eve fell before the sixth day had ended (see §1.3.1). However, this does not change the principal and fundamental fact that we have this inspired testimony about the beginning of human history: "And God saw everything that he had made, and behold, it was *very good*" (Gen. 1:31), apparently including the first pair of humans. This state of integrity possesses a very tangible feature. Before the Fall, "The man and his wife were both naked and were not ashamed" (2:25). But immediately after the Fall, "The eyes of both were opened, and they knew that they were naked. And they sewed fig leaves together and made themselves loincloths" (3:7). Throughout the centuries, the exact meaning of these statements has been widely discussed.[37] The point right now is not so much the outcome of this discussion, but rather our claim that this statement points to a drastic moral difference between prelapsarian and postlapsarian Adam and Eve:

(a) Before the Fall, all of creation was "very good," including Adam and Eve. After the Fall, they were spiritually dead (Gen. 2:17; cf. Eph. 2:1; Col. 2:13). From now on, "the creation was subjected to futility, not willingly,

---

[36] One striking example is Van de Beek (2005; cf. 2011; 2014).

[37] The discussion involved the relation between sexuality and the Fall; see extensively, Ouweneel (2008, §§9.1.2–9.1.3).

but because of him[38] who subjected it," and was in "bondage to corruption," "groaning together in the pains of childbirth" (Rom. 8:20–22).

(b) Before the Fall Adam and Eve knew neither good nor evil (they were good but did not know what this was), because such knowledge was obtained by means of the very act of eating from the forbidden tree. After the Fall they knew both good and evil, but now (before their regeneration) they knew to do only evil, and could no longer do good. This newly experienced sinfulness was evident, for instance, in their evasive answers (Gen. 3:12–13), whereas the new faith of Adam became evident in the name he gave his wife: "mother of all living" (v. 20).

(c) Before the Fall they did not know, but after the Fall they did know the *shame* that sin produces. Their shame was expressed not only in their covering themselves with loincloths of fig leaves but also in their hiding from God (Gen. 3:7–10). Shame is a characteristic result of sin,[39] as we see it, for instance, in Luke's Gospel: in the prodigal son (15:19, 21), in the dishonest manager (16:3), and in Zacchaeus (19:4).

7.2.2 WHAT KNOWLEDGE WAS OBTAINED?

In which way exactly does knowing good and evil relate to eating from the tree of the knowledge of good and evil? At least three views have been propounded.[40]

(a) *Experiential knowledge* (e.g., Wilhelmus à Brakel, Marcus A. van den Oudenrijn). Thus, for instance, the Annotation in the Dutch States' Translation of the Bible (Statenvertaling) on Genesis 2:9 says of the tree of knowledge: "Named this way because through eating of this tree Man would experience (or has experienced) what good he would lose through it, and into what evil he would fall." According to Oosterhoff, the difficulty of this interpretation is that, through eating from the tree, humans would be "like God,"[41] and God does not know evil experientially. Moreover, in this case the tree should be called the tree of the knowledge of evil because Adam and Eve already knew what is good; humans even knew before eating from the tree that the good would lead to life, and evil to death.

I cannot agree with Oosterhoff's objections, though. Before the Fall, Adam and Eve did *not* know what was good. In my view it is essential to see

---

38 Usually, this "him" is thought to refer to God; a few have thought it referred Adam—e.g., Godet (1998, ad loc.)—but this is mistaken: Adam introduced sin into the world, but he did not actively "subject" the world to the power of sin and death.

39 Regarding shame as a result of sin, see extensively, Bonhoeffer (1955, 20–26).

40 Oosterhoff (1972, 142–55); see source references there.

41 Possibly "like gods" (so the Septuagint), but this does not fit Gen. 3:22, said Gispen (1974, 138).

that humans could, and can, really discern what is good only in contrast with evil. Through the Fall, they would learn not only what is evil but, by way of contrast, also what is good. And moreover, God does not have to learn experientially—people do.

(b) *Discerning knowledge* (Gerhard Ch. Aalders, Jan Ridderbos). In this view, expositors argue that, through the tree of knowledge, Adam and Eve were placed before a clear choice between good and evil. Precisely by *not* eating, they would obtain the clear, conscious faculty of discerning between good and evil, obedience and trespass.

This is the opposite of view (a), which claims that Adam and Eve learned to discern between good and evil by eating, whereas according to (b) they learned to discern this by *not* eating. My objections are the same: only through the Fall could Adam and Eve have learned not only what was evil but, by way of contrast, also what was good. We should note the remarkable fact that, after the Fall, God said that, through their eating, Adam and Eve became "like one of us in knowing good and evil" (Gen. 3:22). This is a key verse because it declares that through their eating, Adam and Eve learned not only what is evil but also what is good. Some authors have tried to escape the obvious conclusion from this verse by suggesting that God was speaking with irony here.[42] Others have recognized the essential importance of this verse, and have rejected the idea of "irony."[43]

(c) *Ethically determining knowledge* (John Calvin, Abraham Kuyper, Herman Bavinck, Franz M. Th. Böhl).[44] Here, the key significance of Genesis 3:22 is recognized: the knowledge of good and evil is viewed as a knowledge (in the sense of choosing, determining) that was originally possessed by God alone. Eating is genuine apostasy and rebellion: the first humans wished to determine for themselves what is good and what is evil, as if they were God himself. This is an attempt to emulate God. The central question in Genesis 2–3 is whether Adam and Eve desired to be dependent on God, that is, whether they were willing to be led by God in all matters, especially those of good and evil, and whether they were willing to find life in this attitude, or to follow their own ways. However, for any human being, fashioned to be naturally dependent on the Creator, such so-called "emancipation" entails disaster, misery, and death.

---

42 Sailhamer (1990, 59).

43 Morant (1960, 197); Gispen (1974, 155).

44 See Kuyper (2016, 1:235–44); Bavinck (*RD* 3:32–33). Berkouwer (1971, 271–73), Hamilton (1990, 165–66), Van Genderen (2008, 390 [English translation mistakenly omits the element of "choosing"), and Atkinson (1990, 85) follow this interpretation; also cf. Sikkel (1923, 157–58); Aalders (1981, 101–103); Gispen (1974, 111).

7.2.3 OTHER ASPECTS

That "knowing" can indeed mean "choosing, determining" is evident from several passages, such as Genesis 18:19 ("I have chosen him," lit., "I have known him"); Jeremiah 1:5 ("I knew you," i.e., "I chose you"); Amos 3:2 ("You only I have known," i.e., "chosen"). Children who have "no knowledge of good and evil" (Deut. 1:39) are human beings who are not yet able to *determine* what is good and fair, and what is bad and unfair, and to *choose* to do what is good (cf. 2 Sam. 19:35; Isa. 7:15–16; Jonah 4:11). As Franz M. Th. Böhl rightly said, this knowledge is what little children and the mentally handicapped lack: the faculty of deciding independently one's own destiny and actions, especially in the moral domain. As young Solomon prayed, "Give your servant…an understanding mind…that I may discern between good and evil"; and the LORD answered, "I give you a *wise* and discerning mind" (1 Kings 3:9, 12).

Indeed, this is true *wisdom*: "Behold, the fear of the LORD, that is *wisdom*, and to turn away from evil is understanding" (Job 28:28; cf. Prov. 8:13; 16:6). Paul says, "Among the mature we do impart *wisdom*" (1 Cor. 2:6), and "the mature" are "those who have their powers of discernment trained by constant practice to distinguish good from evil" (Heb. 5:14; Ps. 34:14; 37:27; Amos 5:14–15; see the contrast in Jer. 4:22). Paul also says, "I want you to be *wise* as to what is good and innocent as to what is evil" (Rom. 16:19).[45]

By usurping the right to decide their own destiny and actions, the first humans emancipated themselves from God, so to speak, in sinful conceit. They took their destiny into their own hands, and thereby injured God's honour and love. The emancipation of a minor (being freed from parental control) may be a normal stage in a person's natural development but not in one's religious development (cf. Luke 15:11–12). Instead of childlike dependence, Adam and Eve wished to stand alongside and over against God. They left God's sphere, and this is *the* sin *par excellence*.[46]

It is important to note that the prohibition of Genesis 2:17 ("of the tree of the knowledge of good and evil you shall not eat") had a particular character because it was not a moral but a formal test.[47] That is, this prohibition did not involve some moral commandment ("you shall not murder, not

---

45 Interestingly, in pagan mythology the *serpent* is associated with such wisdom; see Henderson and Oakes (1990). Some Gnostic sects were called Ophites (from Gk. *ophis*, "serpent") because they saw the serpent as the embodiment of wisdom; cf. Couliano (1991); Rasimus (2007).

46 Böhl (1923, 69); cf. Oosterhoff (1972, 175).

47 Cf. Louis Berkhof (1949, 242); to change his formulation a bit: there is nothing injurious in working on Saturday; doing so is not sinful *per se*, for it is not a transgression of the moral law. This means that it would not have been sinful, if God had not said, "Remember the Sabbath day, to keep it holy" (Exod. 20:8).

commit adultery, not steal"), a kind of prohibition that even atheists might have found reasonable, but it involved a test of simple obedience. The Ten Commandments also contained such a test; regarding the Sabbath commandment, there is no moral or logical reason why the seventh day should be a day of rest. For this very reason, the Sabbath was the sign of the Sinaitic covenant (Exod. 31:13, 17), and the Israelites were severely faulted for violating this commandment (2 Chron. 36:21; Ezek. 20:12–13, 16, 20–21, 24). Just as in the case of Adam and Eve, they were ultimately unable to persevere in a test of pure obedience.

Before the Fall, the first humans possessed no knowledge of good and evil, and therefore, strictly speaking, they possessed no morality; they only had to cleave to the Lord. Along the pathway of obedience, the first humans would have found life; the pathway of disobedience led to death. This rule reverberates throughout Scripture: "The fear of the LORD is the beginning of wisdom, and the knowledge of the Holy One is insight. For by me your days will be multiplied, and years will be added to your *life*" (Prov. 9:10–11). At its deepest level, this is what the *Torah* is: the word of the Torah "is no empty word for you, but your very *life*" (Deut. 32:47). The words of the *torah* (here, "instruction") spoken by the father in Proverbs 4:2 "are life to those who find them" (v. 22). "For whoever finds me [i.e., *chokmah*, i.e. *torah*] finds life and obtains favor from the LORD, but he who fails to find me injures himself; all who hate me love death" (8:35–36; see also 3:1–2, 21–22; 4:4; 6:23; 9:6, 11).

7.3 CONSEQUENCES FOR AEH

7.3.1 ADAM AND DEATH

In the previous section, I focused somewhat more extensively on the precise nature of the "knowledge of good and evil." I deemed this necessary because, in my view, it helps us to understand the impossible task with which AEH advocates are confronted. I do not have in view those theologians who do not seem to care very much about the historicity of Adam or of the Fall, such as Peter Enns. But especially Gijsbert van den Brink is someone who wishes to leave room for evolutionary creationism (or whatever he would care to call it), and at the same time seriously believes it is possible to maintain the notions of the historical Adam and the historical Fall. However, in my opinion, his view does not pass the test of biblical accuracy.

First and foremost, the Bible does not speak of life and death in a biological sense only, but especially in an ethical, hamartiological, and soteriological sense.[48] Nobody can escape from the apostle Paul's unambiguous statement: "Sin came into the world through one man, and *death through sin*, and so death spread to all men because all sinned" (Rom. 5:12). Even if we do not think of death here in the most universal biological sense but only in the sense of human death,[49] only one explanation of Paul's statement is possible: *through Adam and Eve's sin, human death entered the world*.

This idea was not Paul's invention; it was nothing but a rewording of Genesis 2:17, "Of the tree of the knowledge of good and evil you shall not eat, for in the day that you eat of it *you shall surely die*." If we interpret the word "day" literally here, then Adam and Eve died spiritually on the day of their fall (cf. Eph. 2:1; Col. 2:13). If we render the phrase "in the day that" as "when" (ESV note), then it referred to the physical death they would one day undergo (Gen. 5:5). Abel was the first human who died (4:8). The frequently repeated "and he died" in Genesis 5 is the outcome of 2:17, and is explained in Romans 5:12: first there was innocence without death, then there was sin, which brought along death, and this death involved Adam and Eve as well as all their descendants. Compare this with Romans 6:23, "the wages of sin is death," and 1 Corinthians 15:21, "by a man [or, by man] came death."

The moment we accept the claim that, by the time the imaginary "Adam" of AEH advocates first fell into sin, many people had already died, we are theologically compromised. This is because, if this were true, human death is nothing more than a natural, biotic phenomenon. Death has lost its ethical and hamartiological dimensions. Death is nothing but a plain fact of life. The deep link that the Bible establishes between sin and physical death disappears; and only some kind of relationship between sin and *spiritual* "death" might be retained. But in the Bible, sin is also explicitly related to physical death: "The soul who sins shall die" (Ezek. 18:4, 20). As the Belgic Confession puts it (Art. 17), "Man had thus thrown himself into physical and spiritual death."[50]

In soteriology, this is of the greatest importance. The life that Jesus Christ has obtained for his people through his death and resurrection (2 Tim. 1:10;

---

[48] Note how van den Brink (2017, 228–45) wrestles with this subject, with a constant "on the one hand" and "yes, but on the other hand." In the end, he does not really solve the problems he himself has raised.

[49] Cf. Moule (1891, ad loc.): "Scripture nowhere says that death in *animals* is due to human sin. Death was the specially threatened penalty to the sole race which was on the one hand created with an animal organism, which could die, and on the other, 'made in the image of God.' The *penal* character of death is essential to St Paul's argument" (http://biblehub.com/commentaries/romans/5-12.htm). Moreover, in order to eat plants (Gen. 1:29), the death of the plant is inevitable.

[50] Dennison (2008, 2:434).

1 John 4:9) has consequences not only for spiritual death but also for physical death. For instance, Jesus taught us that a believer will no longer "die": "Whoever believes in me, though he die, yet shall he live, and everyone who lives and believes in me *shall never die*" (John 11:25–26). Therefore, the New Testament usually describes the death of believers as "falling asleep" (Matt. 27:52; John 11:11; Acts 7:60; 1 Cor. 15:6, 18, 20; 1 Thess. 4:14–15). For them, physical death has lost its sting; it is merely a transition to being "with Christ" (Phil. 1:23). "O death, where is your sting? The sting of death is sin" (1 Cor. 15:55–56). Where the power of sin is broken, the power of death is broken as well.

7.3.2 AEH AND THE FALL

Now let me ask: what happened to all those supposed people during the period before or during the time of "Adam" according to AEH advocates? Did they die like animals, that is, did they cease to exist at death? Did they die to be forever "with God" because sin had not yet entered the world? As I asked earlier, when did hominids reach the level that this wonderful miracle happened: hominids who had previously died like animals (ceasing to exist) evolved into hominids who were now "eternity beings," who would exist forever? Moreover, *how* must we imagine this everlasting existence? Was there a heaven or a hell for them—eternal life or eternal fire (Matt. 25:41, 46)? Or was there some special limbo for them, like the *limbus patrorum* (the place of the Old Testament saints after death) or the *limbus puerorum* (the place of unbaptized children after death) that the ancient church has invented?

I do not wish to be sarcastic about this because the question is serious: What happened to the humans who, according to AEH advocates, died before the Fall? And conversely, what consequences did Adam's fall have for the other people that supposedly lived on the earth? What happened to them when they died after the Fall? Did Adam's sin affect them, or not? If yes, how did it affect them? And especially, how *could* it possibly affect them if they could not be made accountable for what Adam had done? Compare Ezekiel 18:20, "The soul who sins shall die. The son shall not suffer for the iniquity of the father, nor the father suffer for the iniquity of the son. The righteousness of the righteous shall be upon himself, and the wickedness of the wicked shall be upon himself."

"It is appointed for man to die once, and after that comes judgment" (Heb. 9:27).[51] What did this imply for any higher hominid? What did this

---

[51] John Gill comments on this verse: "And as it is appointed unto men once to die.... Not a moral, or what is commonly called a spiritual death, nor an eternal one, but a corporeal one; which does not arise from the constitution of nature, but from the sin of man, and God's decree

imply for the first real human beings? What judgment would they face if God had never warned them? What eternal life would they enjoy if the first humans never heard some gospel? Who might have been the prophets who brought them any gospel? Was this (imaginary) gospel only for those who had "fallen" (no matter the meaning of the word)? Was it also for those early humans who had *not* "fallen"? The questions continue to multiply. Why should we *begin* to take AEH advocates seriously as long as we do not receive any specific answers to questions like these? And we know beforehand that these answers can only be speculative. This is what happens when we discard the simple answers of Scripture.

Moreover, what does it mean according to the AEH view that the first humans (called "Adam and Eve") were pure and innocent? How can we imagine any form of human evolution in which hominids did *not* already kill each other, were *not* promiscuous, did *not* steal from each other, did *not* cheat, and so on, for hundreds of thousands of years before the "fall" supposedly occurred according to AEH advocates? And even if we begin at the miraculous point in this supposed human evolution when hominids became morally responsible, how could we claim that they remained innocent until they had fallen?[52] How could we imagine even that shortest of moments when these hominids *had become* morally responsible but *were still* innocent? This is virtually impossible. Hominids had been killing, stealing, sleeping around, and cheating for an enormous period of time, but then they supposedly reached a point where they—by a miracle of human evolution!—suddenly became morally responsible. And what happened? They simply continued killing, stealing, sleeping around, and cheating, with this difference, that now they were morally responsible for it. *In this picture, there never was a time of innocence, and there never was anything like a fall into sin.*

7.3.3 A FORMAL TEST

One final question for now: According to AEH advocates, what was the precise formal test that God set before Adam and Eve? It was not a moral test but a simple "you shall not eat of that tree." It made the disobedience of the first pair of humans all the more heinous. It was not a matter of killing, stealing, cheating, or sleeping around, but a *formal* commandment, explicitly *revealed* to the first humans: "Don't eat." And from their side, it was a matter of either trusting God or choosing their own way.

---

on account of it" (http://biblehub.com/hebrews/9-27.htm).

[52] Cf. van den Brink (2017, 240), who needs this argument very much to preserve a historical Fall.

Now how does this work according to the AEH view? Or, to put it in a more general form: How exactly do AEH advocates think that God revealed himself to their "Adam"? The Bible is not very specific on this point, but at least it tells us that God "spoke" to the historical Adam and Eve (Gen. 1:28–29; 2:16–17; 3:9–19). What becomes of this in the AEH view? Even if we assume that God selected, from the extant hominids, one pair, or a small population of people, did he speak to them alone (in whatever way), and not to the others? We may call this "election," like God later elected Abraham and Israel;[53] but the point is that, in Genesis, there is not the slightest trace of such an "election" of Adam from among a larger community of humans. We cannot prove anything by suggesting parallels. Adam was created directly by God, after God had created the plants and the animals. And *if* indeed other humans had existed, it could not have been a problem for God to tell us so, perhaps telling us something as simple as: Adam and Eve were graciously and sovereignly chosen from among all the humans who were on the earth at the time.

Moreover, look at Eve. At first, Adam was all alone; he had no wife, although he figured out that all the male animals had their females (cf. 2:18–20). He received his wife in a miraculous way (she was not born of his mother-in-law but taken from his side!), and this first couple became the model for all the millions of couples that descended from them (2:21–24). In the AEH ideology, all of this is understood figuratively. How could we posit an "election" of the first humans from a much larger population of hominids?

I respect van den Brink for trying to undertake such an impossible task. I appreciate his desire to maintain the notions of a historical Adam and a historical Fall. But I can guarantee that a subsequent generation will watch with admiration his on-the-one-hand/on-the-other-hand wrestlings, only then to comfortably dispense with those struggles. Why struggle with the problem of a historical Adam and a historical Fall? Simply to maintain the appearance of some kind of orthodoxy? A new generation of evolutionary creationists will devise newer, more clever and more creative solutions for all the soteriological problems that their view will produce. The first representatives of this new generation have already appeared on the scene.[54]

I can imagine that AEH advocates feel that I am fighting a losing battle "because evolution is true after all." But I honestly feel that AEH advocates themselves are fighting a losing battle. Some of them, coming from orthodox backgrounds, have already come to the point of flatly denying that there ever was a historical Adam and a historical Fall the way the Bible describes

---

[53] van den Brink (2017, 238); cf. Appendix 3.

[54] See Lamoureux (2008; 2009); Enns (2012).

them.[55] Ultimately this will mean that the notion of "sin" involves nothing but the evolutionary shortcomings of *Homo sapiens*; in further evolution ("maturation") lies his true hope of reaching a better world. Why would humans need a new heaven and a new earth if evolution brought them where they are now, after millions of years? Why would human evolution not simply continue (under God's providence, if you like), and, in a few million years, by itself reach a new and better world (cf. §§9.4.2, 9.4.3)? And above all, why would evolving humans need any work of redemption to set them free from their shortcomings? In the end, evolution is inevitably substituted not only for biblical creation but also for biblical redemption.

7.4 THE SERPENT IN EDEN

7.4.1 ITS IDENTITY

We must now pay closer attention to the way the first pair of humans fell. In a book like this, it is not necessary to offer a full exegesis of Genesis 3. But the problem of the serpent is certainly one of the main questions that demand closer examination. How must we view the serpent in the Garden of Eden? In 1926, there was a serious controversy within the Kuyperian Reformed Churches in the Netherlands, which was decided at the Synod of Assen in that year. The conflict involved whether the serpent had talked in a "literal" and "empirically observable" way. The progressive party said no, the conservative party said yes, and in between were a number of church leaders who tried to defend an on-the-one-hand/on-the-other-hand position. Following typical Dutch tradition, the controversy led to a church split.[56]

This is rather strange because, in my view, the identity of the serpent is quite clear (see §4.5.3). Did Eve indeed see an ordinary snake? Oosterhoff claimed, "Almost every exegete has felt that this serpent is not simply an ordinary animal."[57] John H. Sailhamer argued that the best translation of Genesis 3:1 would be: the serpent was "subtle [or crafty, sneaky, clever, cunning; Heb. *arum*] as none other of the beasts," a formulation entirely parallel with exactly the same construction in verse 14: "Cursed [Heb.

[55] See Giberson (2009); Carlson and Longman (2010); Enns (2012).

[56] The progressive party formed the Gereformeerde Kerken in Hersteld Verband (Reformed Churches in Restored Union); in 1926, this small denomination joined the Nederlandse Hervormde Kerk (Dutch Reformed Church).

[57] Oosterhoff (1972, 165). Those who thought otherwise included Proksch (1913, 32); Jacob (1974, 102); Sailhamer (1990, 50). The rabbis did not agree; Obadiah ben Jacob Sforno (d. 1550) said that the serpent was a symbol of Satan, whereas Abraham Ibn Ezra (d. 1167) understood the passage literally, and rejected any symbolism; see Cohen (1983, 12).

arur] are you above all livestock, and above all beasts of the field."[58] Thus, the serpent was crafty,[59] and was later cursed, and the other animals were *none of these*. Sailhamer concluded from this that the serpent was not an ordinary animal, at least not in every way; 2 Corinthians 11:3 tells us that "the serpent deceived Eve by his cunning." Thus, the story clearly leaves room for identifying the serpent with Satan.

In other words, the question is not whether snakes can talk but whether Satan can talk.[60] The answer is: of course he can talk (Job 1:9–11; 2:2–5; Matt. 4:1–11; Luke 4:1–13). His talking is also implied in Zechariah 3:1, Jude 9, and Revelation 20:10.

Victor P. Hamilton pointed to the possible connection between the Hebrew words *nachash* ("serpent") and *nechoshet* ("copper, bronze"), and observed, "This connection with bronze suggests a shiny and luminous appearance, which would arrest Eve's attention."[61] In his view, too, the serpent was hardly an ordinary snake. John Gill spoke of "a beautiful flying serpent, which looked very bright and shining."[62] I may add that such an appearance reminds us of Ezekiel 28, where we find stationed behind the king of Tyre his angelic prince, who was an exalted angel of Satan, if not Satan himself. We read in Ezekiel 28:12b–14:

> You were the signet of perfection, full of wisdom and perfect in beauty. You were in Eden, the garden of God; every precious stone was your covering.... and crafted in gold were your settings and your engravings. On the day that you were created they were prepared. You were an anointed guardian cherub. I placed you; you were on the holy mountain of God; in the midst of the stones of fire you walked.[63]

---

[58] Sailhamer (1990, 50–51); cf. Gen. Rabbah 19.1; the same construction appears in Judg. 5:24.

[59] The Heb. adjective *arom* (plur. *arumim*) means "naked" as well; thus, Gen. 2:25–3:1 says, "And the man and his wife were both naked (Heb. *arumim*) and were not ashamed. And the serpent was more crafty (Heb. *arum*) than all the beasts of the field." Böhl (1923, 68) observes, "The serpent promises to the humans that they would become ʿ*arumim*, that is, astute, and indeed they did become ʿ*arumim*, but in the other meaning of the word: naked!" Cf. Hamilton (1990, 187).

[60] Assuming this sense of the word, Walton (2001, 45–49) can argue that the serpent really talked.

[61] Hamilton (1990, 187). He and others, including Böhl (1923, 68), also point to the connection between, the noun "serpent" (Heb. *nachash*) and the verb "to practice divination, observe signs" (Heb. *nachash*).

[62] http://biblehub.com/genesis/3-1.htm.

[63] See our related comment in chapter 5, note 64.

7.4.2 THE SERPENT UNMASKED

German-Baltic theologian Hellmuth Frey wrote,

> In our chapter, we first see the serpent as an ordinary animal. However, the more actively he intervenes, the deeper his significance becomes and a mysterious, dark reality emerges, which shines through its outer form. Initially, he distinguishes himself from the other beasts of the field only by his cunning (v. 1). Later, he manifests himself more and more: we see a masterly designed plan, the perfect player on the instrument of the human soul (vv. 1–4), his enigmatic, supernatural knowledge concerning the mystery of the trees (v. 5), until he overtly speaks as God's opponent, in order to finally become the great opponent of humanity throughout the centuries (v. 15).[64]

Frey's last point is especially important: in Genesis 3:15, where we read that God puts enmity between the serpent's seed and the woman's seed, the serpent's final mask is removed: he is Satan and nothing else. As Jesus said to his enemies, the religious leaders of Israel, "You are of your father the devil, and your will is to do your father's desires. He was a murderer from the beginning, and does not stand in the truth, because there is no truth in him. When he lies, he speaks out of his own character, for he is a liar and the father of lies" (John 8:44). In the expression "from the beginning," Jesus was apparently referring to Genesis (1–)3 (cf. Mark 10:6, "from the beginning of creation").

German Evangelical theologian Georg Huntemann observed,

> The serpent that is mentioned in the story of the Fall has nothing to do with the serpent as we encounter it in nature. In John's Revelation (20:2; 12:9) we read that the serpent is the dragon, the devil. The evil one himself broke into that domain that had been assigned to humanity for the unfolding of its possibilities.... The serpent does not belong to the domain of the animal world.... It was absurd to depict Satan in the seduction story as a snake[65] such as it occurs today in an animals' lexicon.[66]

---

[64] Frey (1962, 45).

[65] There were exceptions: some artists depicted the "serpent" as a dragon, as in the *Speculum Humanae Salvationis* (1324–1500) (http://publishing.cdlib.org/ucpressebooks/view?docId=ft7v19p1w6; chunk.id=0;doc.view=print).

[66] Huntemann (1977, 31–32).

Indeed, the book of Revelation is decisive here: "And the great dragon was thrown down, that ancient serpent, who is called the devil and Satan, the deceiver of the whole world" (12:9); an angel "seized the dragon, that ancient serpent, who is the devil and Satan" (20:2).

Please note that the approach of Huntemann and others does not consist of surreptitiously turning the story of the Fall into something entirely figurative. A genuinely historical encounter occurred between this being and the first woman. However, it was not a confrontation between an ordinary animal and an ordinary human (cf. Gen. 49:17; Num. 21:6; Deut. 8:15), but between Satan and the first pair of humans. If the first humans were to fall, this would lead to their death. Satan was the one who would lead them to their death for he is the "murderer from the beginning." And he did this through lies, for he is also the "liar from the beginning," so to speak.

7.4.3 THE IDENTITY OF THE SERPENT

Swiss theologian Peter Morant also argued,

> That this serpent is not a member of the Ophidia [snakelike reptiles] from zoology, and that the verdict about it does not simply point to the antipathy that humans naturally feel toward it, is evident from the serpent's speaking, its superhuman knowledge, and its cunning art of seduction.[67]

He went on to sharply oppose theologians such as German theologians Hermann Gunkel and Gerhard von Rad, and Swiss theologian Walter Zimmerli, who saw the serpent merely as a fairy tale figure, a mythical character. He also opposed Philo and ancient theologians such as Clement of Alexandria and Origen, who saw the serpent merely as the symbol of awakening human sensual desire. Instead, he insisted that the serpent was an empirical historical personality; however, it was not an ordinary snake but the representative or even embodiment of evil, Satan himself.

Usually, AEH advocates limit themselves to presenting abstract theological comments about how to salvage the notions of the historical Adam or the historical Fall, without entering too deeply into such unsettling questions as: What about the Garden of Eden? Or the four rivers in that Garden? Or the two special trees? Or (the identity of) the serpent? The more orthodox they wish to appear, the less they tell us about the trees and the serpent because it might become evident that their view is not all that different

---

[67] Morant (1960, 162–63).

from the view of those liberal theologians who overtly characterized these as belonging to fairy tales and myths. AEH defenders themselves would never use a word like "myth." Instead, even though they accept the theory of human evolution, they stress that in good orthodox fashion they believe in some kind of historical Adam or some kind of historical Fall. But they avoid discussing the exegetical details of Genesis 3.

Those who do take a little more seriously the verse-by-verse exegesis of Genesis 3 *must* determine who or what the serpent exactly was. Peter Morant supplied us with three possible explanations, each of which had defenders: (a) Satan employed an ordinary snake, (b) Satan personally appeared in the form of a snake, or (c) Satan was symbolically presented as a snake.

The first explanation seems closest to the text.[68] In my view, however, there are several arguments that make it unlikely that through a kind of "possession" Satan employed an ordinary snake. The Bible nowhere says that, in the Garden of Eden, the devil *employed* a snake; rather, it says explicitly that the serpent *was* the devil. Thus, the apostle John said, "the devil has been sinning from the beginning" (1 John 3:8), and was presumably referring to the Garden of Eden. John also told us that the "ancient serpent" was none other than *the devil himself*, as we saw: "the great dragon ...that ancient serpent, who is called the devil and Satan, the deceiver of the whole world" (Rev. 12:9; cf. 20:2).

This appears also to have been the way Christ understood the serpent. In one breath, he mentioned "serpents" and "all the power of the enemy" (Luke 10:19). When he called the devil "a murderer from the beginning" as well as a "liar" (John 8:44), he apparently meant that the serpent in the Garden of Eden *was* the devil. Oosterhoff stated the matter simply: "To Jesus, the serpent in Paradise was none other than the devil."[69] This was also the way the Jews understood the serpent: "It was the Devil's jealousy that brought death into the world" (Wisdom 2:24). As we have seen, the medieval expositor Rabbi Obadiah ben Jacob Sforno saw the serpent as a symbolic representation of the devil.[70] And in 1 Enoch 69:6–7 we read, "the third [angel] was named Gadreel:[71] he it is who showed the children

---

[68] This was the view of Bavinck (*RD* 3:34); Hoek (1988, 17); see also Heyns (1988, 171), who compares the situation with other occasions where Satan hid himself behind others (Bathsheba in 1 Kings 2:19–24; Job's three friends; the false prophets in Jer. 28; Peter in Matt. 16:23; cf. 2 Cor. 11:14).

[69] Oosterhoff (1972, 165).

[70] Cohen (1983, 12).

[71] In Jewish apocalyptic literature, Gadreel was one of the chiefs of the fallen "watchers"

of men all the blows of death, and he led astray Eve, and showed the shield and the coat of mail, and the sword for battle, and all the weapons of death to the children of men."[72]

7.5 THREE EXPLANATIONS

7.5.1 SATAN EMPLOYED AN ORDINARY SNAKE

As we consider the arguments just mentioned, we discover the irrelevance of asking whether the serpent in Paradise could speak. If donkeys can speak (Num. 22:28–30), why not snakes? But this is not the point. The point is whether the *devil* identified in Genesis 3 could speak. Of course, the answer is Yes (see §7.4.1). He has spoken many times since. He spoke with God (Job 1–2; Zech. 3:1), he spoke with Jesus (Matt. 4:1–11), and he argued with Michael (Jude 10). It remains quite possible that a literal physical snake was employed; but the essence of the matter is that Scripture makes very clear who the figure was behind this snake: it was Satan himself who spoke to Eve.

The claim that a literal physical snake was employed by Satan has some problems. Though not insurmountable, these exegetical difficulties make us think. A literal physical snake must have been an animal that crawled on its belly for the first time after the Fall (Gen. 3:14), and thus, presumably had limbs before the Fall. The Targum of Jonathan, an ancient midrash (Gen. Rabbah 20.5), and the medieval expositor Rashi (Rabbi Shlomoh Yitzchaqi), held this view,[73] as did the Jewish historiographer Flavius Josephus, [74] and Martin Luther.[75] Abraham Kuyper, too, spoke about the post-Fall modification "in the appearance and manner of the serpent's existence."[76] If this was not a literal physical snake but rather a representation

---

(cf. Dan. 4:13, 17, 23) as well as one of the "sons of God" who seduced earthly women (Gen. 6:1–4).

[72] http://wesley.nnu.edu/sermons-essays-books/noncanonical-literature/noncanonical-literature-ot-pseudepigrapha/book-of-enoch/. I am careful with quoting this type of literature, though, because it often contains unbiblical and contradictory statements; cf. Enns (2012, 65–68, 99–103).

[73] Cohen (1983, 15); Ginzberg (1969; https://philologos.org/__eb-lotj/vol1/two.htm#1) points to this ancient Jewish tradition: "Among the animals the serpent was notable. Of all of them he had the most excellent qualities, in some of which he resembled man. Like man he stood upright upon two feet, and in height he was equal to the camel."

[74] *Jewish Antiquities* I.1.50.

[75] Morant (1960, 177); Oosterhoff (1972, 171).

[76] Kuyper (2016, 278).

or personification of the devil—but also if we are AEH advocates—then we do not need such speculative explanations.

Nevertheless, the possibility remains that we are dealing here with a literal physical reptile that, before the Fall, had limbs. I point here to a verse mentioned earlier (Gen. 1:21), which says that God created "the great sea creatures" (Heb. *tanninim*). These are beings that elsewhere in the Old Testament are usually described as "dragons," but sometimes as "serpents" (see §7.6.1). The serpents mentioned in Exodus 7:9–12, Deuteronomy 32:33, and Psalm 91:13b are not described with the Hebrew word *nachash* but *tannin*, like the *tanninim* elsewhere (Job 7:12; Ps. 74:13; Isa. 27:1; 51:9; Jer. 51:34; Ezek. 29:3; 32:2). That is, the meanings "serpent" and "dragon" (sea monster) are flexible. It is not impossible to think of a walking dragon that after the Fall looked more like a snake, going on its belly.[77]

Hardly anyone would understand the statement "dust you shall eat" (Gen. 3:14) literally. Even if a snake eats dust along with its food—which is true for most animals—one could hardly maintain that dust is "the serpent's food" (so Isa. 65:25), or that it "licks the dust" (Mic. 7:17). Incidentally, it strikes me as inconsistent when some conservative supersessionist theologians spiritualize prophetic passages like Isaiah 65 and Micah 7, but vigorously protest when the same approach is applied to Genesis 3. In my view, all these three chapters must be understood literally, although they may contain a lot of figurative language. When such theologians protest against AEH, I protest together with them. But if one insists that, as a test of orthodoxy, we must interpret Genesis 3:14 literally, then one must also explain how the serpent moved around before the Fall, and what it ate after the Fall.

This is an important point. I am sharply criticizing the AEH ideology. But doing so by rushing to the other extreme of defending an extreme literalism is unhelpful (see §4.3). We must proceed carefully in the exegesis of Genesis 1–3, avoiding both extremes of literalist biblicism and of spiritualizing liberalism. We may arrive at diverging views on a number of points. But we should never allow any natural-scientistic theory to influence the way we read these chapters. *Exactly this* is the failure and "fall" of AEH.

7.5.2 SATAN APPEARED AS A SERPENT

This is the second interpretation suggested by Morant concerning the identity of the serpent in the Garden of Eden. We might think here of a parallel with the way angels sometimes assumed the form of a man; good examples are found in Genesis 18:2, 19:1, and Judges 13:3–20. John Calvin wrote very

---

[77] In Isa. 30:7 we read of a dragon "sitting," and in Rev. 12:4 a dragon "standing."

clearly that Moses "described Satan, the prince of unrighteousness, under the form of his servant and instrument," namely, the serpent.[78] Abraham Kuyper, who sometimes strongly emphasized the literalness of the serpent, could nevertheless write that he left the matter undecided as to "whether Satan actually was in the serpent and spoke through him, or whether the word 'serpent' was just a name for Satan himself."[79] Franz Böhl argued, "The assertion: the serpent in Paradise is the devil, is definitely not wrong from a historical viewpoint."[80] Bastiaan Wielenga wondered why the serpent could not be viewed as the form in which Satan appeared to Eve.[81] Berend Oosterhoff spoke of an "incarnation": "The serpent in Paradise is the incarnation of all anti-divine powers, which rise against God and bring humanity to destruction."[82]

If Genesis 3 does not refer to an ordinary, literal, physical snake but to a snake-like incarnation of Satan, one might object that verse 1 says, "the serpent was more crafty than *any other beast* of the field that the LORD God had made"—so it was a "beast" after all. But what does it mean that the serpent was the most "crafty" of all animals? Are not many animals, especially mammals, much more intelligent or cunning than any ordinary snake? Or did the Fall change something in the intelligence or the cunning of the serpent? Of course, this problem does not occur if Sailhamer (§7.3.1) and others are right who say that we should translate the phrase something like: "The serpent was crafty among [or even, in opposition to] all the beasts of the field," where the serpent itself is not necessarily an animal at all, or at least not an ordinary animal. That is to say, among the animals there was one that *presented* itself as an animal—a serpent—but that in reality was a manifestation of Satan, and as such was far more intelligent and cunning than ordinary animals.

Notice in these interpretations that they *all* accept the literal existence of Eve, who was confronted by some literal power, who addressed her in the form of a snake. This entire discussion can scarcely be meaningful for AEH advocates, whose paradigm has no place for such a confrontation between the "first" woman and Satan himself. By contrast, we may well struggle with the precise identification of the serpent, but at least we struggle. For AEH advocates, no struggle exists: there supposedly never was a first,

---

[78] Calvin, *Comm. on Genesis*, ad loc.

[79] Kuyper (1923, 206).

[80] Böhl (1923, 68).

[81] In *Acta* 43–47 (Synode Assen 1926).

[82] Oosterhoff (1972, 170–71). Joh. Francke (1974, 40–53) had many objections against this view, but I find his arguments to be quite dogmatic, and academically less solid than Oosterhoff's.

perfectly innocent pair of humans, in an ideal world, a pair who were literally confronted with Satan, who either employed a snake or adopted the form of a snake.

7.5.3 SATAN SYMBOLICALLY PRESENTED AS A SERPENT

Swiss theologian Theodor Schwegler hesitated between Morant's second and the third option:

> It must therefore involve a being that is particularly hostile to God, one that has assumed the form of a snake, or makes use of a snake as mask and instrument. However, it is also conceivable, even likely, that the snake is merely a symbol of a wicked higher being, one that was perhaps once created good by God, but then ended up in opposition to, and enmity toward, God.[83]

This being later turned out to be Satan, or the devil.

I can imagine that this third option is especially popular among AEH advocates. This is because those who choose this option can easily claim that there never was a *literal*, *historical* encounter between some figure called Satan and the first woman. This was why well-known Reformed theologians in the Netherlands, Jan Ridderbos and Gerhard Ch. Aalders, forcefully rejected this interpretation.[84] German theologian Hans Möller rightly argued that the serpent may never be reduced to the will or the thoughts in the heart of the woman herself;[85] it is of essential importance that the temptation of the "tempter" (Matt. 4:3; 1 Thess. 3:5; cf. 2 Cor. 11:3) came to the woman from the outside.[86] The first pair of humans faced a literal temptation; there is every reason to maintain that in Genesis 3 we are standing on historical ground.

If so, the first option (see §7.5.1) and the second option (see §7.5.2) are far more likely. The first woman had a historical encounter with something evil outside her, namely, with Satan, who either made use of a literal physical snake (or dragon, monster), or appeared to her in the form of a snake (dragon, monster). Surely because of the clear parallels between the temptation of Adam and Eve and the temptations of Christ (see §8.2.2), the encounter between Eve and the serpent was quite likely of the same

---

[83] Schwegler (1960, 105).

[84] Ridderbos (1925, 28); Aalders (1932, 486).

[85] Möller (1977, 36).

[86] Even though his effect was certainly on Eve's *heart*; cf. Acts 5:3, "Ananias, why has Satan filled your heart to lie to the Holy Spirit?" (cf. John 13:27).

nature as the encounter between Christ and Satan in the wilderness (Matt. 4:1–11). At any rate, the former encounter was just as historical as the latter encounter. At the same time, notice these important differences:

(a) Adam and Eve were tempted under the most favourable circumstances, namely, in the pristine Garden of Eden. Jesus was tempted under the most *un*favourable circumstances, namely, in the desolate wilderness.

(b) Adam and Eve had enough food (Gen. 1:29, "Behold, I have given you every plant yielding seed that is on the face of all the earth, and every tree with seed in its fruit. You shall have them for food"), whereas for forty days Jesus had had nothing to eat, and was tempted when he was hungry.

(c) Adam and Eve might have helped each other to resist the temptation, which they didn't. Jesus was all alone; only after the temptations were over did angels come to minister to him (Matt. 4:11).

(d) Adam and Eve had little physical and spiritual power during their temptation. Jesus, however, did have the power to turn stones into bread, and to jump from the temple's pinnacle and land safely, but he did not abuse this power.

(e) Adam and Eve failed in their temptation, and the devil triumphed. Jesus triumphed in his temptation, and the devil "departed from him until an opportune time" (Luke 4:13; probably a reference to Gethsemane and Calvary; cf. 22:53; John 14:30).

At the beginning of the Old Testament, Satan came to tempt the first Adam, and he succeeded. At the beginning of the New Testament, Satan came to tempt the last Adam, and he failed. Each event entailed a turning point in human history. As we will see later, the three temptations of Adam and Jesus, respectively, were basically of the same character (§8.2.2). But what concerns us here is this point: if the encounter between Jesus and the devil was real and historical, so was that between Adam and Eve and the devil. It may be true that we will never know *exactly* what was the relationship between the serpent and Satan.[87] But this uncertainty does in no way affects the historicity of the encounter between Satan and the first pair of humans.

7.6 THE INVISIBLE POWERS

7.6.1 THE SEA MONSTERS

Let me conclude this chapter with this question: Are all "animals" in Genesis 1 the ordinary animals of our zoological handbooks? I remind the

[87] Cf. Sikkel (1923, 184); Popma (1972, 18); Huntemann (1977, 32).

reader here of those fascinating "sea creatures" (Heb. *tanninim*) mentioned in verse 21, even though they are not "beasts of the field." What *are* they? Certainly not the "whales" of the KJV and other translations.[88] I am pleased that, among the Bible translations, the JUB has the courage to translate here the word as it is usually translated in the Old Testament: "dragons." Even in the passages where the word is commonly rendered "serpent," the JUB consistently renders it as "dragon(s)" (Exod. 7:9–12; Deut. 32:33; Ps. 91:13b). Other translations avoid the word "dragon" because they seem to wish to present us with animals that are familiar to the readers. This also explains the tendency to render the terms *Behemoth* in Job 40:15 as "hippopotamus" (CEV),[89] and *Leviathan* in 41:1 as "crocodile" (TLB).[90] Translators should leave untranslated the Hebrew terms *tannin*, *Behemoth*, *Leviathan*, and *Rahab* (e.g., Isa. 51:9); these creatures may appear strange to us,[91] but in the Bible they are very real powers of darkness. As Gordon Wenham wrote, "Gen 1:1–2:3 is a polemic against the mythico-religious concepts of the ancient Orient.... The polemic intent of Genesis is even more clear in its handling of the sea monsters and the astral bodies [say, souls]."[92]

The same is true about the dragons as about the serpent mentioned in Genesis 3: we may struggle with their precise identity, but *they are historically real*. They are *not* dinosaurs, as creationists have sometimes suggested,[93] but neither are they mythical, as liberal theologians have suggested.[94] In my view, they refer to spiritual powers, just as real as Satan and his powers.[95] Whether Satan employed a snake or appeared in the form of a snake, his confrontation with the first woman, Eve, and her husband, Adam, was real and historical. He really spoke to a real first pair of humans, he really referred to the real trees in the Garden, even as he spoke of the reality of God.

I wonder whether especially the term *Rahab* is a name for Satan, or at least for a highly placed Satanic angel. *Rahab* is primarily the name of the chaos-angel, the angelic prince of the worldwide, turbulent "deep,"

---

[88] Incidentally, the word "whale" itself means etymologically nothing but "big (fish)."

[89] Cf. Job 40:17 (NIV), "Its tail sways like a cedar"—ever watched the tail of a hippo?

[90] Cf. Job 41:34 (NIV), "It looks down on all that are haughty"—can you imagine the crocadile "looking down" on certain humans?

[91] But no more strange than the "satyr" (KJV; Heb. *sair*) and the "night monster" (CJB; Heb. *lilit*) in Isa. 34:14, or the phoenix in Job 29:18 (CJB).

[92] Wenham (1987, 37).

[93] As I once did: Ouweneel (1976); see also https://answersingenesis.org/dinosaurs/dinosaurs-and-the-bible/; cf. Paul (2017, 196–201).

[94] See Batto (1992).

[95] Cf. Ouweneel (2018a, §3.1).

or "flood." The word *Rahab* literally means "impetuous" or "reckless." We meet a Rahab for instance in Job 26:12 (cf. 9:13), where the word is parallel with the term for the sea, and apparently can be identified with this. The Talmud and Midrash conclude, also on the basis of this verse, that *Rahab* is the name of the angelic prince of the sea or the primeval ocean.[96] In Psalm 89:11 and Isaiah 51:9, *Rahab* may refer to the primeval ocean (see Gen. 1:2), but also to the Red Sea.[97] In addition to its general meaning as angelic prince of the primeval ocean, *Rahab* also has a more limited meaning, in relation to Egypt (Ps. 87:4).

We should not think of the *Rahab* as a prehistoric or legendary (mythical) monster, in spite of the language in Ezekiel 29:3 and 32:2. It is the monstrous angelic prince, whether in the wider sense of the ocean's ruler, or in the narrower sense of Egypt's prince. Rahab is usually described as referring originally to the feminine chaos monster, comparable with the chaos monster Tiamat of Babylonian mythology. This name Tiamat presumably echoes the Hebrew *tehom*, "primeval ocean, (water)flood, deep," which we encounter in Genesis 1:2.[98] Rahab, then, is the chaos angel, the angelic prince of the original state of emptiness and void, and particularly of the worldwide, uncontrolled "flood." This angelic prince stands under the control of God's power, as is shown by Job 26:12. In Job 38:8–11, too, we clearly see how God has subdued the angelic prince of chaos.

Especially modern form criticism has devoted much attention to the "monster" (angelic prince) of the primeval chaos, or ocean. However, it uses such reflections to assign Genesis 1 to the realm of (ahistorical) mythology.[99] This is why many conservative Bible interpreters shrink from such views. We should not, however, throw out the baby with the bathwater here. If the Old Testament indeed refers to the chaos powers, which played a negative role at the time of God's creation work, we should not suppress these data, but do justice to them. We must acknowledge them, not in order to "dehistorize" or "demythologize" Genesis 1, but, on the contrary, to draw attention to the spiritual warfare in the heavenly realms, which presumably played a role already in the creation story.[100]

---

[96] Baba Bathra 74b; Num.R. 18 (185a).

[97] Cf. Pesakhim 118b and Arakhin 15a, which speak of the angelic prince of the Red Sea; cf. also Ps. 74:13.

[98] See further Gen. 7:11; 8:2; 49:25; Deut. 33:13; Job 28:14; 38:16, 30; 41:23; Ps. 36:7; 42:8; 104:6; Prov. 8:27–28; Isa. 51:10; Ez. 26:19; Am. 7:4; Jon. 2:6; Hab. 3:10.

[99] See more extensively, Ouweneel (2018a, Appendix 3).

[100] See more extensively, Ouweneel (2018a, Appendix 4).

7.6.2 CONFRONTATIONS

In the Old Testament, I perceive at least the following confrontations between the dragon(s) and God (as well as his people).

(a) The implicit conflict in Genesis 1:2, "The earth was without form and void [Heb. *tohu wabohu*], and darkness was over the face of the deep [Heb. *tehom*; see previous section]." A most helpful introduction to this subject remains the work by Dutch-South African theologian Johannes H. Kroeze.[101]

(b) The confrontation between the dragon (Satan) and the first woman (Eve). See chapters 7–10 in this volume.

(c) Expositors have presumed that evil powers were also involved in the Noahic Flood because of the typical vernacular of Genesis 7:11 (cf. 8:2), "on that day all the fountains of the great deep [Heb. *tehom*] burst forth, and the windows of the heavens were opened" (cf. Isa. 24:18).[102]

(d) During Israel's exodus from Egypt, the "gods" of Egypt, that is, its angelic prince with his helpers, were clearly involved (cf. Exod. 12:12; Num. 33:4).[103] *Rahab* is the name of the chaos-angel, the angelic prince of the worldwide, turbulent "deep," or "flood" (see previous section). In Psalm 89:10, *Rahab* refers either to the primeval ocean or to the Red Sea. In Psalm 74:13–14, the Hebrew terms *tanninim* and *Leviathan* may refer to the chaos-powers at the creation, or to the "dragons" of Egypt. In Psalm 87:4, *Rahab* is a reference to Egypt, but this includes Egypt's king, who is viewed as the embodiment of Egypt's main "god," that is, Egypt's angelic prince. In connection with Exodus 15:1 ("the horse and his rider"), a Jewish tradition connects this with the angelic prince of Egypt.[104] Also notice verse 8: "At the blast of your nostrils the waters piled up; the floods stood up in a heap; the deeps [Heb. *tehomot*] congealed in the heart of the sea"; and verse 11: "Who is like you, O Lord, among the gods [Heb. *elim*, the powerful ones]? Who is like you, majestic in holiness, awesome in glorious deeds, doing wonders?"

(e) Another such confrontation seems to be implied in Isaiah 51:9–10:

> Awake, awake, put on strength, O arm of the Lord; awake, as in days of old, the generations of long ago. Was it not you who cut *Rahab* in pieces, who pierced the dragon [Heb. *tannin*]? Was it not you who dried up the sea, the waters of the great deep [Heb. *tehom*], who made the depths of the sea a way for the redeemed to pass over?

---

[101] Kroeze (1962).

[102] Cf. Ouweneel (2018a, Appendix 3).

[103] See extensively, Ouweneel (2018a, §3.1 and Appendix 4).

[104] Mechilta Exod. 15:1 (43b).

This is what the prophet is praying about, as he speaks about the end of Judah's exile in Babylon: LORD, at the exodus you did a great miracle: you defeated the "dragon" (angelic prince) of Egypt, you overcame the *tehom*, and let your people go through the sea. Please, repeat this miracle, LORD! The king of Babylon, as the embodiment of his angelic prince, is a *tannin* (cf. Jer. 51:34 with v. 44). Free your people from his hands, and free them to return to the Holy Land!

In light of this brief summary, we see that the encounter between the dragon and the first woman recorded in Genesis 3 was only one in an entire series of such encounters. In the New Testament this series continues and expands to include the great encounters between Satan and Christ. In the wonderful imagery of the book of Revelation this includes the encounters between the dragon and the Lamb (see especially Rev. 12, 17, 19–20).

7.6.3 CLOSING COMMENT

In light of all the information offered in this chapter, we may begin to understand the cunning of the serpent (the dragon, Rahab, Satan). He is the head of all demonic powers, the epitome of craftiness and deceit. Thus, Genesis 1 may be primarily a treatise about the origin of the flora and fauna as we know them today. But beyond this, it is a treatise concerning the God of gods, who triumphs over the powers. As we saw, sun and moon are no gods in Genesis 1—they are not even mentioned by name (vv. 14–18)—but they are mere creatures (§5.2.2). Even the "sea monsters," Rahab, Behemoth, and Leviathan, are under God's command.

As we saw, the threat of the chaos powers seems to reverberate in the formless void (Heb. *tohu wabohu*) identified in verse 2 (cf. Ps. 42:7; Isa. 45:18; Jer. 4:23). No one can reasonably deny that, already during creation, there existed a confrontation between God and the powers (see §5.2.3). At any rate, this is the message of other passages, which sometimes (also) seem to refer to the exodus from Egypt, but according to many are (also) references to the creation (see Ps. 74:12–14; 89:8–12). In those cases (c), (d), and (e) mentioned in the previous section, the great conflict dominating world history—the one between the woman's offspring and the serpent's offspring—is repeated time and again. It continued in the New Testament, first in connection with Jesus Christ: the temptations in the wilderness, Gethsemane, and Calvary. Now the conflict rages on as Satan and his angels persecute those who belong to Christ (see Rom. 16:20; Eph. 2:2; 4:27; 6:11–12; James 4:7; 1 Pet. 5:8–9).

Finally, let me say this. The very promise that, according to Genesis 3:15, the "offspring" of the woman will ultimately "bruise the head" of the *serpent*, not just the serpent's "offspring," shows very clearly that we are not reading about an ordinary snake in Genesis 3. The *very same* serpent of Genesis 3 is the "ancient serpent, who is called the devil and Satan" (Rev. 12:9; 20:2) who in the end will be "crushed" under the feet of the "woman's offspring" (cf. Rom. 16:20), and will be thrown into the lake of fire and sulphur (Rev. 20:10). One day, the serpent of Genesis 3:9 will be cut in pieces (cf. Isa. 51:9).

8

THE SIGNIFICANCE OF THE FALL

For I desire steadfast love and not sacrifice,
the knowledge of God rather than burnt offerings.
But like Adam they transgressed the covenant;
there they dealt faithlessly with me (Hosea 6:6–7).

THESIS

To understand the biblical gospel it is essential to accept that the redemption in and by Christ presupposes the fall into sin by the first pair of humans; similarly, this fall presupposes the preceding state of being "very good" (Gen. 1:31). The Fall was a real fall, not a kind of maturation process of early humans. Redemption consists of the restoration of the original state of humanity (though on a much higher level, but this does not affect the principle of restoration).

8.1 THE FALL[1]

8.1.1 THE TEMPTATION

In this chapter, we will enter more deeply into the exegetical details of Genesis 3, and their systematic-theological consequences. Without mentioning this each time, I emphasize at the outset that AEH advocates are far more occupied with systematic-theological aspects than with exegetical

[1] For the typological-theological contents of Gen. 3, which are referred to several times in what follows, see Darby (n.d.-1, 19, 63–78, 117–121; n.d.-2, 1, ad loc.); Kelly (1870, 27–33; n.d., 19 and 20); Mackintosh (1972, ad loc.); Grant (1890, 34–37; n.d., 37–52); Coates (1920, 30–42); Bloore (1938, 12–13).

details. This is not surprising, since they do their best to appear to be upholding the notion of a historical Fall, but they can never believe that there was a first pair of humans, created directly by God (Adam from dust, Eve from his side), a pair who were tempted precisely in the way described in Genesis 3. According to AEH advocates, a conversation like the one described in this Bible chapter never occurred. To those, however, who may find a lot of metaphors and anthropomorphisms in these chapters but who basically maintain their literal-historical contents, these details are very important, not only for our exegesis but also for systematic theology, as we will see.

Many of the more conservative expositors have pointed out how cunningly the serpent operated. Compare the other reference in Genesis (49:17), "Dan shall be a serpent in the way, a viper by the path, that bites the horse's heels so that his rider falls backward."[2] Or as Jeremiah warned, "Behold, I am sending among you serpents, adders that cannot be charmed, and they shall bite you" (Jer. 8:17), probably referring to figurative serpents. Psalm 140:3 says of the wicked, "They make their tongue sharp as a serpent's, and under their lips is the venom of asp." Jesus himself used the expression "shrewd as serpents" (Matt. 10:16 HCSB).

Satan chose as his victim not the man but the woman, perhaps first because she was not the head of the human race as Adam was; second, because she did not personally hear the words that God had spoken to Adam (Gen. 2:16–17); and third, because presumably she could most easily reach Adam's heart.[3] Or was there a fourth reason: women are more gullible than men (cf. 2 Cor. 11:3; 1 Tim. 2:14)?[4]

The serpent's introductory question is also shrewd: "Did God actually say, 'You [plur.] shall not eat of any tree in the garden'?"[5] He, and also the woman, avoided God's covenant name YHWH, which could have reminded her of God's faithfulness to his creation, and especially to the first humans.

We may insert here the poignant question: According to AEH advocates, what did the first humans know about, and especially *from*, God? What kind of revelation concerning God did they possess, besides, of course,

---

[2] Hippolytus of Rome, and many after him, linked this verse with Gen. 3, and argued that the Antichrist (1 John 2:18, 22), a "vessel" of Satan, would come from the tribe of Dan (http://www.newadvent.org/fathers/0516.htm).

[3] Berkhof (1949, 243).

[4] Cf. Eccl. 7:28, "One man among a thousand [i.e., a man who represents what a human should be] I found, but a woman among all these I have not found." Sirach 25:19 (RSV), "Any iniquity is insignificant compared to a wife's iniquity; may a sinner's lot befall her!"

[5] This could also be translated: "Has God indeed said, 'You shall not eat of every tree of the garden'?" (NKJV). In this case, Satan might have been referring more directly at the tree of knowledge; cf. Leupold (1942, 144); Oosterhoff (1972, 164); Gispen (1974, 134–35).

God's general revelation in nature (cf. Ps. 19:1; Rom. 1:20)? In the AEH paradigm, was there ever a God who had blessed them and spoken to them (Gen 1:28)? AEH advocates have much to say about what early humans morally could, or could not, know and do, but very little about what divine revelation this pair may or may not have received.

To continue our presentation of the events of Genesis 3: Satan began with sowing doubt in the woman's heart by presenting God's commandment as more severe than God had given it.[6] Moreover, God had given the commandment to Adam before the creation of Eve (2:16–17; on the basis of AEH thinking, this is a nonsensical statement, of course). Therefore, for Eve, God's commandment was learned only secondhand. As a consequence, she was certainly the most vulnerable of that first pair, as Satan apparently realized very well.

8.1.2 EVE'S REPLY

From the woman's reply it is clear that either the "schemes of the devil" struck their target immediately (cf. Eph. 6:11, 16; she could not yet say, "we are not ignorant of his designs," 2 Cor. 2:11), or she was already uncertain in her heart concerning the commandment. In a certain sense, we might say that her fall began at this point, in her own heart. The sinful act never occurs without a prelude; the sinful act always proceeds from a moral preparation in the soul.[7] (As has been said, if we were to remain close enough to God, or to Christ, we would not sin.[8]) Before any sin has been committed there is always the inclination to this sin, the tendency to sin: "Each person is tempted when he is lured and enticed by his own desire. Then desire when it has conceived gives birth to sin, and sin when it is fully grown brings forth death" (James 1:14–15).

Eve answered, "We may eat of the fruit of the trees in the garden, but God said, 'You shall not eat of the fruit of the tree that is in the midst of the garden, neither shall you touch it, lest you die'" (Gen. 3:2–3). I discern here no fewer than five errors.

(a) Eve omitted the word "every" ("You may surely eat of *every* tree of the garden," 2:16), thus diminishing God's generosity.

(b) Eve "suppressed" (to put it in a Freudian way) the actual name of the tree: "the tree of the knowledge of good and evil."

---

6 Cf. Bonhoeffer (1995, 103–110).

7 Cf. 1 Tim. 1:19, "holding faith [i.e., faithful confidence] and a good conscience. By rejecting this, some have made shipwreck of their faith [i.e., Christian truth]."

8 Cf. these lines from a hymn by Carl Brockhaus, "If my eye is fixed upon Jesus, I will neither falter, nor fail" (cf. Matt. 14:30–31).

(c) In Eve's mind, she had not placed the tree of life but the tree of knowledge in the midst of the garden (cf. 2:9), as if this very tree had captured her attention.

(d) Eve made God's commandment more strict, for he had *not* said that Adam and Eve should not "touch" the fruit of the tree of knowledge.

(e) Lastly, Eve weakened the sanction; the Lord had said, "In the day that you eat of it you shall surely die" (Heb. *mot tamut*, "die the death," DRA, GNV, WYC). Eve said, however, "lest you die" (Heb. *pen temutun*). This latter formulation entails uncertainty; we will have to see whether, after the Fall, the fallen humans will die after all.

8.1.3 THE ARROW HITS

After these five errors, which basically suggested the first woman's willingness to surrender to Satan, the latter struck back immediately by now openly saying, "You will not surely die [or, You certainly will not die AMP]. For God knows that when you eat of it your eyes will be opened, and you will be like God, knowing good and evil" (vv. 4–5).

Satan began with an outright lie: Adam and Eve would not die at all.[9] Thus, Satan cast doubt on the *truth(fulness)* of God. The apostle John says in a general way, "Whoever makes a practice of sinning is of the devil, for the devil has been sinning from the beginning" (1 John 3:8). If we may introduce the Ten Commandments at this point (Exod. 20) (under which Satan was never formally placed, of course), we notice that in this brief episode, Satan violated most of them.

First Commandment: "You shall have no other gods before me. You shall not make for yourself a carved image, or any likeness of anything." Satan implicitly presented himself in Genesis 3 as the substitute "god"; since the Fall, he is indeed the "god of this world" (2 Cor. 4:4).

Third Commandment: "You shall not take the name of the LORD your God in vain." Satan offended God by flatly lying about him; the devil "does not stand in the truth, because there is no truth in him. When he lies, he speaks out of his own character, for he is a liar and the father of lies" (John 8:44).

Sixth Commandment: "You shall not murder." Satan caused the first humans to die by claiming that, if they ate, they would *not* die. This is precisely what Jesus meant when he called Satan the "murderer from the beginning" (see again John 8:44).

---

9 Rashi (usually following ancient rabbinical traditions) said that the serpent pushed Eve against the tree and then said, "As you did not die from touching it, so shall you not die from eating thereof"; see Cohen (1983, 13). It is remarkable, however, that the serpent never explicitly invited the woman to eat from the tree; see Bates (n.d., 11).

Seventh Commandment: "You shall not commit adultery." According to one rabbi, the serpent said, "I will kill Adam and marry Eve."[10]

Eighth Commandment: "You shall not steal." Gerrit Berkouwer wrote, "Through the temptation regarding an *alleged* privatio [i.e., being robbed of being like God] the man is seduced into being *robbed* of all of God's gifts."[11] That is, through the suggestion that they were being robbed, and through their subsequent fall, the first humans *were* indeed robbed of all blessings.

Ninth Commandment: "You shall not bear false witness against your neighbor." Belonging to the essence of Satan, and thus the essence of sin, is falsehood (see again John 8:44).

Tenth Commandment: "You shall not covet." "It was the Devil's jealousy that brought death into the world, and those who belong to the Devil are the ones who will die" (Wisdom 2:24 GNT).

After telling his explicit lie, Satan told an implicit lie. He suggested to Eve as it were that God did not wish his human creatures to be like him. God denied them this exalted position, and therefore he had prohibited their access to the fruit of the tree of knowledge. Satan suggested as it were that, as crown and vice-gerents of the creation, positions to which God himself had appointed them, Adam and Eve were *entitled* to all that belonged to creation. However, God did not allow them everything within their own domain because supposedly he did not want them to emulate him. In this way, Satan also cast doubt on the *righteousness* of God.

Moreover, Satan presented the matter in a way that made it seem as if God had withheld something from the first humans. In this way, Satan implicitly cast doubt on the *love* of God: "Here, we hear the demonic lie that God himself, who posits the Torah, that is, the requirement of love toward God and the neighbor, does not know love but is a God full of envy, like Zeus and so many figures in the pagan world of the gods, not a God for people but against people."[12] God had not demanded that humans merely obey blindly but trust in his love. Therefore, this very trust of humans toward God was being tested by Satan. As soon as one does not trust God with all one's heart and without any reserve, mistrust gains ground, and sin enters in: "Trust in the LORD with all your heart, and do not lean on your own understanding" (Prov. 3:5).[13]

---

[10] Talmud: Sotah 9b. See also chapter 7, note 37, on the presumed relationship between sexuality and the Fall

[11] Berkouwer (1971, 270); our translation.

[12] Verkuyl (1992, 113).

[13] Leupold (1942, 147).

After the first humans had sinned, these very attributes of God, doubted by Satan, came to light in such a remarkable way. God vindicated himself, as it were. His *righteousness* was demonstrated in executing the punishment that he had announced. His *holiness* was demonstrated by driving away the first humans from the Garden, and thus from being near to him. However, God's *love* became evident (a) through the prophecy and implicit promise of verse 15, (b) from the death of an innocent animal (v. 21), and thus (c) from his re-acceptance of Adam and Eve. However, this might lead someone to wonder whether, in this way, God's righteousness and holiness were nonetheless shortchanged (even though God drove the first humans from Paradise). These fundamental issues were eventually fully resolved only on and through the cross of Christ. Here, both God's righteousness and holiness as well as God's love receive full satisfaction.[14] Only at Calvary were Satan's lies completely exposed, together with God's perfect truth.

The question may be asked here again: What remains in the position of AEH advocates of this fundamental conversation between Satan and Eve? I address this question especially to those who claim that, to some extent, they wish to maintain the notion of a historical Fall. I cannot conclude anything other than that the details of Genesis 3, along with their far-reaching redemptive and theological implications, are being ignored by these advocates.

8.2 THE NEW "SEEING"

8.2.1 THE THREE DESIRES

Satan had promised Eve: If you will eat of the tree, your eyes will be opened (Gen. 3:5). From that moment, it seemed as if Eve began to "see" things in the tree that she had not seen before—but at the same time, she remained spiritually blind. This is clear from the remarkable contrast with verse 7; verse 6 says, "when the woman saw," but verse 7 says, "Then the eyes of both were opened"—and they began to see things so differently from what Satan had suggested. What opened their eyes spiritually was not Satan's instructions, but the sinful eating of the tree.[15]

Yet, after Satan's double lie there was some form of a new (self-deceiving) "seeing." In the tree of knowledge, the woman perceived three things: she "[a] saw that the tree was good for food, and [b] that it was a delight to the eyes, and [c] that the tree was to be desired to make one wise" (v. 6). This list

[14] See extensively, Ouweneel (2018e, especially chapter 9).
[15] Cf. Morant (1960, 169); Kidner (1967, 68–69).

is not arbitrary: these three things correspond seamlessly with what are still the three moral features of "the world" (the domain of sin and Satan): "All that is in the world—the desires of the flesh and the desires of the eyes and pride of life—is not from the Father but is from the world" (1 John 2:16).

(a) The attraction of the tree's fruit ("good for food") corresponds with the "desires of the flesh."

(b) The fruit was "a delight to the eyes," which corresponds with what the apostle John calls the "desires of the eyes."

(c) The fruit was a desirable object to acquire wisdom (insight, understanding), that is, in fact, nothing but the knowledge of good and evil, which implies one's own choices, autonomy, emancipation, and independence; this is the "pride of life."[16]

"Desire" (Lat. *concupiscentia*) characterizes the person who abandons God and chooses one's anchor in the desirable things of the *world* (cf. 2 Tim. 4:10). "Pride" (Lat. *superbia*) characterizes the person who abandons God and chooses one's anchor in *oneself*. In fact, these two vices are closely related[17] (cf. James 1:14, "each person is tempted when he is lured and enticed by his own desire"; cf. also the lists of sins in 1 Cor. 5:11; 6:9–10; Gal. 5:19–22; Eph. 4:18–19; Col. 3:5; 2 Tim. 3:1–5; Titus 3:3).

Notice the importance of this. Genesis 3 is not just about eating some ordinary fruit, as has been often said. Rather, it is about "all that is in the world," that is, about the attractiveness of the "world," a term referring to the totality of all that belongs to the domain of sin and Satan. In her heart, the woman had already fallen before she ate of the fruit.[18] When God is forsaken, basically two idols take his place: one outside the person (the "world"), and one inside the person (one's Ego; see the previous paragraph). It has often been said that the same thing caused both Satan and Adam to fall: pride. Regarding Satan's pride, see 1 Timothy 3:6 (if we read the text correctly): a new overseer "must not be a recent convert, or he may become puffed up with conceit and fall into the condemnation of the devil."[19] Regarding the pride of the first humans, they fell because they desired to be "like God" (cf. Gen. 3:5, 22).

As the wise Solomon said, "When pride comes, then comes disgrace" (Prov. 11:2). "Pride goes before destruction, and a haughty spirit before

---

[16] Cf. Leupold (1942, 151); see more extensively, Ouweneel (2008, §9.2.2).

[17] Berkhof (1986, 190).

[18] Leupold (1942, 147).

[19] Some read this as if the devil himself was condemned for pride (cf. Isa. 14:12–15; Ezek. 28:1, 12–19), others understand that it is the devil who condemns the conceited overseer; see Towner (2006, 257–60).

a fall" (16:18). "Before destruction a man's heart is haughty" (18:12). "One's pride will bring him low" (29:23). Adam and Eve provide the model for this pattern: first the "pride of life," then the fall, which is humiliation. For a comparison, consider Isaiah 14:14–15 (though the context is different): "'I will ascend above the heights of the clouds; I will make myself like the Most High.' But you are brought down to Sheol, to the far reaches of the pit." And Ezekiel 28:2, 13, 17 (again a different context, but the underlying principle is the same): "Because your heart is proud, and you have said, 'I am a god, I sit in the seat of the gods...,' yet you are but a man, and no god, though you make your heart like the heart of a god.... You were in Eden, the garden of God.... Your heart was proud."

8.2.2 THE FIRST AND THE LAST ADAM

It is remarkable that the three worldly temptations that are mentioned here also seamlessly correspond with the three tests to which Satan subjected Jesus in the wilderness (Matt. 4:1–11; Luke 4:1–12).[20] The first Adam and his wife were tempted under the most favourable circumstances, in an abundant Garden, where they lacked nothing. The last Adam was tempted under the most *un*favourable circumstances, in a barren wilderness, after he had had nothing to eat for forty days (see §7.5.3). Moreover, the first Adam considered it a desirable "robbery" to be equal with God (cf. Gen. 3:5, 22), whereas the last Adam did not (Phil. 2:6).[21]

(a) The first temptation of Jesus involved *eating*. Just as the woman saw that the tree was good for food (cf. the "desires of the flesh"), so it must have been objectively attractive to Jesus to turn stones into bread (only objectively, because subjectively it did not really appeal to Jesus[22]), thereby to satiate his tremendous hunger. Eve did fall, but Jesus answered the devil, "Man shall not live by bread alone, but by every word that comes from the mouth of God" (Matt. 4:4, a reference to Deut. 8:3). That is, I will never act independently of God. Serving God is always better, even if it means I cannot satisfy my hunger.

(b) The second temptation (according to the numbering in Luke 4) involved the "delight to the eyes," for Satan showed Jesus "all the kingdoms of the world"—"*and their glory*," as Matthew 4:8 adds. To Jesus, the idea of being able to acquire all power over these glorious kingdoms so easily (simply by kneeling down before Satan) must have been objectively

---

[20] Cf. Morris (1976, 113–14); Ouweneel (2007b, §11.3.3 note 74).

[21] Ouweneel (2007b, 324).

[22] Ouweneel (2007b, §§10.4–10.6).

attractive again. Eve did fall—she followed the "desires of the eyes"—but Jesus answered the devil, "You shall worship the Lord your God, and him only shall you serve" (Luke 4:8, a reference to Deut. 6:13). Moses said this to restrain the people from serving demons (cf. Deut. 32:16–17); Jesus repeated it to indicate that he would not serve the prince of the demons, Satan.

(c) The third temptation involved acquiring autonomy with respect to God (obtaining wisdom apart from God, Gen. 3:6; the "pride of life," 1 John 2:16). Thus, the devil placed before Jesus the temptation of "putting God to the test" (see Jesus' quote from Deut. 6:16), that is, taking the initiative and "forcing" God to fulfill his promise, namely, to protect him during his jump (cf. Ps. 91:11–12, quoted by the devil). This temptation resembled the actions of the Israelites, who "tested" the Lord at Massah and Meribah by trying to exploit God for their own ends (Exod. 17:1–7; Ps. 95:8–9). Trying to subject God to one's own purposes is pride.

Expositors have often observed that all three of Jesus' quotations came from Deuteronomy (chapters 6 and 8). These citations linked him with Moses, but also with the promised land, and God's promises given with a view to this land. These promises had to be (re-)fulfilled in Jesus' day. Whereas Adam and Eve had lost their vice-gerency through their fall, Jesus looked forward to the establishment of God's kingdom in glory and majesty, and the final restoration of Israel (cf. Deut. 30:1–10). Israel would be driven out of the land, just as the first Adam had been driven out of Eden. But one day, Israel would return to the Lord, and in the last Adam "all things" would be "restored" (cf. Acts 3:21).

Through the parallels and contrasts between the first and the last Adam, the essential historicity of Genesis 3 is underscored.

(a) What happened here in the Garden of Eden finds its counterpart in what happened to Jesus in the wilderness.
(b) The Fall finds its counterpart in redemption and final restoration.
(c) The first Adam finds his counterpart in the last Adam.
(d) The first Eve finds her counterpart in the "last Eve," so to speak (the bride of Christ; see next section).
(e) *Paradise Lost* finds its counterpart in *Paradise Regained*, as poet John Milton taught us.

The one set of realities is just as historical as the other—not just in some vague, general sense, as some AEH advocates portray them, but in a detailed sense, to such an extent that even the temptations substantively correspond.

8.2.3 "SHE ATE…HE ATE"

Without realizing the full consequences of what she was doing, Eve chose for the world of sin (by eating) and Satan (by listening to him), and thus *against* God. Adam followed her in this evil: "She took of its fruit and ate, and she also gave some to her husband who was with her, and he ate" (Gen. 3:6b). Notice these words, "who was with her": Adam had been present right from the beginning of the conversation, and he had done nothing to intervene and stop it.

Again we are reminded here of 1 Timothy 2:14, "Adam was not deceived, but the woman was deceived and became a transgressor." Eve was deceived by Satan, and Adam witnessed it. In my view, this does not make Adam's sin less grave. Rather, it is the very opposite: Eve was deceived (seduced, persuaded, talked into it), and fell; but apparently, Adam consciously followed her into evil. We could even say that he did not "fall" but "leaped" after his wife. In this sense, his deed was *more rebellious* than that of his wife: she was persuaded to sin, but Adam *chose* to sin.[23]

Interestingly, in Romans 5 and 1 Corinthians 15 the apostle Paul does not mention Eve at all (as he does in 2 Cor. 11:3 and 1 Tim. 2:13, and implicitly in 1 Cor. 11:8–12), but only Adam. We might be inclined to think that Eve's sin was worse, for she ate first (cf. Sirach 25:24 GNT, "Sin began with a woman, and we must all die because of her"). But Adam's sin was in fact far worse because of his position of responsibility.[24] Yet, of course, like Eve, neither did Adam fully realize what he was doing. There is another reason why Adam but not Eve is mentioned in Romans 5 and 1 Corinthians 15: as covenant head, the first Adam stands in contrast to the last Adam. This does not at all minimize the reality that, just as there was a "first Eve," so to speak, there is also a "last Eve," which is the bride of Christ (see 2 Cor. 11:2, "I betrothed you to one husband," and the link with Eve in v. 3).

Incidentally, notice that all such theological connections that we have been discussing are meaningless within the context of the AEH paradigm: there never was a first woman (wife) who was seduced, and a first man (husband) who deliberately followed her in her fall. On this basis, there cannot be any discussion about who was more responsible: Adam or Eve. I cannot repeat often enough that, in my view, those who believe in human evolution, and yet in some way wish to maintain some form of a historical Fall, suffer from a serious form of self-deception. That is, in no way can

---

[23] Cf. Rice (1975, 127–30), who in this regard views Adam as a type of Christ, which in my view should rather be an *anti*-type; cf. Leupold (1942, 153); Chafer (1983, 2:211); Morris (1976, 114–15), De Haan (1995, ad loc.).

[24] Morant (1960, 171–72).

they do any justice to the careful exegesis of so many details in Genesis 2 and 3. They hide behind their systematic-theological explanations and formulations, but pervasively neglect biblical exegesis. This has been one of the besetting sins of theologians: they often overemphasize systematic theology at the expense of biblical exegesis.[25]

8.3 CONSEQUENCES OF THE FALL

8.3.1 THE HUMAN CONSCIENCE

The attentive reader is necessarily impressed by the riches that the first humans possessed before the Fall, by the blessings and privileges that a loving God had granted them, by the blessed rule under which God had placed them, *and* the rule over creation that he had entrusted to them. All the more deep, then, was the foolishness of the first pair of humans, how deep their fall, and how great the misery into which they had plunged the world.[26]

Yet, we are not allowed, in deference to the claims of AEH advocates, to make the gap between the pre-Fall and post-Fall states larger than it really is. For instance, we can hardly say that Adam and Eve were holy and righteous before the Fall, because this would imply that, before eating of the tree of knowledge, they possessed a certain "knowledge of good and evil." Originally, the first humans did not know what evil was, and therefore could not know what was good, righteous, and holy. For this reason, strictly speaking we cannot say that prelapsarian humans possessed a conscience. As a consequence, we can hardly assert that their conscience was the criterion for their (dis)obedience; they needed merely to obey a simple but clear (inherently non-moral) commandment. It was the eating itself that *gave* the first humans a conscience. This helps us to understand what happened in Genesis 3:8, "They heard the sound of the LORD God walking in the garden in the cool of the day, and the man and his wife hid themselves from the presence of the LORD God among the trees of the garden."

This conscience was basically the same as what they had acquired through their eating, namely, "knowledge of good and evil." It was not an absolute knowledge—only God possesses this—but from now on, they did know the relative difference between good and evil. They now began to realize the good that they came from, and the evil in which they had landed. This knowledge led to shame, a matter always closely linked with the human

[25] Cf. Ouweneel (2015).
[26] Berkouwer (1971, 269, 274–75).

conscience (cf. §7.2.1). Animals possess neither a conscience (true moral awareness), nor shame—only fear (although I realize that anthropomorphizing scientists readily ascribe such human aspects also to animals).

Imagine how AEH advocates explain this. For them, it is inconceivable that early humans developed a conscience instantaneously *through* falling. Their conviction is rather the opposite: during their evolution, the early humans *developed* a conscience, or a moral consciousness, just as all their mental abilities, including so-called "religious awareness,"[27] were the product of evolutionary development. Again, we encounter here in AEH thinking the very opposite of what we find in Genesis 3. Let us examine further.[28]

On the one hand, this is the AEH picture: hominids are neither good nor bad (just like all animals). At a certain stage in their evolution, they developed something that one might call a conscience (basic moral awareness) in some form or other. Now they began to understand, on the most elementary level, the distinction between good and bad. Their consciences began immediately "to bear witness" (cf. Rom. 2:15; 9:1). There never was a "time of innocence"[29] during which they did only the good. That is, as soon as they began to understand the difference between good and bad they continued doing both good and bad things, as they had been doing for hundreds of thousands of years. The only difference was that they were now aware of what this entailed, and hence they began to bear responsibility.

On the other hand, this is what I see as the biblical picture: there was no human evolution, and hence no evolutionary development of a conscience in the sense of some basic moral awareness. There was a "time of innocence" (between the creation and fall of the first humans—a period perhaps of no more than one or two days) during which the first pair of humans strictly speaking had no conscience, that is, had no knowledge of good and evil. When they fell, they instantaneously acquired a conscience, that is, the knowledge of good and evil. From now on, they were aware of good things, but they could do only bad things (at least according to God's standards). For the first time, they realized what *happiness* they had come from (they could never know before what happiness means, because the term "unhappy" did not yet make sense to them[30]). After their fall, it was only through regeneration (and the power of the Holy Spirit) that they received the renewed capacity for doing things pleasing to God.

---

[27] Cf. van den Brink (2017, 224).

[28] For a critical evaluation, see Van Woudenberg (2007).

[29] *Contra* van den Brink (2017, 239–41).

[30] Cf. the impossibility of explaining to children before they have reached puberty the meaning of sexual pleasure.

8.3.2 LOINCLOTHS OF FIG LEAVES

Originally the first humans were "very good" (cf. Gen. 1:31), but after their fall they had become evil. They had now learned to know evil, but this evil thoroughly characterized their new condition. In contrast with this evil, they now also really knew what was good—but the good had become an unattainable ideal. In a certain sense, through eating of the tree of knowledge their moral insight and ability had increased. But without spiritual life from God, this ability could be only a dreadful capacity. They could almost say with the apostle: "I *know* that nothing good dwells in me, that is, in my flesh. For I have the desire to do what is right, but not the ability to carry it out. For I do not do the good I want, but the evil I do not want is what I keep on doing" (Rom. 7:18–19).

In the same chapter Paul says, "Wretched man that I am! Who will deliver me from this body of death? Thanks be to God through Jesus Christ our Lord!" (vv. 24–25). Paul knew the answer; Adam and Eve still had to learn it (see the next sections). At first, they tried to find deliverance through their own action. Their first act after the Fall was trying to undo their sin, or at least to cover it up. As Solomon said, "Whoever conceals his transgressions will not prosper, but he who confesses and forsakes them will obtain mercy" (Prov. 28:13). Perhaps Adam's act is what Job was referring to (Job 31:33 KJV): "If I covered my transgressions as Adam, by hiding mine iniquity in my bosom." But others prefer to render the Hebrew expression *ke-Adam* here as "like other man" or "as people do," or something similar.[31]

Adam and Eve had eaten, they had expected wonderful results, they waited—and what happened was the emergence of a feeling of shame.[32] The loincloths of fig leaves that Adam and Eve sewed together were intended to cover not only their bodily nakedness, but also their spiritual nakedness. It did not have the desired result: they were, and remained, "wretched, pitiable, poor, blind, and naked" (Rev. 3:17).[33] Just as the Preacher in the book of Ecclesiastes, they sought wisdom and insight (1:13, 17; see Gen. 3:6), but they found only vanity and mischief.[34] In a sense, this is the essence of all false religion, that is, every religion that arises from the demonic world and champions human self-righteousness (cf. Rom. 10:3). Every such religion

---

[31] Remarkably, the only references outside Genesis (and 1 Chron. 1:1) to Adam as a person (namely, Job 31:33 and Hos. 6:7) are uncertain. David Kimchi understood "your first father" (Isa. 43:27) as referring to Adam (see Slotki [1983, 211]); but this is unlikely, since the reference is probably to Abraham or Jacob.

[32] Proksch (1913, 154).

[33] Grant (1956, 24).

[34] Sailhamer (1990, 52).

is a fig leaf, an attempt to cover personal "nakedness" with the products of personal meritoriousness (cf. Isa. 64:6, "We have all become like one who is unclean, and all our righteous [not just all our *un*righteous!] deeds are like a polluted garment [and nothing more]").

Notice the difference: before the Fall, Adam and Eve had been naked, but without fear and shame. After the Fall, they were ashamed and covered their nakedness, but the fear and shame remained. They hid from the presence of him before whom "no creature is hidden" but "all are naked and exposed to the eyes of him to whom we must give account" (Heb. 4:13). He is the One whose Word "discern[s] the thoughts and intentions of the heart" (v. 12).

8.3.3 SIN AND SACRIFICE

It was not the handmade loincloths of fig leaves that made possible restored fellowship between God and fallen humans. Rather, it was the garments of skins that *God himself* prepared for them (Gen. 3:21). This presupposes that one or more animals had to die for this.[35] The first humans could now begin to discover how their helplessness supplied them with the opportunity to begin learning God's tender mercies. These mercies could be bestowed only through the death of *another*, a substitute that itself was innocent of the wrong Adam and Eve had committed. This vicarious death supplied them with a covering that corresponded to God's way of thinking about sin. The first humans could do nothing but accept God's grace and the covering that he, in his grace, had supplied them. In principle, it is no different for people today, who must entrust themselves to God's grace and accept the "covering"[36] that he offers them in the person and work of Christ.

Lewis Sperry Chafer wrote, "Much truth...lies hidden in the facts that they [i.e., Adam and Eve] attempted to clothe themselves, which clothing was of no value; and that God clothed them with skins, which meant the shedding of blood."[37] It is not people's own covering that saves them, but God's covering.

Francis A. Schaeffer maintained—and I agree with him—that here in Genesis 3:21 we find the roots of the entire Old Testament sacrificial system. The verse looks forward, as it were, to the fulfillment of verse 15:

---

35 Ouweneel (2018e, §6.6.1); see this entire volume for an answer to the vital question *why* sin demands a vicarious sacrifice.

36 Notice that the common Heb. word for "to atone" (*kapar*) probably has the root meaning "to cover" (see Gen. 6:14).

37 Chafer (1983, 2:218).

the coming of him who would bruise the head of Satan.[38] Redemption is not only about atonement for sins but also about defeating the "tempter" (Matt. 4:3; 1 Thess. 3:5), the one who incited, and incites, humans to sin. In the consummation of the ages, death is taken away, as well as what caused death, sin, along with what caused sin, the devil.

Henry M. Morris suggested that Adam and Eve perhaps sadly watched as God selected two of their animal friends—possibly two sheep—and killed them before their eyes, pouring out the animals' innocent blood before their eyes. In a typological form they learned that a true atonement or covering of human spiritual nakedness can be supplied by God alone.[39] Interestingly, Calvin speculated that God left the slaughtering to Adam and Eve: "Since animals had before been destined for their use, being now impelled by a new necessity, they put some to death, in order to cover themselves with their skins, having been divinely directed to adopt this counsel; therefore Moses calls God the Author of it" (see further in §4.5.2).[40]

C. John Collins wrote, not directly in reference to the garments of sin but to the relationship between the Fall and the sacrificial system in general:

> In the biblical story sin is an alien intruder; it disturbs God's good creation order. This comes through clearly in the way that the Levitical sacrifices deal with sin: they treat it as a defiling element, which ruins human existence and renders people unworthy to be in God's presence—and that is dangerous. The sacrifices work "atonement," "redemption," and "ransom," addressing sin as a defiling intruder that incurs God's displeasure (e.g., Lev. 16).[41]

Given this standpoint, it is natural for us to expect in Genesis 3 a first hint about the sacrificial system. To put it more strongly: in God's counsel the sacrifice had already been prepared before the first humans actually fell: the Lamb "without blemish or spot...was foreknown before the foundation of the world" (1 Pet. 1:19–20).[42]

---

[38] Schaeffer (1982, 2:75); cf. Ouweneel (2008, §10.2.1).

[39] Morris (1976, 130); sheep are an obvious suggestion because in the Torah these are the most common sacrificial animals (Gen. 4:4; 22:7, 13; Exod. 12:3–6; 29:38–41); moreover, their skins are suitable for clothing.

[40] Calvin, *Comm. Genesis* (ad loc.; http://www.ccel.org/ccel/calvin/calcom01.ix.i.html).

[41] In Barrett and Caneday (2013, 160).

[42] Some (e.g., Rabbi Moshe Alshich) thought here of Eccl. 10:11 (depending on the translation), "If the snake bites before it's been charmed, what's the point in then sending for the charmer?" (MSG; cf. AMPC, CJB, LEB): before the serpent "bit," the charming had already occurred (e.g., John Gill's comments on this verse at http://biblehub.com/ecclesiastes/10-8.htm).

Again we ask, what happens to the loincloths of fig leaves and the garments of skin in the AEH portrait of early human history? These items vanish. They are entirely out of place, even in those portraits that seek to uphold a historical Fall. If there are no garments of skin, then there was no primeval sacrifice, no primeval atonement for Adam and Eve. The implicit promise of Genesis 3:15 evaporates, for if there is no serpent, there is no serpent's head to be bruised. From those early chapters of Genesis, AEH advocates can proclaim to us nothing but the most meagre gospel imaginable.

8.3.4 THE EVIL ONE

No human ever invented sin; Satan did, probably through deceit (if we read 1 Tim. 3:7 correctly). The Hebrew word *satan* means "enemy, adversary, opponent."[43] He is the devil, from Greek *diabolos*, that is, "agitator, provoker, detractor." He is also Beelzebul (Mark 3:22)[44] or Belial/Beliar (2 Cor. 6:15).[45] Among the consequences of the Fall belonged the reality that, by listening to the serpent's temptation, the first humans came under the power of Satan.[46] Ever since the Fall—we cannot point to any later moment of origin—"the whole world lies in the power of the evil one [or "in evil," Gk. *en tōi ponēroi*]" (1 John 5:19), and Satan is the "ruler of the/this world [Gk. *cosmos*]" (John 12:31; 14:30; 16:11). He is the "god of this world [or age, Gk. *aiōn*]" (2 Cor. 4:4), the "prince of the power of the air, the spirit that is now at work in the sons of disobedience" (Eph. 2:2).

Too often, Genesis 3 has been considered from the viewpoint of only sin. But it is equally important that, through their fall, the first humans inadvertently surrendered their dominion over the world into the hands of the devil. Therefore, Satan could say that the power over the kingdoms of the world "has been *delivered* to me, and I give it to whom I will" (Luke 4:6)—a fact that Jesus did not deny, and could not have denied. Before the Fall, the first humans were God's vice-gerents over the earth (Gen. 1:26,

---

[43] It has been argued that, in passages such as Job 1–2 and Zech. 3, *satan* ("adversary") fulfills the role of a prosecutor in court, which is not necessarily negative (although it *is* here, in my view); however, in 1 Chron. 21:1 his role is clearly obstructive.

[44] Derived from *Baal-Zebul*, "lord of the height," which Israel mockingly changed into Baal-Zebub, "lord of the flies" (2 Kings 1:1–6, 16). Some versions read "Beelzebub" in Mark 3:22 (cf. KJV).

[45] Derived from *Beli-ya'al*, "worthlessness"; in the Old Testament, "sons/children of Belial" or "men of Belial" are worthless men (e.g., 1 Kings 21:10). Paul used the word as a name for Satan, as does the pseudepigraphical book Ascension of Isaiah (2:4, "the angel of lawlessness, who is the ruler of this world, is Beliar").

[46] See extensively, Ouweneel (2018a); cf. Chafer (1983, 2:217–18); Heyns (1988, 166–69).

28); through the Fall, it was the usurper, Satan, who was enabled to exercise control over the world (of course, under God's providential guidance). Jesus overtly recognized the existence of this "kingdom of Satan":

> If Satan casts out Satan, he is divided against himself. How then will *his kingdom* stand? And if I cast out demons by Beelzebul, by whom do your sons cast them out? Therefore they will be your judges. But if it is by the Spirit of God that I cast out demons, then the *kingdom of God* has come upon you (Matt. 12:26–28).

This is an important point. What can AEH advocates tell us about this "kingdom of Satan," which began at a well-defined point in time? Adam and Eve were the first rulers of the world under God (Gen. 1:26, 28)—Satan was the second ruler (see §10.6.2). His kingdom began in the Garden of Eden at the moment when Adam and Eve fell into sin. What do AEH advocates know about Satan, and about the beginning of his dominion? I am not asking these advocates how Satan himself fell into sin; all of us can do little more than speculate about the answer to this question.[47] But I am asking AEH advocates (a) what exactly the dominion of the first humans entailed (how was it given to them, and in what sense?), (b) how the confrontation between them and Satan occurred, and (c) how they ultimately surrendered their dominion to Satan by succumbing to the latter's temptation.

The matter is soteriologically important because we read that people who come to faith "turn from darkness to light and from the power of Satan to God" (Acts 26:18). The Father "has delivered us from the domain of darkness and transferred us to the kingdom of his beloved Son" (Col. 1:13). Peter speaks of "him who called you out of darkness into his marvelous light" (1 Pet. 2:9). No soteriology that wishes to underscore people's deliverance from the power of Satan can afford to ignore how people came under the power of—not only sin and death but—Satan in the first place. Genesis 3 has a clear-cut answer to this question. Do AEH advocates?

---

[47] Cf. Luke 10:18; Rev. 12:9; see also 2 Enoch 29:3–4, "And one from the order of angels, having turned away with the order that was under him, conceived an impossible thought, to place his throne higher than the clouds above the earth, that he might become equal in rank to my [i.e., God's] power. And I threw him out from the height with his angels" (cf. 2 Enoch 31, where he is identified as the Satan of Eden; https://archive.org/stream/AllTheBooksOfEnochenoch1Enoch2Enoch3/AllBooksOfEnoch #page/n71/mode/2up/search/satan).

8.4 THE RESTORATION OF THE FIRST HUMANS

8.4.1 GOD'S PURSUING LOVE

It is remarkable that the new era after Jesus' resurrection also began in a garden (John 19:41–20:17), namely, the garden of Joseph, where he had cut a tomb out of the rock (Mark 15:46). Notice the differences.

(a) In the former garden, the first Adam died his spiritual death; in the latter garden the last Adam arose from the dead.
(b) In the former garden, the sad, fallen Eve hid among the trees; in the latter garden, the sad Mary Magdalene wandered about as the "new Eve."[48]
(c) In the former garden, God asked the woman, "What is this that you have done?" (Gen. 3:13). In the latter garden, Jesus asked, "Woman, why are you weeping? Whom are you seeking?" (John 20:15).
(d) In the former garden, God revealed himself in his forgiving grace toward humanity; in the latter garden, the risen Jesus revealed himself to Mary Magdalene.
(e) When the first humans had been driven out of the Garden of Eden, a cherub guarded the entrance to the garden (Gen. 3:24). In the garden of Joseph, there were two angels in the empty and open tomb, who indicated that there was no longer any reason to weep (John 20:12–13).

Look at the Fall again: what was its core, its essence? Not that God rejected the first humans on account of something as trifling as forbidden fruit, but that the first humans rejected *God* for something as trifling as the forbidden fruit. Adam's sin drove him away from God, and in response to this, God drove him away from the Garden. However, this expulsion was not God's first act. His first act was his *pursuing love*. To be sure, this love could not be manifested immediately; the sinner's miserable condition first had to be exposed.[49] This God does by asking questions. These questions have the character of a judicial cross-examination (cf. Gen. 4:6–10; Num. 22:9; 1 Sam. 13:11; 15:14), but this does not necessarily conflict with the motive of love.

---

[48] One ancient tradition calls Mary Magdalene the "new Eve"; others reserved this title for the Virgin Mary; see extensively, Ouweneel (1998, 97–101).

[49] Pannenberg (2010, 179) suggested that talking about human *misery* describes in a more encompassing way the lost condition of humanity, who is far away from God, than does talking about the theological doctrine of sin.

God addressed one question to Adam, and one to Eve. The first is: "Where are you?" (v. 9). This question concerns the human spiritual *condition*. The second question, addressed to Eve, is: "What is this that you have done?" (v. 13). This question concerns human evil *actions*. This is the same difference between sin as an evil power and sin(s) as evil action(s) that we read about in the letter to the Romans (e.g., "sins" in Rom. 3:25; 4:7, and "sin" in 5:12–13, 21). These are probing questions; they bring to light the entire situation into which the first humans had fallen. But at the same time these questions demonstrate God's pursuing loving-kindness. "Walking in the garden in the cool of the day," God came to the first humans as their interlocutor: Man, woman, what happened to you? The underlying thought is: I feel sorry for you; how can I get you out of this mess?

That these questions reflect God's character is evident from the mention in Genesis 4 of the same type of questions in God's relationship with Cain. God's question to him, "Why are you angry, and why has your face fallen?" (v. 6) was not necessarily an indictment, but an invitation as well as a warning, for God continued, "If you do well, will you not be accepted? And if you do not do well, sin is crouching at the door. Its desire is contrary to [or, toward] you, but you must rule over it" (v. 7). After the fratricide, this pursuit is even stronger: just as God had asked Adam, "Where are you?" so he asked Cain, "Where is Abel your brother?" (v. 9). The first question means, Where are *you*? The second question means, Where is your *brother*? These questions reverberate throughout history: Human being, what is your individual condition? But also: What did you do to your neighbour? In both respects, humanity has sinned against *God*.

In these and similar questions, we hear the anger of God but an anger that can never be severed from his pursuing love.[50] It is true that God's anger sometimes assumes such forms that his love is hidden behind his hatred (Ps. 5:5; 11:5; Jer. 12:8; Hos. 9:15; Mal. 1:3). However, behind this hatred, his pursuing love always remains hidden.[51] We find a beautiful example of this in Mark 3:5, where Jesus was angry with his adversaries, but at the same time he was "grieved at their hardness of heart." His anger and his loving-kindness were not mutually exclusive.

8.4.2 THE *PROTOEVANGELIUM*

At first, the hearts of Adam and Eve seemed hardly reachable. Adam blamed his wife ("the woman whom you gave to be with me," Gen. 3:12; so actually

[50] Verkuyl (1992, 132–33).

[51] Erickson (2007, 619–20).

it was all *God's* fault!), and Eve blamed the serpent: "The serpent deceived me, and I ate" (v. 13). These replies demonstrated their sinful condition: God was at fault, or the serpent/devil was at fault, not they. It is not easy for fallen human beings to say to God, "I have sinned" (cf. Luke 15:18, 21), and "God, be merciful to me, a sinner!" (18:13). For the time being, God went along with these replies to the extent that after Adam's response he addressed Eve, and after Eve's response he addressed the serpent.

After the curse upon the serpent (Gen. 3:14; cf. Isa. 65:25), we find the well-known verse 15: "I will put enmity between you and the woman, and between your offspring [lit., seed] and her offspring [seed]; he[52] shall bruise your head, and you shall bruise his heel." It is not quite correct to call this statement a "promise" or *prot[o]evangelium* (Gk. *prot[o]evangelion*, "first" or "primeval gospel," a term found first among church fathers like Irenaeus[53]). It is neither a promise, nor a "*good* message" (Gk. *evangelion*) for humans; in fact, it is only a *bad* message for Satan. Humans can derive from this a promise for themselves only indirectly.[54] Humans, fallen into sin, have forfeited every right to a promise of blessing. To put it more strongly, strictly speaking, even indirectly this was not a promise for Adam because he did not belong to the "woman's offspring." Nor was it a promise for Eve but only for her offspring. It is like the sign for Ahaz in Isaiah 7:13–14, which was not for Ahaz himself but for the "house of David," a sign for a very distant future, since Ahaz had personally forfeited every right to a sign.

For the "woman's offspring" there is this indirect prophecy: "you [i.e., Satan] shall bruise his [i.e., the woman's offspring's] heel."[55] This is a heavy blow, but it does not imply a definitive ending. There is also the "serpent's offspring," which are the "sons of the evil one" (Matt. 13:38; cf. John 8:44;

---

52 The old Vulgate reads *ipsa* ("she"), which led many expositors to think of Mary; see Ouweneel (1998, 173–74). Even in the dogma of Mary's immaculate conception (1954), an appeal was still made to this false rendering of Gen. 3:15; see extensively Morant (1960, 182–89). The Nova Vulgata corrected this error and reads *ipsum* ("it"), i.e., the woman's seed.

53 Not to be confused with the Prot[o]evangelium of James, also called the Infancy Gospel of James, which deals with Jesus' youth.

54 Morant (1960, 180) rightly opposed the interpretation of Gunkel (1997, 20–21), Zimmerli (1943, ad loc.), Von Rad (1972, 89–90), and Westermann (1984, ad loc.), who understand v. 15 to refer exclusively to a literal physical snake and its offspring, and reject every Christological exegesis. Schelhaas (1932, 27–50) chose the opposite extreme and viewed the "directly Messianic" exegesis as the only tenable interpretation. See Hamilton (1990, 197–200).

55 Only God can predict the future; this is one way to read Isa. 44:7, "And who, as I…shall declare it, since I appointed the ancient people [Heb. *am-olam*]?" That is, who knows the future since the earliest people? Cf. David Kimchi; see Slotki (1983, 213).

Eph. 2:2–3). We should not think of demons here but of humans, not in the sense of possessed people, but of the wicked in general.[56] Note carefully the following:

(a) the enmity exists between the serpent (Satan) and the woman (Eve), and
(b) between the serpent's offspring (i.e., as I read it, the wicked) and the woman's offspring (i.e., the righteous);
(c) the woman's *offspring* (at the deepest level, Christ) bruises the head of the *serpent* (Satan; not the serpent's *offspring*), and
(d) the *serpent* (Satan; not the serpent's *offspring*) bruises the heel of the woman's *offspring* (at the deepest level, Christ).[57]

Thus, the ultimate announcement of judgment does not concern the serpent's offspring but the serpent itself. On Calvary, he (through his children) may have bruised Christ's heel, but in the end "it [i.e., the woman's offspring, Christ] will bruise your [i.e., the serpent's, Satan's] head"—which implies the definitive ending of the serpent (Rom. 16:20; Heb. 2:14; 1 John 3:8; Rev. 20:10).[58]

What does all this mean in terms of the AEH position? Very little, I am afraid. According to AEH advocates, there was no first woman, there is no human race descended from this one mother ("the mother of all living," Gen. 3:20),[59] there is no serpent, nor any serpent's seed (regardless of how this is identified). There is no meaningful *protoevangelium*. AEH defenders leave us with an impoverished and empty gospel, at least to the degree that the gospel is to be rooted in Genesis 3.

8.4.3 THE LAST BATTLE

Indeed, the ultimate conflict will occur between the person of the "serpent" (in which the "serpent's offspring" is thus individualized) and the woman's *offspring*. Therefore, it seems obvious, in the spirit of Galatians 3:16 ("It does not say, 'And to offsprings,' referring to many, but referring to one, 'And to your offspring,' who is Christ"), that we should understand the woman's offspring to be one particular person as well, namely, Jesus Christ. In the light of the New Testament, this is the deepest significance of the *protoevangelium*:

---

[56] *Contra* Morant (1960, 181).
[57] Leupold (1942, 166–68).
[58] See also Ouweneel (2007b, §5.2.1).
[59] In this context of origins as well as of the "woman's seed" (Gen. 3:15), "mother" clearly means ancestral mother, *contra* van den Brink (2017, 226).

the hope for fallen humans lies in the One who will ultimately destroy Satan. However, this interpretation does not change that, first and foremost, the serpent's offspring and the woman's offspring each refers to a collective entity, as, for instance, in Genesis 9:9 and 12:7. (For the individualized meaning of "offspring/seed" as "son," see, e.g., Gen. 4:25; 21:13.)

Cain and Abel were the first people who, spiritually speaking, represented the serpent's offspring and the woman's offspring, respectively (cf. 1 John 3:8–12); as Johan Verkuyl put it, "Cain has the fever of the serpent's poison in his blood."[60] One Jewish tradition understood this very literally:

> After the fall of Eve, Satan, in the guise of the serpent, approached her, and the fruit of their union was Cain, the ancestor of all the impious generations that were rebellious toward God, and rose up against Him. Cain's descent from Satan, who is the angel Samael, was revealed in his seraphic appearance. At his birth, the exclamation was wrung from Eve, 'I have gotten a man through an angel of the Lord.'[61]

Abel was the first one whose "heel" was "bruised," a reference to physical death, whereby he became a type of Christ (cf. Heb. 12:24). Thus, throughout human history the woman's offspring (the *tsaddiqim*, the "righteous") and the serpent's offspring (the *reshaim*, the "wicked") have existed side by side.[62] In John 8, Abraham's children (v. 39, children in the true spiritual sense) are opposed to the devil's children (vv. 38, 44), who may have descended physically from Abraham but who descended spiritually from Satan. The devil's children bruised the heel of Christ, who is the woman's offspring *par excellence*. So from Genesis 3:15 onward, the Bible portrays these two lines all the way through redemptive history until the consummation of the ages. In this way, God gives us a remarkable view of the end, the triumph of the woman's offspring: not only Christ but all those who are of his "race": "The God of peace will soon crush Satan under *your* feet" (Rom. 16:20).

I repeat, there is no promise for the first Adam; in him, all are lost (Rom. 5:15–19; 1 Cor. 15:22). In the second Man, the last Adam, all those who belong to *him* are saved. Already in Genesis 3, strictly speaking, God did not restore the "first man," but accepted Adam and Eve in the second Man, without these humans being aware of this wonder at the time. In the last

---

[60] Verkuyl (1992, 111; cf. more extensively, 135).

[61] Ginzberg (1969; https://philologos.org/__eb-lotj/vol1/three.htm#1).

[62] This word pair appears frequently in the book of Proverbs, as in 3:33, "The LORD's curse is on the house of the wicked [Heb. *rasha*], but he blesses the dwelling of the righteous [Heb. *tsaddiqim*]."

Adam, "all the promises of God find their Yes" (2 Cor. 1:20). What was lost in the first Adam, was and is regained in the last Adam, and much more than that: much more was and is and will be obtained than was ever lost.[63]

Again we notice that in terms of AEH, all this makes little sense. There may be a last Adam and a future Paradise, but the AEH view has no "Paradise Lost" in any literal sense (one can invent virtually any figurative scenario). It has no first Adam in any truly biblical sense. It also hardly makes sense in terms of AEH to say that, from the beginning of humanity, there were two kinds of people: the sons of the Living (cf. the Heb. name *Chawwa*, "Eve") and the sons of the serpent, that is, Satan, the regenerate and the unregenerate, the *tsaddiqim* and the *reshaim*. According to AEH, God did not speak *to* the first pair of humans, nor did he speak *through* the first pair of humans (at least I have not read of this in any AEH literature). Where and when, in AEH thinking, did God first speak to humanity, either directly (cf. Gen. 1:26, 28–30; 2:16–17) or through a prophet? Enoch "walked with God" (Gen. 5:24)—how did he know this God if not through God's revelation to his seventh forefather, Adam? What room exists in AEH thinking for any kind of "redemptive history" before the time of the patriarchs? This remains an open and unanswered question.

8.5 THE RE-ACCEPTANCE OF THE FIRST HUMANS

8.5.1 SINNING INADVERTENTLY

We have seen that, before the Fall, the first pair of humans did not know what evil was, and therefore neither did they know what was good, righteous, or holy. If Adam had possessed this knowledge, no restoration would have been possible, according to the principle of Hebrews 10:26–27, "For if we go on sinning *deliberately* after receiving the *knowledge* of the truth, there no longer remains a sacrifice for sins, but a fearful expectation of judgment." Of course, this must be not interpreted as an absolute statement, because, unfortunately, every human sins knowingly on occasion ("letting oneself go"); the passage in Hebrews refers to sin characterized by outright rebellion against God, like blasphemy against the Holy Spirit (Matt. 12:31–32).[64]

This principle is found throughout the Bible: "Declare me innocent from hidden faults" (Ps. 19:13), sins that I have inadvertently and unknowingly

[63] See extensively, Ouweneel (2012a).

[64] Blasphemy against the Holy Spirit is the example *par excellence* of a *conscious* sin of rebellion; cf. Hoek (1988, 24–26); Ouweneel (2007a, §6.2.2).

committed. The Torah of Moses speaks often of sins that have been committed unintentionally (or inadvertently; see Lev. 4:2, 13, 22, 27; 5:15, 18; 22:14); atoning sacrifices were prescribed for those sins alone. When a person sinned "with a high hand" (Num. 35:20), in conscious and deliberate rebellion against God (presumptuously, defiantly), the death penalty was administered immediately.

When Jesus was hanging on the cross, he prayed, "Father, forgive them, for they *know not* what they do" (Luke 23:34; cf. Acts 17:30, "times of *ignorance*"). After the mass of the Jewish nation had rejected its Messiah, the gospel was offered to them with the argument: "And now, brothers, I know that you acted in *ignorance*, as did also your rulers" (Acts 3:17). The apostle Paul called himself "formerly a blasphemer, persecutor, and insolent opponent. But I received mercy because I had acted *ignorantly* in unbelief" (1 Tim. 1:13). This might be one reason why salvation is not for fallen angels: they are too close with God to ever call them ignorant (cf. Isa. 14:13–14; 1 Tim. 3:6; Jude 6). Indeed, Adam was not in the heavenly regions, in the direct presence of God's glory, as Satan was.

Earlier, I made a distinction between Eve being deceived and Adam deliberately following her in her fall. Does this mean that Eve might be eligible for forgiveness, but Adam not? Apparently, this was not the case. I think this was because although Adam followed his wife intentionally, he nevertheless did not really know what he was doing (see again Jesus' prayer on the cross on behalf of his enemies). He was disobedient, but he could not possibly have realized the consequences of what he did. I do not mean this as an excuse, but only wish to point out that, despite the difference between Adam and Eve, they both sinned inadvertently, in ignorance. This opened for them the possibility of forgiveness.

8.5.2 WISDOM'S SALVATION

Indeed, there was re-acceptance for Adam and Eve: "Wisdom protected the father of the world [i.e., Adam], the first man that was ever formed [!], when he alone had been created. She [i.e., Lady Wisdom] saved him from his own sinful act" (Wisdom 10:1 GNT). I understand this to mean that God's wisdom surrounded the very first man on earth, and after he had fallen, God's wisdom took care of his restoration. While, on the one hand, God mapped out the future course of all redemptive history until its consummation in Christ (Gen. 3:15), on the other hand, he had a message for those two people who had just fallen: they had to bear the consequences of their sin. There was the implicit promise of the "woman's seed," but the woman would experience the pains of bringing forth this very offspring as well as the (sinful!) rule of the man over her (v. 16),[65] whereas the man would experience the troubles of laboring to sustain life in a cursed creation (vv. 17–19).[66]

It was not pregnancy and labour as such that were consequences of sin; before the Fall, the first humans had been called upon to multiply (1:28) and to work (2:5, 15). It was the intensifying of the *pains* of pregnancy and labour that belonged to the consequences of the Fall. The intention of this curse was to make clear that the broken relationship between God and humanity would bring about brokenness in the rest of the creation, too: between people, between the sexes, between nations, between races, between humans and nature. Fallen humanity encounters the cracks and crevices of this brokenness on every side.

Again I must ask: what remains of all this in the AEH position? Did evolved women give birth to their children more easily before than after the Fall (whatever the nature of this "Fall")? Did the men have less (oppressive) dominion over their women before than after the Fall? Was the ground cursed after but not before the Fall? Did men expend more sweat after than before the Fall? Did thorns and thistles exist after but not before the Fall (or more broadly, was agricultural work much easier before than after the Fall)?

Of course, AEH advocates will not easily answer these questions in the affirmative. In their position, Genesis 3 must be understood largely or entirely figuratively. Please note, even those who claim that they wish to retain the notion of a historical Fall cannot do so with fairness. In their view, all the things just mentioned must necessarily be understood figuratively. These advocates have no option. Time and again I must emphasize this point: *you cannot accept both human evolution and a historical Adam and a historical Fall*—at least not an Adam and a Fall that even remotely resemble what the Bible tells us about these things.

8.6 DEATH AND LIFE

8.6.1 ADAM'S FAITH

After announcing the temporal curses that the first humans had brought upon themselves, the Lord announced that they would eventually undergo

[65] This verse can never be quoted as describing God's "creation order" with respect to the relationship between man and woman, more specifically, between husband and wife, as has been done (cf. W. O. Einwechter, http://darashpress.com/articles/men-and-women-and-creation-order). Genesis 3:16b describes a consequence of sin, not God's normative order.

[66] The emphasis is on "eating" (v. 17): because Adam had eaten what he was not allowed to eat, he will have to extract food from the ground in drudgery; see Sailhamer (1990, 57).

the physical death with which he had threatened them earlier: "By the sweat of your face you shall eat bread, till you return to the ground, for out of it you were taken; for you are dust, and to dust you shall return" (Gen. 3:19; cf. 2:17). This is not necessarily speaking of eternal death; God does not refer here to the eternal destination of the human personality but to the "dust" that returns to the "dust" (cf. 2:7; Eccl. 12:7). He speaks of the *consequences of*, not the *punishment for*, sin. These two matters must be carefully distinguished. The *punishment for sin* is the "second" or eternal death, to which those are destined who are not only sinners but stubbornly refuse to repent from their sins (cf. Matt. 10:28; 25:41, 46; 2 Thess. 1:9; Jude 7; Rev. 20:14–15). It was impossible to escape the *consequences* of the Fall; thank God, it *was*, and *is*, possible to escape eternal punishment.

Where lay the initiative to the re-acceptance of Adam and Eve? In a certain sense it lay with God, as indicated in his implicit promise of verse 15. In a certain sense it also lay with these two humans, because after God's pronouncements we read, "The man called his wife's name Eve, because she was the mother of all living" (v. 20). This was an utterance of a new nature from the fallen but reborn Adam, an utterance of *faith* in God's words. Only then did the woman receive the name Eve (Heb. *Chawwah*), that is, she who produces "life" (Heb. *mechawwah*).[67] In this way, Adam pronounced that, despite the Fall and despite the announcement of physical death, "life" would sprout from Eve; not just physical offspring, but children in whom *God's* "life" would be manifested (the *tsaddiqim* mentioned earlier). Adam saw that he himself would not bring forth this "life";[68] therefore, he called his *wife* "life." From this point onward, children would be born in whom God would kindle the true life. With pain and grief, his wife—and all women after her—would give birth to these children; this is part of the inevitable consequences of the Fall. But many (not all) of those children would be born for *life*; these would be the recipients of God's lasting mercy, also after the Fall.

Adam's faith taught him that what has the last word is not death but rather the life that God grants, namely, in the woman's offspring. There *is* a "serpent's offspring"—which, incidentally, is also born literally and physically of earthly women—but in the spiritual sense, there is also a "woman's offspring," in which the true life is manifested. All children are born of

---

[67] Thus Böhl (1923, 71); Leupold (1942, 178) calls *chawwah* an archaic cognate of *chayyah*, "life"; cf. Hamilton (1990, 205).

[68] Something of this may be read in John: "the children of God, who were born, not of blood nor of the will of the flesh *nor of the will of man*, but of God" (John 1:12–13); the true life is not through Adam, but through Eve, in her offspring *par excellence*, Christ.

mothers; and siblings are born from the same mothers. But some of Eve's children are spiritually "children of the devil," and remain such because they refuse to repent; others of these children become "children of God" spiritually through regeneration and faith. The former belong to the "serpent's offspring," the latter to the "woman's offspring"—the woman viewed here as the mother of those who really *live*. It is the contrast between the spiritually dead and the spiritually living, between the *reshaim* and the *tsaddiqim*, the wicked ones and the righteous ones.

Whether Adam was fully aware of all these things or not is not relevant. He probably was not. Nevertheless, the true life that he received through regeneration and faith was founded upon the One who would also be born of Eve, and who, in the end, would bruise the serpent's head.[69]

8.6.2 BEARING THE CONSEQUENCES, YET SAVED

To Adam's act of faith the Lord responded immediately by preparing garments of animal skins for the first humans (v. 21). This is the gospel at its deepest core—so deep that we can perceive it only in the light of all subsequent divine revelation: true life is based upon an innocent, vicarious sacrifice. In Abel's case this was clearer (4:4), in Noah's case still clearer (8:20–21), in Abraham's case very clear: "Abraham went and took the ram and offered it up as a burnt offering instead of his son" (22:13). God could not forgive Adam and Eve, or anybody, without further action on his part; he cannot turn a blind eye to sin. God is "of purer eyes than to see evil and cannot look at wrong" (Hab. 1:13). Therefore, it says, "It is the blood that makes atonement by the life" (Lev. 17:11); "without the shedding of blood there is no forgiveness of sins" (Heb. 9:22). Stated more broadly, without an innocent, vicarious sacrifice there is no forgiveness of sins.

God drove Adam and Eve from the Garden only after he had covered their nakedness with garments of skins. Their expulsion belonged to the inevitable consequences of their sin; their clothing evidenced the sovereign mercy of God. They had to wander outside the Garden—but they did so covered with the tokens of God's grace.

Until the consummation of the ages, humans must continually suffer the consequences of their sins, but at the same time they may enjoy the proofs of God's love. This is not a contradiction; on the contrary, it is an important principle for understanding God's ways with humanity, including with

---

[69] This is why on December 24, the Eastern Orthodox Church honours Adam and Eve as saints; the Western church does not venerate them in this way. They were the first converts in world history.

believers. We see it illustrated in the life of king David. After his sin with Bathsheba, he came to true repentance and confession, as we see in Psalm 51. Therefore, he was fully restored as God's servant. Yet, he was never freed from the *consequences* of his sins (both adultery and murder). The prophet Nathan told him, "The sword shall never depart from your house" (2 Sam. 12:10). After Nathan's story about the rich man who stole a poor man's lamb, David had exclaimed, "He shall restore the lamb fourfold" (v. 6). This is precisely what happened to himself: as a consequence of his sins, he lost four children, three of them by the sword: Bathsheba's first child (v. 16), Amnon (13:28–29), Absalom (18:14), and Adonijah (1 Kings 2:24–25).

Adam and Eve had to bear the consequences of their sin: painful pregnancies and perspiring labour, and eventually physical death. At the same time, God re-accepted them and blessed them. The serpent had overtly called into question God's love; but God demonstrated his love, as well as his righteousness, by sacrificing an innocent animal.[70] His *righteousness* demanded satisfaction; therefore, an animal had to die. His *love* bestowed grace on the sinner by executing judgment not on him but on this substitute. This is exactly what happened at the cross: God's *righteousness* demanded satisfaction; therefore, the full judgment fell upon Christ. God's *love* bestowed grace on the sinner by executing judgment not on the sinner but on Christ. Thus, in the end, judgment involves every human being: either through Jesus Christ, who takes one's place under judgment, or the unrepentant sinner, who rejects him as one's substitute.

The question we must ask AEH advocates here is obvious. I would almost do it with the words of Isaac, "Where is the lamb for a burnt offering?" (Gen. 22:7). Where do AEH advocates explain the vicarious sacrifice rendered on behalf of the historical Adam and Eve? If some, like Gijsbert van den Brink, believe in human evolution, yet wish to maintain some form of a historical Fall, how do they explain the substitute provided on behalf of the early humans who had sinned? In the Bible, the very first human beings fell into sin, a fact that had dramatic consequences for their offspring (there were no other human beings on earth at the time[71]). However, from the very outset, from Genesis 3 onward, history bears witness not only to sin but also to vicarious sacrifice for sinners. We read on the opening pages of the Bible, particularly when we also take Abel and Noah into account,

---

[70] So Sikkel (1923, 201–02); Morris (1976, 130); *contra* Kidner (1967, 73), who finds the notion of atonement here to be an exaggeration; Leupold (1942, 179) hesitates.

[71] *Contra* van den Brink (2017, 255): "We can assume...that it all began with the sin of one man; this sin led to a chain of reactions as a consequence of which sin spread rapidly among [then extant] humanity."

about both the universal presence of sin and the universal provision of atonement and forgiveness.

This is the core of the gospel. As far as I can tell, this core disappears in any AEH narrative about origins. Their story of struggling and maturing hominids has no gospel. They may defend a kind of New Testament gospel, and I take their attempt at face value. But my heartfelt and urgent question is this: What gospel was proclaimed to the first humans who had evolved so far that they could be held responsible, and then sinned? How can we take AEH advocates seriously if they do not provide serious answers to these and many similar questions? If the gospel in Genesis 3 disappears, what becomes of the gospel in the New Testament? Can we surrender the former and hope to retain the latter?

8.6.3 THE FALL AND REVELATION

Let me conclude this chapter with a few comments about the book of Revelation. C. John Collins pointed to several authors who have claimed that, in the New Testament, the apostle Paul "is the only writer to appeal to the story of Adam, Eve, and the serpent."[72] Collins showed that these authors were mistaken. He could have drawn attention here especially to the apostle John, who in his Gospel (8:44) reports what Jesus told his opponents, "your father the devil...was a murderer from the beginning... he is a liar and the father of lies"—words that according to many expositors clearly refer to the serpent's role in the Fall.

Collins used a different, somewhat less obvious example from the Gospels: in Matthew 19:3–9 Jesus explained God's design for the husband and wife relationship "from the beginning." However, in Deuteronomy 24:1–4, additional legislation had become necessary because of the Israelites' "hardness of hearts." Apparently, there had been some change of circumstances since the beginning—namely, the Fall.

Another non-Pauline part of the New Testament is the book of Revelation, to which Collins refers. Here, again, there is an implicit but clear reference to the Fall: the "dragon" is described as "that ancient serpent, who is called the devil and Satan, the deceiver of the whole world" (12:9; cf. 20:2). Some elements of Genesis 2–3 return in the description of Paradise (2:7, "the tree of life, which is in the paradise of God") as well as that of the New Jerusalem (Rev. 22:1–5), where we hear about the "river of the water of life," and "on either side of the river, the tree of life with its twelve kinds of fruit,

---

[72] In Barrett and Caneday (2013, 161–63), he pointed to Harlow (2010, 189); cf. Barr (1993, 4).

yielding its fruit each month. The leaves of the tree were for the healing of the nations." In opposition to the curses pronounced in Genesis 3:14 and 17, we read, "No longer will there be anything accursed, but the throne of God and of the Lamb will be in it, and his servants will worship him. They will see his face, and his name will be on their foreheads" (vv. 3–4).

Adam and Eve were expelled from access to Eden and the tree of life because they had become defiled by sin. In Revelation 22:14–15 we find the reverse: "Blessed are those who wash their robes, so that they may have *the right to the tree of life* and that they may enter the city by the gates. Outside are the dogs and sorcerers and the sexually immoral and murderers and idolaters, and everyone who loves and practices falsehood." Both in Genesis 3 and in Revelation 22, there is an inside and an outside. Inside are those who have not fallen into sin, as well as those have washed their robes, and thus are clean once again, respectively. Outside are the wicked, that is, those who in Adam have fallen into sin *and* refuse to "wash their robes," that is, to repent and to humbly return to God. In the end, those who fell in Adam but repented will return to Paradise, or the New Jerusalem.

I think it is impossible to read these references in Revelation in any other way than against the background of Genesis 2–3.

9

ORIGINAL SIN

Behold, I was brought forth in iniquity,
and in sin did my mother conceive me.
Behold, you delight in truth in the inward being,
and you teach me wisdom in the secret heart (Psalm 51:5–6).

THESIS
The biblical subject of original sin presupposes a very first (original) pair of humans, Adam and Eve, who fell into sin (this is the origin of sin), and passed on their now sinful nature to all their descendants. It also presupposes the view that there are no other humans in the world than those who descended from them.

9.1 WHERE ARE WE?

9.1.1 AEH OR ENNS?

In any review of AEH, the topic of original sin must occupy a central place. AEH defenders have clearly understood this.[1] What is original sin? Is the doctrine tenable within the AEH ideology? Where (with whom) did original sin start? Did a pair of humans, or a small population of early humans, ever fall into sin? What effects did this have on their descendants? And especially, what effect did it have on the other hominids/humans who were

[1] See C. Deane-Drummond in Cavanaugh and Smith (2017, 23–47); J. B. Green in ibid., 98–116; van den Brink (2017, 245–262); among recent AEH opponents, see C. R. Trueman in Phillips (2015, 183–210).

supposedly living on earth at the time of the supposed Fall? What about *their* offspring? It is my thesis that the AEH ideology cannot properly uphold a biblical doctrine of original sin.

Van den Brink mentions as one of the aspects of the problem of sin the following assumption: "The tendency to sin is not part of our original constitution, but a consequence of the first sin that occurred at the beginning of human history. Thus, it is in fact a corruption."[2] In his entire argument concerning the problem of original sin, he struggles especially with this feature, and understandably so.[3] If no first pair of humans existed, who were created directly by God, who fell into sin, and who passed on their sinful nature to all their descendants, then the AEH position has a major problem when it comes to the matter of original sin. Then we invent such totally unbiblical ideas as "that it all began with the sin of one man [whoever this was]; this sin led to a chain of reactions as a consequence of which sin spread rapidly among [then extant] humanity."[4] Sin spread not only among "Adam's" offspring but also among the other humans supposedly living in "Adam's" time! Profound gullibility is required to imagine that this portrayal has anything to do with the biblical picture.

Gijsbert van den Brink needs to learn here from Peter Enns, who is a collaborator of BioLogos (see §1.6.2) but takes a far more liberal—but, in my view, also far more consistent—position than most other AEH advocates:

> We do not reflect Paul's thinking when we say, for example, that Adam need not be the first created human but can be understood as a representative "head" of humanity. Such a head could have been a hominid chosen by God somewhere in the evolutionary process [this is the AEH position], whose actions were taken by God as representative of all other hominids living at the time and would come to exist. In other words, the act of this "Adam" has affected the entire human race not because all humans are necessarily descended from him but because God chose to hold all humans accountable for this one act. *But this would hardly have occurred to Paul, and posing such an "Adam" does not preserve Paul's theology.*[5]

---

[2] Van den Brink (2017, 247).
[3] Ibid., 255–57.
[4] Ibid., 255.
[5] Enns (2012, 120); italics added.

9.1.2 CONSEQUENCES

Notice what is happening here. Peter Enns and I agree on this important point: the AEH position is untenable, because its "Adam" can hardly correspond with the Adam whom the apostle Paul took for granted. AEH advocates such as N. T. Wright, James K. A. Smith, and Gijsbert van den Brink are attempting to defend an impossible theory, which does justice neither to current evolutionary thought nor to Paul's teaching.

However, please note that Peter Enns and I draw opposite conclusions from this state of affairs. The AEH view is a kind of middle position, which in the long run is counter-productive. Therefore, we can do one of two things: (a) Either we simply abandon our fruitless attempt to preserve a kind of historical Adam and historical Fall that in fact have little to do with the biblical data. This is Enns' position (and that of many others); it is nothing but a variation of liberal theology, as we have known it especially since the time of the Enlightenment. (b) Or we disallow evolutionary thought from any longer governing our theology of creation, Fall, and redemption. This is my position (and that of many others); one may call it the traditional or conservative position, if one so wishes.

Peter Enns has stated the matter accurately: "By saying that Paul's Adam is not the historical first man, we are leaving behind Paul's understanding of the *cause* of the universal plight of sin and death. But this is the burden of anyone who wishes to bring evolution and Christianity together—the only question is how that will be done."[6] Exactly. If you are wishing to reconcile evolution and Christianity (Smith's objective; see §§1.4.2 and 3.6.2), you may try the AEH position for a while, as Wright, Smith, and van den Brink are doing. You may maintain such a position for some time, and thus (in my view) unintentionally delude orthodox Christians. As long as some kind of historical Adam and historical Fall can be salvaged, orthodoxy may be rescued (at least, this is the impression created). But this doesn't work, says Enns (who is "orthodox" in his own way, for he maintains the resurrection of Christ). Enns is right; it doesn't work. In the end, AEH advocates (or at least their followers) will end up where Enns has already arrived: defending a thoroughly liberal understanding of Genesis 1–3, and thus to some extent also of Romans 5. But AEH advocates have the other option open to them: return to the thoroughly biblical position.

The present chapter is not the place for an extensive historical survey of the doctrine of original sin,[7] nor for a comprehensive systematic-theological

---

6 Enns (2012, 123).

7 See recently Maduemer and Reeves (2014), especially the contributions by P. Sanlon, R. Kalb, D. Macleod, T. McCall, and C. Trueman.

treatment of the same.[8] I will focus on those matters that are of special relevance to our investigations concerning the historical Adam and the historical Fall.

9.1.3 THE RESURRECTION

To the superficial reader it might almost have seemed that I was defending Peter Enns in the previous section. Of course, this was not the case; I was merely appreciating his criticism of the AEH position. AEH advocates face an impossible dilemma, and this reality is sharply exposed by Enns. The problem, however, is that he, too, faces an impossible dilemma. Before we move on with our discussion of original sin, we must illuminate Enns' dilemma as well. In my view, there are only two consistent positions: the liberal one and the conservative one. Liberal means: I accept from the Bible only what fits my own preconceived ideas, which have largely been formed by modern science. Conservative means: I believe that content of the Bible that, broadly speaking, centuries of Christian expositors have believed; this belief was formed by what the Bible itself had to say, not by preconceived ideas that are imposed upon it. (Of course, this picture is rather idealistic, but it will serve my present purpose.)

Between the liberal and the conservative positions we now find at least two intermediate positions, which are both inconsistent, and are thus untenable.

(a) The AEH position: orthodox on the resurrection of Christ, semi-conservative on the historical Adam and the historical Fall ("semi," because the AEH view of Adam and the Fall only vaguely resembles the biblical picture), liberal on creation (Gen. 1–3 are understood in evolutionary terms).

(b) The position of theologians like Peter Enns and Denis Lamoureux:[9] orthodox on the resurrection of Christ, liberal on the historical Adam and the historical Fall (these are basically denied), liberal on creation (Gen. 1–3 are understood in evolutionary terms).

Both positions are inconsistent because they are partly conservative and partly liberal. In this sense, we may say that Enns' accusation against the AEH position returns like a boomerang: his view is just as inconsistent with the Pauline picture as the AEH view, if not more so. Let me turn Enns' conclusion (see the end of §9.1.1) against his own view: positing a kind of "semi-Adam" does not preserve Paul's theology, but a kind of "non-Adam" does so even less.

---

8 See some well-known systematic-theological handbooks: Van Genderen (2008, chapter 9); Grudem (1994, chapter 24); Geisler (2011, 741–89), Thiselton (2015, 148–53).

9 Enns (2012); Lamoureux (2008; 2009); see also the latter's contributions to Barrett and Caneday (2013).

This is not a matter of "taking one's pick." The *same* chapter that presents to us Paul's faith in the resurrection of Christ also presents to us the historical Adam (1 Cor. 15). It is this chapter that explains to us "the gospel by which you are being saved" (vv. 1–2). The facts that Christ died, was buried, and was raised (vv. 3–4) are presented as *historical truth*. These are accepted as such by conservative Christians, and also by AEH advocates, along with Enns and Lamoureux. The crucial difference is this: conservative Christians also accept the rest of 1 Corinthians 15 as historical truth; I am referring now especially to verses 45–49.[10] Both the AEH camp and the Enns/Lamoureux camp begin their strategy of picking and choosing right here: Enns and Lamoureux believe in the last Adam but not in the first Adam, while AEH advocates believe in both the first and the last Adam—but then a first Adam of their own design. Which position is more objectionable? If Enns *rightly* faults AEH advocates, then I would reply to him: "Physician, heal yourself" (Luke 4:23), and: "Why do you see the speck that is in your brother's eye, but do not notice the log that is in your own eye?" (Matt. 7:3).

AEH advocates believe in some kind of historical Adam, but then one of their own design. Enns rejects this "semi-Adam"—and rightly so—but believes in a notion of creation that is his own design. The followers of both camps want to hold on to the resurrection of Christ, but in the end this might be nothing but a resurrection of their own design—a resurrection that will satisfy the demands of modern science (e.g., the belief "that the resurrection has already happened," 2 Tim. 2:18; a resurrection entirely of a spiritual nature; cf. Eph. 2:6; Col. 2:12).

9.2 BIBLICAL STARTING POINTS

9.2.1 WHAT IS ORIGINAL SIN?

Let us now delve more deeply into the subject of original sin, which is of the utmost importance in our understanding of AEH (as well as of Peter Enns and thinkers congenial to his view). The expression "original sin" is the English rendering of what the Western church fathers called *peccatum originale*.[11] Strictly speaking, the term refers to the original sin that Adam and Eve committed in the Garden of Eden: the *peccatum originale originans* ("original, originated sin") refers to Adam's original sin originating misery

[10] I could mention vv. 52–53 as well: why believe in the last trumpet but not in the first Adam? Why should that last trumpet be more acceptable to modern science than the first Adam?

[11] For the views of various church fathers and later expositors, see extensively, Smith (1994, chapters 1–5). The subsequent Latin phrases are explained by Bavinck (*RD* 3:101, 106–110).

in all his descendants. But the term came to include the consequences of this original sin in their descendants: the *peccatum originale originata* (inherent sin, punishment) refers to the misery that original sin originated in all Adam's descendants. In summary, the fathers used this term to refer to the origin of sin, which involved:[12]

(a) The Fall of the first pair of humans; this is the collective-historical aspect.[13]
(b) The sinful beginning of each human life, which soon begins to lead to personal sins; this is the *individual-historical* aspect.[14]
(c) Sin in an abstract-moral sense, as the corrupt root that produces perpetually new corrupt shoots; this is what many call the "sinful nature," which characterizes each individual (cf. [b]), and was inherited by every human being from fallen Adam (cf. [a]). To use the language of the Heidelberg Catechism, it is a "corrupt" nature.[15]

Thus, the term "original sin" refers not only to the first sinful act itself, but also to the sinful nature and condition of unregenerate humans. However, I must warn immediately that theologians have differed widely on the subject of original sin. Dutch dogmatician Herman Bavinck wrote, "The doctrine of original sin is one of the weightiest but also one of the most difficult subjects in the field of dogmatics,"[16] with a reference to French philosopher Blaise Pascal, who spoke of the transmission of sin throughout the human race as "the mystery furthest from our understanding,"[17] and also to French philosopher Jean-Jacques Rousseau, who said that original sin explained everything except itself.[18]

9.2.2 ROMAN CATHOLIC VIEW

It is interesting to consider here the attitude of the Roman Catholic Church since Pope Pius XII.[19] In his encyclical *Humani Generis* (1950), the pope

[12] Cf. Hoek (1988, 73).
[13] Enns (2012, 125) doubts whether this aspect has any biblical support.
[14] Murphy (2006) distinguishes between (a) and (b) by calling the former "original sin" in the stricter sense, and the latter "sin of origin."
[15] Dennison (2008, 772): "[Q/A] 7. *From where, then, does this depraved nature of man come?* From the fall and disobedience of our first parents, Adam and Eve, in Paradise (Genesis 3; Rom. 5:12, 18–19), whereby our nature became so corrupt that we are all conceived and born in sin (Ps. 51:5)." Cf. Ouweneel (2016, ad loc.).
[16] Bavinck (*RD* 3:100).
[17] Ibid., 3:101.
[18] Ibid., n87.
[19] Cf. A. Riches in Cavanaugh and Smith (2017, 117).

and subsequent Catholic leaders have emphasized that there is no conflict between the theory of evolution and the Bible. However, on one specific point the church stood firm: it refused to surrender the historical Adam, and especially his historical fall into sin. As Pius XII stated the matter,

> The faithful cannot embrace that opinion which maintains that either after Adam there existed on this earth true men who did not take their origin through natural generation from him as from the first parent of all, or that Adam represents a certain number of first parents. Now it is in no way apparent how such an opinion can be reconciled with that which the sources of revealed truth and the documents of the Teaching Authority of the Church propose with regard to original sin [*peccatum originale*], which proceeds from a sin actually committed by an individual Adam and which, through generation, is passed on to all and is in everyone as his own.[20]

Similarly, the *Catechism of the Catholic Church*, promulgated by Pope John Paul II in 1992, declares, with reference to *Humani Generis*, "The account of the fall in *Genesis* 3 uses figurative language, but affirms a primeval event, a deed that took place *at the beginning of the history of man.*"[21]

Thus, we may conclude that the view of Roman Catholic Church is characteristically in agreement with that of AEH. Both are wrestling with the problem occupying us in this book: how can those who feel obliged to embrace what they see as the results of modern science harmonize those results with the traditional and unrelinquishable view that sin came into this world "by an individual [historical] Adam," who fell into sin? Aaron Riches claimed that this "agonizing tension of the church's double commitment" (to modern science and to Christian teaching) "is possible only if the center of unity is Jesus Christ, the 'Paradox of paradoxes,'" referring to the French Roman Catholic theologian Henri de Lubac.[22] In other words, according to Riches we must learn to live with the paradox. However, this solution is rather superficial and evasive. This problem is not at all a paradox. A paradox is a *seeming* contradiction; but Riches is describing a *real* contradiction.

In opposition to this, we argue that such an amalgamation as the one Riches is defending simply does not work. We do not *need* to live with the

[20] *Humani Generis* 37 (http://w2.vatican.va/content/pius-xii/en/encyclicals/documents/hf_p-xii_enc_12081950_humani-generis.html).

[21] *Catechism of the Catholic Church* 390 (available at http://www.vatican.va/archive/ccc_css/archive/ catechism/p1s2c1p7.htm).

[22] A. Riches in Cavanaugh and Smith (2017, 117–18).

paradox, or rather, the contradiction: we need only surrender either the theory of evolution or Christian teaching. You simply cannot have both: you cannot have Christian teaching blended with the so-called "results of modern science." So then, let us look a little more closely at the problem of original sin.

9.2.3 THE CONFUSING TERM "NATURE"

Adam became a sinner because he sinned; among his descendants it is the other way round: Adamites are not sinners because they sin, but they sin because they are sinners, that is, because they have a sinful nature. As the prophet says, "Can the Ethiopian change his skin or the leopard his spots? Then also you can do good who are accustomed to do evil" (Jer. 13:23). You cannot be an Ethiopian without having dark skin; you cannot be a leopard without having spots. Moreover, this cannot be changed: Ethiopians cannot get rid of their darkness, nor leopards of their spots. Similarly, the sinner cannot be "accustomed to do evil," and then suddenly begin doing good. A sinner requires a fundamental *transformation* for this to happen: the Ethiopian must be transformed into, for instance, a Caucasian, or the leopard into a lion. Or the sinner must be transformed into a "new self, created after the likeness of God in true righteousness and holiness" (Eph. 4:24).

A sinner cannot but sin; even his best deeds are only "dead works" (Heb. 6:1; 9:14), and his righteous deeds are "like a polluted garment" (Isa. 64:6). Sinfulness is his *nature*. In the letter to the Romans, Paul usually refers to this sinful nature as the "flesh" (Gk. *sarx*), a rather confusing term because the terms can refer also to the human body in the neutral sense (e.g., the body of Christ, Luke 24:39; 2 Cor. 5:16; 1 Tim. 3:16; 1 John 4:2). Therefore, various translations (like the GW) render that Greek word as "nature": "While we were living under the influence of our corrupt nature, sinful passions were at work throughout our bodies.... I have a corrupt nature, sold as a slave to sin.... nothing good lives in my corrupt nature.... I am obedient to sin's standards with my corrupt nature" (Rom. 7:5, 14, 18, 25). We "do not live by our corrupt nature but by our spiritual nature.... Those who live by the corrupt nature have the corrupt nature's attitude. But those who live by the spiritual nature have the spiritual nature's attitude. The corrupt nature's attitude leads to death. But the spiritual nature's attitude leads to life and peace" (8:4–6; cf. 6:19; 8:7–9, 12–13; Gal. 5:13, 16–17, 19, 24).

The Bible speaks many times of this "natural" sinfulness of humanity: "The wickedness of man was great in the earth, and every intention of the thoughts of his heart was only evil continually" (Gen. 6:5); "the intention of man's heart is evil from his youth" (8:21). "Who can make him clean that is conceived of unclean seed?" (Job 14:4 DRA).[23] "What is man, that he can be pure? Or he who is born of a woman, that he can be righteous?" (15:14). "The fool says in his heart, 'There is no God.' They are corrupt, they do abominable deeds; there is none who does good. The LORD looks down from heaven on the children of man, to see if there are any who understand, who seek after God. They have all turned aside; together they have become corrupt; there is none who does good, not even one" (Ps. 14:1–3; 53:1–3; cf. Rom. 3:10). "Behold, I was brought forth in iniquity, and in sin did my mother conceive me" (Ps. 51:5). "The wicked are estranged from the womb; they go astray from birth, speaking lies" (58:4). "If you then, who are evil" (Matt. 7:11). "That which is born of the flesh is flesh" (John 3:6).[24] We "were by nature children of wrath, like the rest of mankind" (Eph. 2:3; cf. Rom. 2:5).

In the latter passage, the word "nature" speaks of what we are "by nature," that is, through our birth. We must note here that this does not involve the human *creaturely* nature but *fallen* nature.[25] If we say, "By nature I am a thinking being," we are referring to our creaturely nature; if we say, "By nature I am inclined to deceit," we are referring to our fallen nature. The two must be carefully distinguished. We might put it this way: when we speak of the creaturely nature, we are referring to the "first man" (1 Cor. 15:45, 47); when we speak of the fallen nature, we are referring to the "old man" (Rom. 6:6; Eph. 4:22; Col. 3:9 KJV). In many of us, much of the "first man" still remains, since otherwise we could not be called "human" anymore. This has to do with the two dimensions mentioned earlier (§3.2.1): structure and direction. The notion of "first man" involves structure (our creaturely make-up), whereas the notion of "old man" involves direction (directed toward God or idols).

Here is an illustration. When Jesus spoke with the rich young man, he *loved* him (Mark 10:21). This is a strong expression! Jesus loved him because—if I may put it this way—he saw something of the splendour (especially the sincerity) of the "first man" remaining in him. But that the young man turned away from Jesus was due to the "old man" (i.e., the person as dominated by the "old" [i.e., sinful] nature).

---

[23] Perhaps the sense is rather, "Who can bring purity out of an impure person?" (NLT).

[24] Perhaps the sense is rather, "The only life people get from their human parents is physical" (ERV).

[25] Hughes (1989, 127).

9.3 THE THREE "NATURES"

9.3.1 CREATION—FALL—REDEMPTION

Indeed, "nature" is a confusing term insofar as it might suggest that human sinfulness is part of human creatureliness. What sinners do "by nature" (because they are born sinners) is not what they originally (i.e., before the Fall) did as creatures. To put it more strongly, what they do now as fallen human beings goes *against* their originally created nature. Paul makes this very clear in Romans 1:

> Claiming to be wise, they became fools, and exchanged the glory of the immortal God for images resembling mortal man and birds and animals and creeping things. Therefore God gave them up in the lusts of their hearts to impurity, to the dishonoring of their bodies among themselves, because they exchanged the truth about God for a lie and worshiped and served the creature rather than the Creator, who is blessed forever! Amen. For this reason God gave them up to dishonorable passions. For their women exchanged *natural* relations for those that are *contrary to nature* [Gk. *para physin*]; and the men likewise gave up *natural* relations with women and were consumed with passion for one another, men committing shameless acts with men and receiving in themselves the due penalty for their error (vv. 22–27).

In summary, we may distinguish three "natures," where the term "nature" refers to the true character of a thing, creature, or person.

(a) *Creation*. First there is the original creaturely nature of Adam and Eve, which they had between their creation and their fall. The sins that sinners commit are against this original (pre-Fall) nature (Gk. *para physin*). "Nature" is here also a reference to the Ego, the personality, that which a person is. To ask, "What nature do you have as a person?" is the same as asking, "What kind of person are you?"

(b) *Fall*. Second there is the sinful nature of the fallen Adam and Eve, and hence the sinful nature of all their descendants. "By nature" they are all sinners, but this is not their original creaturely nature. Rather, it is the nature they acquired through the Fall, the nature that each of them has since his/her conception. There is correspondence with (a) because a person can still say, "By nature I am a human being." At the same time, there is contrast with (a) because a person can no longer say, "By nature I am a good, innocent being." The adjective "human" involves structure, whereas the adjectives "good, innocent" involve direction.

(c) *Redemption*. Third there is the nature of the regenerated person. Without using the word "nature," Jesus expressed the matter this way: "That which is born of the flesh is flesh [not only physically but also morally![26]], and that which is born of the Spirit is spirit" (John 3:6). That is, the nature of the reborn person is derived from the nature of the Holy Spirit, just as the nature that a human has by birth is derived from the nature of fallen Adam.

The matter of natures has been expressed this way: through *rebirth* we participate in the blessings that Christ has brought about through his work of redemption. As a parallel of this, it is through *birth* that we participate in the terrible effects that Adam has brought about through the Fall.[27] As Johan Heyns put it,

> Adam's actual sin [*peccatum actuale*] became his personal sin [*peccatum personale*], and his personal sin[fulness], of course together with that of Eve, became the natural sin[fulness] [*peccatum naturale*] of the entire human race, and the natural sin[fulness]of the human race became the actual sin [*peccatum actuale*] of each individual.[28]

It began with an *act*, which became Adam's *nature*. This nature subsequently became each descendant's *nature*, and in all descendants, this sinful nature led and leads to their personal sinful *acts*.

9.3.2 GENERATION, REGENERATION

The matter we are discussing right now was expressed by the Council of Carthage (418) as follows: what was obtained through *generation* (begetting, that is here, obtained through inheriting from Adam) is through *regeneration* (rebirth) blotted out and cleansed.[29] Though this was wrongly connected with infant baptism, the principle is correct. Church father Augustine said something similar: "What is contracted through generation [i.e., begetting, birth], is saved through regeneration" (Lat. *Qui generatione contrahitur, regeneratione salvatur*).[30]

---

[26] Cf. John 6:63, "It is the Spirit who gives life; the flesh is no help at all."

[27] Robinson (1982, 298).

[28] Heyns (1988, 190).

[29] Canon 2 states that infants "derive from Adam" "original sin, which needs to be removed by the laver of *regeneration* [i.e., infant baptism]...infants, who could have committed as yet no sin themselves, therefore are truly baptized for the remission of sins, in order that what in them is the result of *generation* may be cleansed by *regeneration*" (italics added) (http://www.earlychurchtexts.com/public/carthage_canons_on_sin_and_grace.htm).

[30] *De dono perseverantiae* II.4.

The Bible frequently tells us that all people are sinners: "There is no one who does not sin" (1 Kings 8:46; 2 Chron. 6:36); "There is none who does good, not even one" (Ps. 14:3; 53:3; cf. Rom. 3:10–12); "Surely there is not a righteous man on earth who does good and never sins" (Eccl. 7:20; cf. John 8:7, "him who is without sin among you"); "For by works of the law no human being will be justified in his sight…for all have sinned and fall short of the glory of God" (Rom. 3:20, 23); "by the one man's disobedience the many were made sinners" (Rom. 5:19); "I am of the flesh, sold under sin. For I do not understand my own actions. For I do not do what I want, but I do the very thing I hate.… I see in my members another law waging war against the law of my mind and making me captive to the law of sin that dwells in my members" (Rom. 7:14–15, 23); "For the mind that is set on the flesh is hostile to God, for it does not submit to God's law; indeed, it cannot" (Rom. 8:7); "If we say we have no sin, we deceive ourselves, and the truth is not in us.… If we say we have not sinned, we make him a liar, and his word is not in us" (1 John 1:8, 10); "we all stumble in many ways" (James 3:2).

This is the result of "generation" or natural begetting; every human born of an earthly father and an earthly mother is by nature impure, unrighteous, sinful, a captive of one's sinful nature, "completely incapable of any good and prone to all evil."[31] This "corrupt nature" (cf. Rom. 8:21; Eph. 4:22, 29; 2 Tim. 3:8; 2 Pet. 1:4; 2:19; Rev. 19:2) cannot be mended, improved, or repaired. There is only one solution: a bad "generation" must be followed by a good (divine) *re*generation; the person must be made all over again, so to speak ("born again," John 3:3, 7). This is basically the same as saying that the only good solution for a bad "generation" is *death*, but then the death of Christ that is reckoned to us, followed by our being made *alive* in and with Christ (Rom. 6:4–6; Col. 2:11–12).

In God's sight, humans are "impure" beings: "Behold, even the moon is not bright, and the stars are not pure in his eyes; how much less man, who is a maggot, and the son of man, who is a worm!" (Job 25:5–6; cf. 4:18–19; 15:15). "Who can say, 'I have made my heart pure; I am clean from my sin'?" (Prov. 20:9; cf. v. 6, "a faithful man who can find?"). "To the pure, all things are pure, but to the defiled and unbelieving, nothing is pure; but both their minds and their consciences are defiled" (Titus 1:15). "For we ourselves were once foolish, disobedient, led astray, slaves to various passions and pleasures, passing our days in malice and envy, hated by others and hating

---

[31] Dennison (2008, 2:772); this is part of Q. 8 of the Heidelberg Catechism; cf. Ouweneel (2016, ad loc.).

one another" (3:3). "We know that…the whole world lies in the power of the evil one" (1 John 5:19).

Each human stands guilty before God: "If you, O LORD, should mark iniquities, O LORD, who could stand?" (Ps. 130:3); "no one living is righteous before you" (143:2). The Gentiles "are darkened in their understanding, alienated from the life of God because of the ignorance that is in them, due to their hardness of heart" (Eph. 4:18; cf. Rom. 1:18–32). There is not one person who could honestly say that he does not need the prayer, "O LORD, be gracious to me; heal me, for I have sinned against you!" (Ps. 41:4), or, "Forgive us our debts" (Matt. 6:12), or, "God, be merciful to me, a sinner!" (Luke 18:13).

9.4 THE SINFUL NATURE

9.4.1 SINFUL FROM CONCEPTION

Why are all people sinners? This is the central question, central because it involves the view that is being undermined by AEH advocates. It is the traditional—and I believe, biblical—view that all people are sinners because they descend from the first pair of humans, who fell into sin and passed on their sinful nature to all their descendants. The AEH paradigm *has* no first pair of humans, directly created by God, who fell into sin.

Again, why are all people sinners? It must be in their nature. Is the question comparable with this one: Why do (virtually) all people have ears? Yes and no. People have ears because they are humans; people are sinners because they are humans; it is in their nature. Yet there is an enormous difference, as we saw. Before the Fall, the first humans had ears[32] (they could hear both the LORD and the serpent); their having ears has nothing to do with the Fall. But it is not like this with human sinfulness. After the Fall, we can definitely say that humans are sinners "by nature":

> You were dead in the trespasses and sins in which you once walked, following the course of this world, following the prince of the power of the air, the spirit that is now at work in the sons of disobedience—among whom we all once lived in the passions of our flesh, carrying out the desires of the body and the mind, and were *by nature* children of wrath, like the rest of mankind (Eph. 2:1–3).

---

[32] Cf. Ps. 94:9, "God made our ears" (ERV).

Again, "by nature" we have ears, and "by nature" we are sinful—but the first nature is our creaturely nature; it concerns what we have and are as creatures of God, irrespective of the Fall. However, the second "by nature" refers to our *fallen* nature, to what we are since and through the Fall. As a creature, I was designed by God to have ears, but I was *not* designed by God to be sinful. My having ears is due to divine creation; my sinfulness is due neither to creation nor to some evolutionary imperfection. It is due to the Fall, to a historical event at a certain time and in a certain place.

Before this historical event, called the Fall, humans were "very good," but after this Fall, as long as they are not born again, humans are "very bad" ("the intention of man's heart is evil from his youth," Gen. 8:21). Notice the difference with the AEH view: initially we were less morally mature, later we became more morally mature. AEH advocates cannot possibly uphold the notion of a "very good" humanity, which then, through one single event, became a thoroughly sinful humanity. As Greg Haslam claims, "Adam fell downwards, not upwards."[33] His fall was not development but degeneration, not progress but regress.

If, since and through the Fall of Adam and Eve, we are all sinners, this is what we call "inheritance."[34] It reminds us of 1 Peter 1:18, which speaks of "the futile ways inherited from your forefathers." This is not inheritance in some biological (genetic) sense of the word but in a spiritual sense, even if we cannot describe what "spiritual inheritance" is; we cannot determine exactly how it works (cf. §9.5.3). We are sinners by birth because we descend from sinful ancestors, in whatever way. We maintain this *contra* (semi-)Pelagians, Socinians, some Remonstrants, and more recently, theologians like Hendrikus Berkhof, who denied the notion of original sin.[35] But even these people cannot deny that each human is a sinner, and is so from early youth: "Behold, I was brought forth in iniquity, and in sin did my mother conceive me" (Ps. 51:5). "The wicked are estranged from the womb; they go astray from birth, speaking lies" (58:3).

Such passages clearly teach that people are born sinners. Some rabbis wondered whether a human can sin already in the mother's womb. Jesus' disciples seemed to believe so; otherwise, their question concerning the man who was born blind is hardly understandable: "Rabbi, who sinned, this

---

33 In Nevin (2009, 58).

34 Cf. Belgic Confession (Art. 15): "We believe that through the disobedience of Adam original sin is extended to all mankind; which is a corruption of the whole nature and a *hereditary* disease, wherewith even infants in their mother's womb are infected, and which produces in man all sorts of sin" (Dennison [2008, 2:433]; italics added).

35 Berkhof (1986, 203–10).

man or his parents, that he was born blind?" (John 9:2). If you ask whether someone was *born* thus-and-so, and whether this condition might be due to the person's sin, then this sin apparently preceded the birth. If humans are sinners from conception—because that is what inheriting involves, even if the inheriting must not be understood in a precise biological sense—then the idea of sinning in one's mother's womb (that is, long before the child has come to the age of discernment) is not necessarily objectionable.

About Genesis 25:22 ("The children struggled together within her," i.e., Rebekah), Rashi commented that, when Rebekah passed a school conducted by Shem and Eber (cf. 10:21)—that is, a godly school—Jacob inclined to it in her womb, but when she passed an idolatrous temple, Esau was inclined to it.[36] In this way, Rashi indicated that some people are already righteous or wicked before their birth. In principle, this is not the Christian view. *All* people are born sinners; it is only through conversion and regeneration that they become godly and righteous (Heb. *tsaddiqim*). We must admit, however, that some of the Lord's servants were apparently called already in their mothers' womb: Jeremiah (Jer. 1:5), John the Baptist (Luke 1:15),[37] and Saul of Tarsus (Gal. 1:15).

But this observation is merely incidental to our main point; what is essential to understand is that humans possess a *nature* that is sinful from their very conception, even if it may take some time before this nature begins to produce bad fruits. For instance, young children are sinners, but do those "who do not know their right hand from their left" (Jonah 4:11) commit specific *sins*? Although most children begin to talk and to walk in their second year of life, they were *born* with the basic ability and inclination to talk and to walk. Similarly, even if children begin to sin at a certain age of discernment, we may say that they were *born* with the basic ability and inclination to sin.

At any rate, we remember that by nature we are not sinners because we sin, but we sin (from whatever age this may be; theologians may differ on this point) because we have a sinful nature. A bad tree is bad before the bad fruits begin to be manifested (cf. Matt. 7:17–19; 12:33). A tree is not bad because it produces bad fruits, but it produces bad fruits because it is a bad tree.

---

36 Cohen (1983, 141); cf. De Graaff (1987, 329).

37 This explains the Spirit-inspired joy with which John leaped in his mother's womb when she greeted Mary (Luke 1:41), says John Gill (http://biblehub.com/luke/1-15.htm).

9.4.2 HUMAN IMPERFECTION

If sin were inherent to human nature from the moment of human origin (whether divine creation or natural evolution), we must realize what this means. In this case, human sinfulness would be due not to a historical Fall, at a certain point in time and space, but to the imperfection of a still evolving humanity. An example of where such an approach may lead is an article by Daniel C. Harlow.[38] Bryan Estelle gives the following comment on it:

> If Adam is not the responsible agent for casting the human race into a condition of sin and misery, then at whose feet should we place the blame for our human predicament? Does it not follow, if one removes the historicity of Adam from the equation and if our historical forefather Adam is not responsible for our condition of sin and misery, that someone else must bear that responsibility? It seems to this author that the necessary consequence is to make God responsible for the evil we observe in the world. A careful reading of Harlow's article, which was previously referenced, will demonstrate that this is the case. These recent suggestions that Adam is merely a literary construct, without any external historical reference to real situations, are not without serious consequences for our theology.[39]

Earlier, Phil Hills and Norman Nevin had arrived at a similar conclusion: the traditional biblical view "makes the sin of man responsible for the dysfunction in our world. A theology that denies a significant fall and denies that physical death is a result of humanity's sin makes God responsible for the suffering in our world."[40] If Adam *fell* by his own fault, *he* is to be blamed for his sinfulness. If Adam was *created* sinful (read: immature, imperfect, as a consequence of incomplete evolution), then the *Creator* is to be blamed. In this case, the creature would have a right to ask, "Why have you made me like this?" (Rom. 9:20), which would mean here: Why have you made me so imperfect, so incomplete?

Also C. John Collins concluded:

> If we abandon the conventional way of telling the Christian story, with its components of a good creation marred by the Fall, redemption as

---

[38] Harlow (2010).

[39] Estelle (2012, 20).

[40] In Nevin (2009, 214).

> God's ongoing work to restore the creatures to their proper functioning, and the consummation in which the restoration will be complete and confirmed, then we really give up all chance of understanding the world. Specifically, if we deny that all people have a common source that was originally good, but through which sin came into the world, then the existence of sin becomes God's fault, or even something that God could not avoid. In either case, there is little reason to be confident that any relief is headed our way.[41]

Of course, not all AEH advocates go as far as this. Some wish to allow for the theory of evolution but are clearly aware of some of the theological consequences of doing so. Therefore, they try by all means to uphold some form of a historical Adam and a historical Fall. Ultimately, all such attempts necessarily and inevitably lead to denying the historical Adam and the historical Fall as these are revealed in the Bible.

Once people have reached that point, the truth will come out. They either believe in a human being who was initially "very good," and through his fall became very bad, or they believe in a human being who was imperfect from the beginning. This is what Estelle is trying to say: in the traditional view, humanity itself is to blame for its sinfulness because people fell by their own decision. When the AEH ideology is simply followed to its logical outcome, as several authors have pointed out, the Creator is to blame for human moral imperfection. In the traditional view, humans fell, but they can be redeemed by God in and through Christ. Following AEH to its logical conclusion, human imperfection can be solved only by further evolution. These advocates may involve the indispensable help of God or Christ in this, but their position requires that further evolution (a *process*), not redemption (an *event*), is the needed solution.

In James 1 we find an interesting reason why God cannot be blamed for human sinfulness:

> Desire when it has conceived gives birth to sin, and sin when it is fully grown brings forth death. Do not be deceived, my beloved brothers. Every good gift and every perfect gift is from above, coming down from the Father of lights, with whom there is no variation or shadow due to change. Of his own will he brought us forth by the word of truth, that we should be a kind of firstfruits of his creatures (vv. 15–18).[42]

---

[41] Collins (2010, 161–62).

[42] Cf. J. B. Green in Cavanaugh and Smith (2017, 112)

Here James implies that the Creator can never be made the cause of human sin. Sin is darkness; God is the "Father of light"; he is light, "and in him is no darkness at all" (1 John 1:5). No darkness can ever spring forth from this source of light.[43]

9.4.3 EVOLUTION AND ESCHATOLOGY[44]

We would pose the following provocative question to any AEH advocate, moderate or consistent: Does he or she believe in *ongoing* evolution? If not, why not? Why would human evolution have stopped somewhere along the way? Indeed, Peter Enns confidently speaks of "the continued evolution of life on this planet."[45] If we accept this view, what room for further evolution does the AEH advocate see within the framework of redemptive history? The next question, following from the previous one, is this: How does continued human evolution relate to Christian eschatology? Is it plausible to accept the Big Bang theory, which supposedly occurred some 13.8 billion years ago, while at the same time believing that only the last 6,000–10,000 years of these 13.8 billion years constitute "redemptive history," and one day God will transform the present cosmos into an entirely new cosmos (cf. §4.4.3)?

When it comes to the *end* of the present cosmos, van den Brink seems to take the Bible literally.[46] But why do that if he refuses to take literally what the Bible has to say about the *beginnings* of the present cosmos? If he allows evolutionary science to govern his exposition of Genesis 1–3, why does he not allow it to govern his eschatology? The same modern science that tells us that the world began 13.8 billion years ago also tells us that, in 5.4 billion years from now, the sun will turn into a "red giant," and at that time will kill all life on earth. The former number is based on extrapolation backward, the latter number on extrapolation forward; there is no basic difference. The same science that can predict solar eclipses thousands of years from now is able to predict the future of the sun. Why not take this seriously if one also takes science seriously when it comes to the beginnings of our cosmos? If there never was a historical Adam in any biblical sense, why would we believe in a literal return of Christ on the clouds of heaven?

Greg Haslam states the matter clearly:

---

43 When the LORD says, "I form light and create darkness" (Isa. 45:7), this is a different matter; he is telling us, as the context suggests, that not only benefits but also trouble may come from him.

44 Regarding Gen. 1–3 and eschatology, see Fesko (2007); Barcellos (2013).

45 Enns (2012, 147).

46 van den Brink (2017, 334–35).

> If God's creative work was as slow, imperfect, slipshod and dangerous to life and limb as [the theory of] evolution asserts it was, then the creation story of Genesis 1–3 is wrong and this throws into question the final consummation accounts in Revelation too. If God didn't say what he meant in Genesis, why should we trust him anywhere else? If God couldn't get creation right first time in the beginning, and was forced to use a "hit and miss," "road-kill" method involving eons of death and destruction, then how long might we have to wait for God to re-fashion the world at the end? It could be a long wait for our final resurrection and new residence! Yet both are said to be instant outcomes of Christ's return in the Bible. Why not at the beginning also?[47]

These are fair questions. Three possible views underlie this discussion.

(a) The Bible says that *one day* it all began ("the day that the LORD God made the earth and the heavens," Gen. 2:4). The Bible also says that *one day* it will all end (the "last day," John 11:24; cf. Acts 17:31; 2 Thess. 1:10; 2:2–3 etc.),[48] and conservative Christians understand both statements literally.
(b) The Bible says that *one day* it all began and *one day* it will all end, but this must all be understood figuratively. We have billions of years behind us, and billions of years ahead of us. It all began extremely slowly, and it will all end extremely slowly.
(c) AEH advocates take a kind of middle position: according to them, we have billions of years of gradual development behind us, but *one day* it will all suddenly end.

In my view, both (a) and (b) are perfectly consistent, whereas (c) is perfectly inconsistent. Basically the discussion turns on this question: Where is AEH going to stop? How consistent will AEH advocates really be? Why is modern science allowed to speak with authority about an inaccessible past, and not about the coming future? Remember the camel: if you allow its nose into your tent, and then allow other parts of its body to come in as well, in the end the camel will be entirely inside and will refuse to leave (see §3.6.3). To the left of the conservative viewpoint we now have the AEH position. To the left of the AEH position we have Peter Enns, Denis

---

47 In Nevin (2009, 72).

48 The question remains open as to whether this day is the day of Jesus' second coming, or the day, a thousand years later, when the new heavens and earth will arrive; see extensively, Ouweneel (2012a).

Lamoureux, and others. To the left of Peter Enns we have full-fledged liberal theology. This is the course of thinking from the moment you allow the theory of evolution to govern your understanding of beginnings. It won't be long before the camel occupies your entire tent; modern science will push out the Bible's teaching altogether.

Pierre Teilhard de Chardin, the Roman Catholic theologian and paleontologist, was consistent in this respect.[49] He referred to a Point Omega in the future, which would be the culmination point of human evolution as well as that of Christian eschatology.[50] His only misfortune was that the majority of neither biologists (Christian or not) nor Christians (Catholic or Protestant) were enamoured of his ideas.

9.5 THE INHERITANCE OF EVIL

9.5.1 FROM VERY GOOD TO VERY BAD

It may seem that not all people are sinners because the Bible describes some as righteous and blameless. If evil is inherited, they appear to have missed out. Noah was such a person: "a righteous man, blameless in his generation. Noah walked with God" (Gen. 6:9). Yet, he turned out not to be sinless at all: after the Flood he fell into drunkenness and exhibitionism (9:21). Abraham was called the "friend of God" (2 Chron. 20:7; Isa. 41:8; James 2:23). Yet, he lied twice about his wife (12:11–19; 20:2–5). Another such person was Moses, "very meek, more than all people who were on the face of the earth" (Num. 12:3). Yet, despite his age and wisdom he fell into thoughtless behaviour (20:10–11). Another righteous man was Job, "blameless and upright, one who feared God and turned away from evil" (Job 1:1, 8; 2:3). Yet, after a time of trial he had to say, "I despise myself, and repent in dust and ashes" (42:6).

One of the most impressive biblical examples is King David. After he had died, God spoke of "my servant David, who kept my commandments and followed me with all his heart, doing only that which was right in my eyes" (1 Kings 14:8). This illustrates a wonderful truth: "I [the Lord] will remember their sin no more" (Jer. 31:34). Yet, David had been an adulterer and a murderer (2 Sam. 11). About King Solomon we read, "He was wiser than all other men" (1 Kings 4:31). Yet, we read about him, "When Solomon

---

[49] See especially Teilhard de Chardin (1959).

[50] Actually, based on what we know about the impact of mutations, the "culmination point of human evolution" is rather extinction in the not-too-distant future.

was old his wives turned away his heart after other gods, and his heart was not wholly true to the LORD his God, as was the heart of David his father" (11:4). Zechariah and Elizabeth "were both righteous before God, walking blamelessly in all the commandments and statutes of the Lord" (Luke 1:6). Yet, at the word of the angel, Zechariah was afflicted with muteness because of his unbelief (v. 20).

We could continue. Virtually every person whose history is described more extensively in the historical Bible books turns out to have fallen into serious sins. Joseph may seem to be an exception, but we read in Genesis 48:17 of a personal weakness of Joseph, and some would take 37:2b as sinful. The Bible's testimony is that "there is none who does good" (Ps. 14:1, 3; 53:1, 3). On the one hand, we discover this great biblical truth: even the *greatest* sinners can repent and be saved (see, e.g., 1 Tim. 1:15–16). On the other hand, even the *best* people turn out to be profound sinners; even "*the* teacher of Israel" (Nicodemus) had to be born again (John 3:3–10).

Our point is this: the Bible claims, and illustrates with many examples, that there are no people who are not sinners by nature, who did not inherit evil. "There is no distinction: for all have sinned and fall short of the glory of God" (Rom. 3:22–23). Many AEH advocates will readily admit this. But the Bible is equally clear in teaching that, before the Fall, humans were *not* like this. In addition to the clear testimony of Genesis 1–3, we hear King Solomon say, "See, this alone I found, that God made man upright [this was before the Fall], but they have sought out many schemes [this was after the Fall]" (Eccl. 7:29).[51] Before the Fall, the first humans were "very good" (Gen. 1:31); after the Fall, humans were very bad, except those who were regenerated and redeemed. Any model, AEH or otherwise, that cannot account for this transition from very good to very bad must be rejected. And please note, in the Bible this transition occurred very quickly; there is no hint that it was part of some evolutionary process (or process of degeneration, for that matter).

In an orthodox soteriology, it is highly important to properly understand the transition (regeneration, salvation) from being a great sinner to being a beloved child of God, yes, someone who no longer "makes a practice of sinning, for God's seed abides in him; and he cannot keep on sinning, because he has been born of God" (1 John 3:9). However, to properly understand this transition, it is essential to also understand that earlier quick transition (the Fall) from being a "very good" person to being a condemnable sinner. If we

[51] Cf. CEV: "We were completely honest when God created us, but now we have twisted minds." Cf. Jewish expositor Yehudah Kiel (1997, on Gen. 3:10), who sees Eccl. 7:27 as a description of the Fall.

lose the historical Fall—historical in the biblical sense of Genesis 1–3—we will also lose redemption in the biblical sense. Salvation is not the perfection of one's evolutionary shortcomings, but the restoration (and much more than that) of what Adam and Eve were before the Fall. Without a proper understanding of the Fall, there can be no proper understanding of salvation.

Richard B. Gaffin has written pointedly about this.

> By now it should be clear that questioning or denying the descent of all humanity from Adam as the first human being has far-reaching implications for the Christian faith. It radically alters the understanding of *sin*, particularly concerning the origin and nature of human depravity, with the corresponding abandonment of any meaningful notion of the *guilt* of sin. It radically alters the understanding of *salvation*, especially in eclipsing or even denying Christ's death as a substitutionary atonement that propitiates God's just and holy wrath against sin. And it radically alters the understanding of the *Savior*, by stressing his humanity, especially the exemplary aspects of his person and work, to the extent of minimizing or even denying his deity.[52]

Or, as Andy McIntosh formulates it, "That mankind knew no physical death in the beginning is integral with the gospel. Remove this and the gospel itself is weakened. Why did Christ die physically if the wages of sin is not physical death?"[53] In other words, if human death is a purely natural phenomenon, then it belonged to the world that God had declared to be "very good" (Gen. 1:31). Why would Christ die to eliminate what from the beginning was part of a "very good" world?

It is simply not true that Christ died only to free us from spiritual death, as, for instance, Denis Alexander has asserted.[54] The book of Revelation clearly distinguishes between "death," which is physical death (9:6; 12:11), and the "second death," which is the everlasting continuation of spiritual death in the lake of fire (cf. 2:10–11; 20:6, 14; 21:8). One day, the former death will be abolished (20:13–14; 21:4, "death shall be no more"; cf. 1 Cor. 15:26, "the last enemy to be destroyed is death"; see 2 Tim. 1:10), but the "second death" shall last forever (14:11; 19:3; 20:10). Only one reason explains why, one day, human physical death will be annulled: death is *not* a natural human phenomenon, but a consequence of the historical Fall.

---

[52] Gaffin (2012, 5).

[53] McIntosh (2008); cf. also his chapter in Nevin (2009, chapter 9D).

[54] Alexander (2014, 265).

9.5.2 ADAM'S NATURE IS INHERITED

In order to obtain a clear picture of this point of conflict with AEH advocates, we must enter a little more deeply into the matter of original sin. There are many views among those who do agree on the concept of original sin but differ on its exact meaning. For instance, some claim that humans are sinners because they supposedly sinned "in and with" Adam (so-called realism). The Bible does not explicitly say this, but perhaps there is an element of truth in this view if we accept the parallel with Levi being present in the loins of Abraham (see §§4.6.3 and 10.5.1).

Another view involves the notion that humans are sinners because what their "covenant head" did is "imputed" to them (federalism).[55] I am not aware of any biblical basis for the claim that Adam's sin is in some way or another our sin. We did not inherit his *sin*, or his *guilt*, but we inherited his *sinful nature*.[56] This is an enormous difference. The essence of the matter is not that humans sinned "in and with Adam," so that in some way or another Adam's sin is their sin. The point is rather that they sin every day in the way that Adam sinned because of the sinful nature that they inherited from him.

What is imputed to them is not Adam's sin—they do not inherit his *guilt*—but only their own sins, the sins that they themselves commit because of the sinful nature that they inherited from Adam. By nature, humans are "under sin" (Rom. 3:9; 7:14; Gal. 3:22). This is so, first, because of the sinful nature they inherited from their parents, and second, because they themselves committed sins from the time of their earliest consciousness and responsibility. The Bible states very clearly that the "soul who sins shall die. The son shall not suffer for the iniquity of the father, nor the father suffer for the iniquity of the son. The righteousness of the righteous shall be upon himself, and the wickedness of the wicked shall be upon himself" (Ezek. 18:20). In other words, no human being can and will ever be punished for the sin that Adam committed. People are condemned for their *own* sins: "We will all stand before the judgment seat of God...each of us will give an account of himself to God" (Rom. 14:10, 12). "Let each one test his own work, and then his reason to boast will be in himself alone and not in his neighbor. For each will have to bear his own load" (Gal. 6:4–5; see more examples in the next section).

I am convinced that only in this way can and must the notion of "primordial sin" be understood: what people inherit is not the sin that

---

[55] Elsewhere I have dealt extensively with this conflict between realism and federalism; see Ouweneel (2008, §§11.3.2–11.3.3).

[56] King Rehoboam did not inherit the *sins* of his father Solomon, but certainly his inclination to sinful behaviour (1 Kings 12:10, 14).

Adam committed, nor the guilt that he incurred as a result; they do not inherit the responsibility for his guilt. What they do inherit is the sinful nature that Adam possessed since the Fall, and which he passed on to all his descendants. They are not even responsible for the sinful nature that they inherited from their parents, just as they are not responsible for the blue or brown eyes that their parents transmitted to them. They are responsible—and this is in itself serious enough—only for what they *do* with this sinful nature. They are responsible only for the sins that they commit as a consequence of this sinful nature. Nothing of what Adam committed is imputed to them; they do not bear his guilt. What connects them with Adam's Fall is only the sinful nature that they inherited from him—and that is bad enough.

Incidentally, behind the idea of imputation lies a pervasive misunderstanding of what the Greek verb *logizomai* means. This error also infects the Reformed idea that the "righteousness of Christ," that is, his Torah-obedience, has supposedly been imputed to (i.e., transferred to the account of) believers, and that this belongs to the essence of biblical justification. The Greek verb *logizomai* never means "to transfer from one account to another," but rather "to count (someone) as (sinful, righteous or whatever)." This is very clear from Romans 4:8, "Blessed is the man against whom the Lord will not count [Gk. *logisētai*] his sin." The idea that Christ's righteousness is transferred to the believer is just as wrong as the idea that Adam's guilt is transferred to his descendants.[57]

9.5.3 MISUNDERSTANDINGS

Let's face it: the phrase "original sin" is not very lucid and the Dutch term (*erfzonde*) and German term (*Erbsünde*) are not must better.[58] At least three misunderstandings may arise here.

First, the phrase as well as some interpretations of it easily lead to the misunderstanding that the sinful *act* is inherited[59] rather than only the inherited sinful *nature*. The Dutch Calvinist term *erfschuld*, "inherited guilt," confuses the matter still further. When Louis Berkhof says, "The guilt of Adam's sin, committed by him as the federal head of the human race, is imputed to all his descendants,"[60] I cannot agree with this (see the previous section). But I

57 See extensively Ouweneel (2018d, chapter 4).

58 Cf. Brunner (1952, 103–117); Berkouwer (1971, 531); Van de Beek (1984, 140–41); Wentsel (1987, 724–26); Hoek (1988, 73); Van Genderen (2008, 403–404).

59 "It does not make sense to think of sins as being inherited!" exclaims Velema (in Van Genderen [2008, 411).

60 Berkhof (1949, 270); Klaas Schilder agrees with him (http://www.dbnl.org/tekst/schi-008curs02_01/ schi008curs02_01_0005.php).

do agree with him when he continues, "This is evident from the fact that, as the Bible teaches, death as the punishment of sin passes on from Adam to all his descendants." I believe that Berkhof has drawn the wrong conclusion from this biblical fact. The conclusion is not that young infants who die prove that Adam's guilt was imputed to them. Scripture never says such a thing. What it does prove is that all humans, including infants, are subject to death because they have inherited Adam's sinful nature (cf. 1 Pet. 1:18).

Second, terms like *erfzonde* and *Erbsünde* (literally, "inherited sin") could easily suggest that the matter somehow involves *genetic* inheritance, which is not the case at all. There is no sin gene, or sin chromosome. Some authors have argued that original sin is inherited through male semen but not through the female ovum, and have thus tried to prove biologically how it is possible that Jesus, who had no earthly father, was sinless.[61] American physician Albert S. Anderson argued that Jesus' virgin birth demanded three divine steps: (a) removal of Mary's "sin mutation" from her ovum,[62] (b) addition of a newly created Y-chromosome, (c) duplication of the entirely sin-free somatic chromosomes in order to attain a full set of chromosomes.[63] The weakest, if not most deplorable, aspect of this kind of argument is, of course, the presupposition of something like a "sin mutation" in the biological sense, and more broadly, the view that the inheritance of the sinful nature would be a genetic matter. There is no such thing as sinful DNA.

This kind of misunderstanding is aggravated if the doctrine of original sin is troubled by the ancient debate between *traducianism* (the doctrine that the soul of the child is transmitted through the father's semen and the mother's ovum) and *psycho-creationism*, the doctrine that the child's body comes from the parents but the so-called "logical soul" (Lat. *anima rationalis*, an Aristotelian notion) comes from outside by a special creational act of God (cf. the popes mentioned in §5.1.2).[64] If psycho-creationism were true, how can a human being be a sinner by nature? Is this sinful nature biologically inherited from the parents, without the God-given soul being affected by it? Either the soul is pure, or God creates sinful souls. Traducianism has its own problems because it requires us to presuppose the body–soul dualism.

Third, the doctrine of original sin is sometimes understood to imply that responsible humans are arbitrarily rejected because of their nature. It is certainly true that the wrath of God rests on an unconverted person from

[61] See Ouweneel (2007b, §10.3.2).

[62] Did Jesus' conception involve an ovum from Mary? See extensively, Ouweneel (2007b, § 10.2.2).

[63] Anderson (1976, 104).

[64] See extensively, Ouweneel (2008, §§7.2.1, 7.3.1–7.3.2, 8.2.1–8.2.2, 8.3.2).

the moment of that person's conception—but at the same time the love of God reaches out to this person from the very beginning.[65] Moreover, when a person has come to the age of responsibility one is not only, and not even primarily, condemned for one's sinful nature but for one's sinful actions.[66] As Scripture teaches, God "will render to each one according to his *works*" (Rom. 2:6). "For we must all appear before the judgment seat of Christ, so that each one may receive what is due for what he has *done* in the body, whether good or evil" (2 Cor. 5:10). "I will give to each of you according to your *works*" (Rev. 2:23). "And the dead were judged by what was written in the books, according to what they had *done*. And the sea gave up the dead who were in it, Death and Hades gave up the dead who were in them, and they were judged, each one of them, according to what they had *done*" (20:12–13). "Behold, I am coming soon, bringing my recompense with me, to repay each one for what he has *done*" (22:12; cf. John 3:20–21).

The term "responsible" is quite important here. Those who might be inclined to complain that God is unrighteous if he rejects people because of original sin are the very people who bear enough responsibility for having added to their inherited sinful nature their own many actual sins. Before the judgment seat of God they will not have to give account of their original sin but of their own actual sins. And let those who complain that, as a consequence of original sin, they *could* do nothing other than sin, repent quickly and receive new life from God and the power of the Holy Spirit, in order to be rich and diligent in good works (1 Tim. 6:18; Titus 2:14; cf. Matt. 5:16; Eph. 2:10; 4:28; Titus 3:8, 14; Heb. 10:24).

9.6 TOTAL DEPRAVITY[67]

9.6.1 THE AEH VIEW

As we have seen before, those theologians who believe in human evolution—or allow room for it—will have great difficulty in formulating a theological anthropology that will do justice to the biblical data. Recall this observation from chapter 6: animals are temporal, humans are eternal because they have been created in the image and after the likeness of God. Even the highest animals are strictly bound to what I have called the sensitive structural layer or organizational level (feelings, drives, instincts),

[65] See Ouweneel (2008, §10.3).

[66] Cf. Barth (1956, IV/1:544–45).

[67] The Lat. terminology is *vitiositas, corruptio totalis*; for Calvin's view of it, see extensively, Torrance (1957, chapters 7–8); see also Demarest (1997, 73–75).

whereas humans have what I have called the mental structural layer or organizational level (thinking, imagination, creativity, volition). But this is not an exhaustive description of the human mystery. If it were, many biologists and psychologists would hasten to convince us of how much thinking, imagination, creativity, and volition certain animals possess. Therefore, I have added in chapter 6 the biblical notion of the *transcendent* heart as the eternal concentration point of all *temporal* functions and structural layers of human beings.

In their hearts, human beings surpass their bodily mode of existence, and stand in a transcendent-religious relationship with God (or, since the Fall, with idols). All varieties of AEH are necessarily united on this point: in their ideas of human evolution there is no room for the rise of eternity beings, created in the image and after the likeness of God, people whose very essence lies in their transcendent relationship to God or idols. If AEH advocates attempt to solve this problem by appealing to God (as a *deus ex machina*) as the One who transformed certain hominids into eternity beings, images of God, they would become the laughing-stock of ordinary evolutionists. Either AEH advocates must adopt the theory of evolution in full, or they must smuggle in some creative acts of God, as the Roman Catholic Church does with its teaching of psycho-creationism. This is because the problem remains: how did *temporal* hominids ever evolve into eternity beings? I have not yet found an AEH publication that has dealt with this question.

The notion of the "heart" is quite important for us to better understand what happened at the Fall. When AEH advocates pretend to believe in some kind of a "Fall," they may write about the weaknesses, errors, mistakes, shortcomings, and imperfections—*sins* if you like—of evolving hominids. What else could they do? However, lower hominids had no "heart" in the theological-anthropological sense of the term; they were animals. Strictly speaking, they therefore could not sin, just as gibbons and gorillas cannot sin. This is the enormous challenge for AEH advocates: at what point in time did the transition from animal to human occur, the transition from temporal being to eternity being, from non-images to images of God, from beings without a "heart" to beings with a "heart"?

A human being has a heart (in the transcendent-spiritual sense of the term), whereas an animal does not.

Jesus said, "From within, out of the heart of man, come evil thoughts, sexual immorality, theft, murder, adultery, coveting, wickedness, deceit, sensuality, envy, slander, pride, foolishness. All these evil things come from within, and they defile a person" (Mark 7:21–23). Please note that animals, including the supposed early hominids, *do not have evil thoughts, are not sexually immoral, are not wicked, not proud*—simply because they have no "heart" from which these things can spring forth. They are not responsible for their behaviour; they do not have to give account. Humans are responsible and do have to given an account. Where and how in the course of human evolution did humans receive a "heart" from which all these evil things could, and do, come forth?

We will never reconcile the Bible's portrait of the Fall with any attempt to defend the evolution of human beings from hominids, because the Fall was a matter of the human heart. Relatively soon after the Fall we read about fallen man that "every intention of the thoughts of his *heart* was only evil continually" (Gen. 6:5); "the intention of man's *heart* is evil from his youth" (8:21).[68]

9.6.2 THE CORRUPT HEART

Let us look a little more closely at this fallen human heart. That the Fall involves not only the physical-bodily aspect of a human being, but the entire person, is seen most clearly from sin being localized in the person's *heart*. In §§6.4 and 6.5, we observed that the heart is the root and the centre of the human mode of existence. This heart is a person's deepest inner being, the most genuine and essential point in the human personality, from where "flow the springs of life" (Prov. 4:23). Negatively this means that the heart is the place where sin is devised (Gen. 6:5; 8:21; Num. 15:39; Ps. 28:3; 58:2; 66:18; 140:2; Prov. 6:14, 18; 12:20; 19:3; Jer. 7:24; Zech. 7:10; 8:17); the heart is the source of so much wickedness: of false prophecy (where what humans devise stands in opposition to what God devises; Jer. 14:14; 23:26); of human pride (Gk. *hybris*, the anti-divine arrogance; Exod. 4:21; 7:13–14, 22; 8:15, 19, 32; 9:7, 34–35; Deut. 8:14; 2 Chron. 25:19; 26:16; Ps. 10:6, 11, 13; 14:1; 36:1; 41:6; 78:18; Prov. 18:12; 21:4; Isa. 9:9; 46:12; Dan. 5:20); of (the inclination to) apostasy (Num. 32:7, 9; Deut. 8:2; 11:16; 29:18; 30:17; 1 Kings 11:2–4, 9; 2 Chron. 36:13; Job 31:7, 27; Ps. 44:18; Ezek. 6:9); of self-deification (Isa. 14:13; Ezek. 28:2, 6); of sinful self-confidence (Prov. 28:26; Isa. 47:10; Ob. 1:3; Zeph. 2:15), and much more. "The heart is deceitful above all things, and desperately sick; who can understand it?" (Jer. 17:9).

The heart is the deepest inner being, that which represents the Ego, the personality (Col. 2:2; 1 John 3:19–20), the source of sin. Jesus himself

[68] Interestingly, the second time we read in Genesis about the "heart" is this: "And the Lord regretted that he had made man on the earth, and it grieved him to his *heart*" (6:6), and the third time: "When the Lord smelled the pleasing aroma, the Lord said in his *heart*" (8:21). The meaning of humans as images of God involves this feature as well: both God and humans have a "heart" (in this specific sense), animals do not.

taught his disciples that "from within, out of the heart of man," all evil comes forth, "from within, and they defile a person" (Mark 7:21–23; see previous section). Therefore, the Bible speaks of the "uncircumcised heart" (Lev. 26:41; Jer. 9:26; Ezek. 44:7, 9; cf. Deut. 30:6; Jer. 4:4), and God puts the "hearts" to the test (1 Chron. 29:17; 2 Chron. 32:31; Ps. 7:9; 17:3; 26:2; 139:23; Jer. 11:20; 17:10; 20:12). "Heart" and "flesh" are not opposed to each other but stand alongside each other: "My heart and flesh sing for joy to the living God" (Ps. 84:2). Also compare Psalm 63:1, where the meaning of "soul" more or less corresponds with that of "heart": "O God, you are my God; earnestly I seek you; my soul thirsts for you; my flesh faints for you, as in a dry and weary land where there is no water."

In the positive sense the heart—under the indispensable guidance of the Holy Spirit—is the organ for seeking God (e.g., 2 Chron. 11:16; 12:14) and for the fellowship with God (e.g., 1 Kings 8:61; 14:8; 15:14), and so on for the fear of the Lord, the service of God, the walk with God, and so on. The New Testament also teaches that the heart is the point of contact for divine working (Acts 15:9; Rom. 2:15; Gal. 4:6; Eph. 3:17). Jesus taught us: "You shall love the Lord your God with all your *heart* and with all your *soul* and with all your *strength* and with all your *mind*, and your neighbor as yourself" (Luke 10:27; cf. Deut. 6:5; Lev. 19:18b).

The general conclusion of all this is that both the Fall and spiritual rebirth are radical, that is, they involve the *radix*, the root, of our human existence.[69] The same "heart" that was thoroughly affected by sin at the Fall is the "heart" in and with which a believer serves the Lord. Animals, including early hominids, have a muscle that pumps blood; but they have no "heart" in the sense just described. And it is inconceivable how beings *with* such a "heart" could ever have evolved from beings *without* such a "heart." No AEH advocate that has yet attempted to explain this.

9.6.3 RADICAL CONSEQUENCES

The inevitable result of belittling the seriousness and impact of the Fall, as AEH advocates do (presumably unintentionally), will be to downplay the profound impact of salvation, which involves the very same *radix* ("heart") of our human existence. Belief in human evolution blurs both one's anthropology (doctrine of humanity) and one's hamartiology (doctrine of sin). Their radical character and their transcendent significance are lost. To accept a theory in which humans are no more than evolved, ennobled primates, is to surrender both the transcendent and the eternal qualities of

---

[69] Cf. Noordegraaf (1990, 66).

being human. Where this happens, the radical character and transcendent significance of salvation will also inevitably be blurred. Our claim is that the AEH position affects soteriology and Christology (see chapter 10), and, as we saw, it also threatens eschatology. Remove the foundation of the Christian edifice, and the entire building will eventually collapse.

The argument of AEH advocates is this: let's assume that humans were created by God through the process of human evolution from earlier hominids. Does Christianity then collapse? Not at all; look, the rest of the edifice remains standing intact. This argument is self-deception. It is analogous to attempting to replace part of the foundation of the St. Peter's Basilica in Rome with paper. Could that be done? Perhaps. But soon thereafter the St. Peter's Basilica won't be standing.

Theological revision cannot stop with origins, because the entire theological building rests upon the bedrock of biblical creation and the historical Fall. As Dutch theologian Johannes P. Versteeg put it, "If an evolutionary view leaves no place for Adam as a historical person and has a place for Adam only as a teaching model, that has direct consequences so far as its view of Christ is concerned."[70] We may state it this way: an improper view of the "first Adam" will endanger our view of the "last Adam." William VanDoodewaard has brought this out very clearly in the title of his book; just as Albert Schweitzer wrote his *The Quest for the Historical Jesus*,[71] VanDoodewaard felt it had become equally necessary to write about *The Quest for the Historical Adam*.[72] If we lose the biblical Adam, we will lose the biblical Christ.

The theory of evolution is a diseased tree that some want to transplant into the orchard of Christian theology, and this tree can yield nothing but diseased fruit: "Every healthy tree bears good fruit, but the diseased tree bears bad fruit. A healthy tree cannot bear bad fruit, nor can a diseased tree bear good fruit. Every tree that does not bear good fruit is cut down and thrown into the fire. Thus you will recognize them by their fruits" (Matt. 7:17–20).

In the Song of Solomon (2:15), the bride clearly perceives what threatens the relationship to her bridegroom, urging him: "Catch the foxes for us, the little foxes that spoil the vineyards, for our vineyards are in blossom." Bad doctrine is a little fox that spoils the vineyards. Or to use another image: to those who boast about reconciling their theology with modern science, I say:

---

70 Versteeg (2012, 67).
71 Schweitzer (2005).
72 VanDoodewaard (2016).

Your boasting is not good. Do you not know that a little leaven leavens the whole lump? Cleanse out the old leaven that you may be a new lump, as you really are unleavened. For Christ, our Passover lamb, has been sacrificed. Let us therefore celebrate the festival, not with the old leaven, the leaven of malice and evil, but with the unleavened bread of sincerity and truth (1 Cor. 5:6–8; cf. Gal. 5:9).

10

THE NEW TESTAMENT ON GENESIS 1–3

Therefore, just as sin came into the world through one man, and death through sin, and so death spread to all men because all sinned—for sin indeed was in the world before the law was given, but sin is not counted where there is no law. Yet death reigned from Adam to Moses, even over those whose sinning was not like the transgression of Adam, who was a type of the one who was to come (Romans 5:12–14).

THESIS

The way that Jesus and the New Testament writers (preeminently Paul) understood Genesis 1–3 as describing historical events is decisive for the way we ourselves ought to read these chapters. Paul has been regarded as a man of antiquity, with an outdated worldview, not knowing any better. However, this cannot be true about the Son of God, who worked with precisely the same worldview as Paul did. Because of this, the issue of the historical Adam ultimately involves Christological dimensions. This confirms the thesis that the historical Adam and the historical Fall are essentially and vitally important for Christianity.

10.1 JESUS ON GENESIS

10.1.1 JESUS ON CREATION

In §4.1.1, I mentioned the statements that Jesus made about the first chapters of Genesis, and I will turn now to comment on them further. I begin with this introductory remark. Both AEH advocates and those to the left of AEH advocates (e.g., Peter Enns, Denis Lamoureux[1]) have sidelined the apostle Paul as a "man of antiquity," who remained enmeshed in the ancient worldview, and therefore understood Genesis 1–3 as fully historical. Of course, these authors go on to assure us that Paul's outdated views are no obstacle for properly understanding Romans 5, for instance. What strikes me, however, that none of these authors has made similar claims about Jesus. It is quite obvious that Jesus had the same worldview as Paul, and the same views of Genesis 1–3 as Paul. Why is *he* not sidelined as someone who belongs among those supposedly ignorant men of antiquity? Is this to spare the feelings of orthodox readers?

Let us consider the following. Jesus placed Adam and Eve at the "beginning of creation" (Mark 10:6). Apparently, for Jesus there was no enormous time lapse between Genesis 1:1 (the beginning of the world) and verses 26–28 (the creation of Adam and Eve). The "beginning of creation" coincided more or less with the beginning of human history; the span was only five days, not billions of years. This is a well-known argument of young-earth creationists,[2] the power of which is hard to evade. The only counter-argument would be that we should not read too much into Jesus' words because he is not dealing here with the subject of origins, but with the subject of marriage and divorce (in Matt. 19:8, he refers simply to "the beginning," not to "the beginning of creation"). Yet, we may wonder whether Jesus ever said anything casually. Could it not be that Jesus simply held the same view of Genesis 1 that Jews and Christians have entertained for centuries?

Jesus also spoke about Satan and his role during the fall of the first humans; he said to his opponents, "You are of your father the devil, and your will is to do your father's desires. He was a murderer from the beginning, and does not stand in the truth, because there is no truth in him. When he lies, he speaks out of his own character, for he is a liar and the father of lies" (John 8:44). Satan is called a murderer and a liar because with his lies he induced Adam and Eve's spiritual death (cf. Gen. 2:17; 3:4). As Wisdom 2:24 (GNT) says, "It was the Devil's jealousy that brought death into the world, and those who belong to the Devil are the ones who will die." Notice in John 8:44 again the expression "the beginning": from the beginning of history, Satan manifested himself as a murderer and a liar. In other words, Adam's fall occurred at the beginning of human history, or even at the beginning of the world's history. Apparently, for Jesus this was self-evident; there

---

[1] See Enns (2012); D. O. Lamoureux in Barrett and Caneday (2013, passim).

[2] See https://creation.com/age-earth-matters.

were not billions of years between Genesis 1 and Genesis 3, i.e., between the creation of the world and the Fall of humanity—not even a few years.

10.1.2 JESUS ON ABEL

When Jesus spoke of "the blood of the righteous Abel" (Matt. 23:35; cf. Luke 11:50–51), he was presenting him as a historical figure. The Bible refers to no person whom we can claim was only a literary figure, and not a historical figure.[3] Moreover, notice the completion of the sentence: "so that on you may come all the righteous blood shed on earth, from the blood of righteous Abel to the blood of Zechariah [the son of Berechiah], whom you murdered between the sanctuary and the altar." Some manuscripts omit the words "the son of Berechiah," understandably because the person in view was Zechariah, the son of Jehoiada and this Zechariah was the last martyr mentioned in the Hebrew Bible (which ends with 2 Chronicles; 2 Chron. 24:22).

How many of Jesus' listeners, as well as Bible readers, have doubted that this Zechariah was a historical person? So then, how could we expect, when Jesus said, "from Abel to Zechariah," that Zechariah was historical, but Abel was not? There can be no reasonable doubt that Jesus accepted Abel as a historical person, and every orthodox Bible reader must consider the consequences of Jesus' statement. However, in the AEH paradigm, doubting Abel's historicity seems a rather obvious conclusion; who could Abel possibly be if Adam were an early hominid individual or from an early hominid population? "By faith Abel offered to God a more acceptable sacrifice than Cain, through which he was commended as righteous" (Heb. 11:4; cf. 12:24; Jude 11). "We should not be like Cain, who was of the evil one and murdered his brother. And why did he murder him? Because his own deeds were evil and his brother's righteous" (1 John 3:12). What could possibly be the historical sense of such statements if Adam were an early hominid, or if there were no historical Adam (in any biblical sense) at all? What can AEH advocates, especially those who assure us that they wish to retain the historical Adam, tell us about his sons Cain and Abel?

In addition, according to liberal theologians as well as AEH advocates, more people must have inhabited the earth because Cain went to "the land of Nod" and there had intercourse with his wife (Gen. 4:16–17).[4] Such a claim overlooks a few obvious facts. First, the Hebrew word *nod* means "wanderer"; we find the same word earlier in verse 12. Apparently, the

3 Supposed counter-examples are Job (Ezek. 14:14, 20; James 5:11) and Jonah (2 Kings 14:25; Matt. 12:39–41; 16:4; Luke 11:29–32), whose historical existence has been doubted by certain expositors.

4 See Haarsma and Haarsma (2011, 255).

land received its name when the wanderer Cain settled there.[5] Second, as to Cain's wife, Adam and Eve were parents not only of sons but also of daughters (5:4). By the time Cain slew Abel, Adam and Eve may already have had a considerable family; this explains why Cain could say, "Whoever finds me will kill me" (4:14). Thus, Cain's wife was simply his sister (or niece, or great-niece, etc.). Notice that still in the time of Abraham it was not uncommon for someone to marry his (half-)sister (Gen. 20:12).

According to a midrash, Cain indeed had a twin sister, and Abel had two twin sisters (more precisely, they were triplets); each married such a sister.[6] It is always difficult to say whether this is pious imagination, or whether such stories are rooted in ancient traditions. One such tradition says, "To ensure the propagation of the human race, a girl, destined to be his wife, was born together with each of the sons of Adam. Abel's twin sister was of exquisite beauty, and Cain desired her. Therefore he was constantly brooding over ways and means of ridding himself of his brother."[7]

10.1.3 JESUS ON NOAH

Jesus spoke also about Noah as a historical figure, and about his vicissitudes as real historical events: "For as were the days of Noah, so will be the coming of the Son of Man. For as in those days before the flood they were eating and drinking, marrying and giving in marriage, until the day when Noah entered the ark, and they were unaware until the flood came and swept them all away, so will be the coming of the Son of Man" (Matt. 24:37–39; cf. Luke 17:26–27). Similar questions must be asked here as we asked about Abel: if Adam were not a historical figure in the biblical sense, why would we take Noah to be a historical figure? But if Noah was *not*, how could Jesus make the comparison he made? If those "days of Noah" never existed, what about the coming of the Son of Man? Why would we believe in the second coming of Christ—because it is orthodox?—and not in those "days of Noah"?

Again, would AEH advocates claim that Jesus did not know any better? They might hesitate to make that claim because they wish to maintain some semblance of orthodoxy. But why do they freely make such claims about Paul while ignoring the fact that Jesus claimed similar things?

The apostle Peter also wrote about Noah as a historical person and about the Flood as a historical event (1 Pet. 3:20; 2 Pet. 2:5; 3:5–6). However, if

---

[5] A land can be identified by a name that it received for the first time many years later; cf. the name "Dan" in Gen. 14:14.

[6] Gen. Rabbah 22.2; see Jewish expositors: Cassuto (1961, 229); Kiel (1997) on Gen. 4:17.

[7] Ginzberg (1969) (https://philologos.org/__eb-lotj/vol1/three.htm#1).

our traditional understanding of the historical character of Genesis 1–3 must be heavily adapted to fit with modern science, what about the story of Noah's Flood? Or, to remain a little closer to the opening chapters of Genesis, Luke wrote about Enoch as a historical figure (3:37), and the author of Hebrews did the same (11:5). Jude 14 said that Enoch was "the seventh from Adam," entirely in accord with Genesis 5:1–8 (cf. §4.6.1). However, given the AEH paradigm, what sense do such statements make? If no historical Adam (in some biblical sense) existed, what about Cain and Abel, Enoch and Noah? In other words, on the basis of AEH, what remains of the genealogy of Genesis 5?

Why do we hear so little from AEH advocates about implications like these? Where is their *exegesis* of Genesis 1–11? And above all: those who have so much to say about the ignorance of the apostle Paul, who ostensibly was just a man of antiquity, what do they have to say about Jesus, who did not seem to differ from Paul in his views on Genesis 1–11? For additional related arguments, see §10.4.

10.2 PAUL ON GENESIS

10.2.1 PAUL ON CREATION

In Acts 17:24–27, the apostle Paul told the people of Athens,

> The God who made the world and everything in it, being Lord of heaven and earth…made from one man [lit., one blood] every nation of mankind to live on all the face of the earth, having determined allotted periods and the boundaries of their dwelling place, that they should seek God, and perhaps feel their way toward him and find him.

Paul clearly states here that all the nations of the world are blood relatives because all humans who have ever lived have one common ancestor (or, more precisely, one common pair of ancestors). They are all of one stock, one origin.

Some suggest that the "one blood" of which Paul speaks here refers to Noah.[8] However, from a strict AEH position, this suggestion is problematic: what room does the AEH position have for a historical Noah from whom all modern humans have descended?

---

8 For references, see chapter 1, note 9.

It may be interesting to note that the idea of all people descending from one pair of humans was the conviction of all orthodox Jews; this follows from the tradition written down later in the Talmud and related books. Thus we read in the tract Sanhedrin,

> Why was only a single specimen of man created first? ...in order that no race or class may claim a nobler ancestry, saying, "Our father was born first"; and, finally, to give testimony to the greatness of the Lord, who caused the wonderful diversity of mankind to emanate from one type. And why was Adam created last of all beings? To teach him humility; for if he be overbearing, let him remember that the little fly preceded him in the order of creation.[9]

For the rabbis, Genesis 1 spoke a clear language: Adam was created as the first of all humans, and as the last of all creatures.

In 1 Corinthians 11:7–12, the apostle Paul says

> A man ought not to cover his head, since he is the image and glory of God, but woman is the glory of man. For man was not made from woman, but woman from man. Neither was man created for woman, but woman for man. That is why a wife ought to have a symbol of authority on her head, because of the angels. Nevertheless, in the Lord woman is not independent of man nor man of woman; for as woman was made from man, so man is now born of woman. And all things are from God.

This entire passage (vv. 1–16) is undoubtedly one of the most complicated in the New Testament, in part because we do not know precisely the clothing habits of those days, the Corinthian clothing fashions relating to public worship gatherings, including possible head coverings, and Paul's objections to their habits. This need not concern us right now; we are concerned only with what Paul had to say about men and women in general, and especially about the first man and the first woman. Nor will we concern ourselves here about another interpretive challenge, namely, why Paul says that the man is the image and glory of God but the woman is this only in a secondary sense; we are not sure what he means here by "glory."[10] But what we do understand—and this is enough for now—is this:

---

[9] As corrected in the Tosefta: Sanhedrin 8:4–9; see http://www.jewishencyclopedia.com/articles/758-adam.

[10] See the commentaries (ad loc.); I recommend especially Fee (1987) and Thiselton (2000).

(a) "Man is not from woman, but woman from man" (v. 8 literally; cf. NKJV). Here, Paul clearly implies that the first man, Adam, was *not* born of a woman, unlike every man *after* Adam was and is. Notice this: whatever view AEH advocates may present concerning Adam, all of them present Adam as having been born of a hominid female. Paul implicitly denies this, entirely in line with Genesis 1–3. Was Paul an ignorant "man of antiquity" simply because he believed in the historical truth of Genesis 1–3?

(b) "Woman [is] from man" (vv. 8, 12): this clearly refers to the remarkable way Eve was created: "The LORD God caused a deep sleep to fall upon the man, and while he slept took one of his ribs and closed up its place with flesh. And the rib that the LORD God had taken from the man he made into a woman" (Gen. 2:21–22). Apparently, Paul understood this literally. AEH advocates cannot possibly do the same; in their view of human evolution, no female hominid ever came out of a male's side;[11] all males and females have been and are born of other females. For Paul, the biblical portrait constitutes a key to his entire argument: every person who ever lived came from a mother—except Adam: he came directly from dust. And every woman who ever lived came from a mother—except Eve: she came from her husband's side. Why did Paul tell us this if he were not firmly convinced of the historicity of Genesis 1–3 *and* of the relevance of these chapters for his argument? Although *we* may be uncertain about the meaning of 1 Corinthians 11, we may know assuredly that *Paul* was certain about the historicity of Genesis 1–3.

In 1 Corinthians 15:47, Paul states as a plain fact, "The first man was from the earth [Gk. *gēs*], a man of dust [Gk. *choïkos*]." This verse shows that Paul understood Genesis 2:7 literally and historically: "The LORD God formed the man of dust from the ground."[12] To Paul, the literal truth of this Genesis verse was important because of the comparison with Christ that he wished to make: "The first man was from the earth, a man of dust; the second man is from heaven. As was the man of dust, so also are those who are of the dust, and as is the man of heaven, so also are those who are of heaven" (1 Cor. 15:47–48). What Paul argues is this: Adam is from the earth, Christ is from heaven. Natural humans resemble their first father: he was a man of dust, and so they are people of dust, too. Believers resemble Christ: he is from heaven, and so they, in the spiritual sense, are people of heaven, too. How could Paul make such statements unless he believed that Adam was indeed

[11] I prefer "side" to "rib" because elsewhere, the Heb. word *tsela* is always rendered "side" (sometimes "side chamber"). The Septuagint has the Gk. word *pleura*, which also means "side."

[12] At this point the Septuagint has basically the same Gk. words that Paul uses: *chous* ("dust") and *gē* ("earth").

just as historical as Christ, and that the description of his creation was to be understood in an essentially literal way? This question will become still more urgent when we now turn to some key passages in Paul's letters.

10.2.2 PAUL ON THE FALL

The rest of the Old Testament seems to pay little attention to the fall of the first pair of humans. This does not mean very much because we do not find many Old Testament references to the book of Genesis anyway. We might conclude that the persons of Noah (1 Chron. 1:4; Isa. 54:9; Ezek. 14:14, 20) and Joseph (no reference to his person, only to his "house," e.g., Zech. 10:6), and even Abraham (e.g., Ps. 105:9, 42; Isa. 51:2; Ezek. 33:24), are not very important because there are few or no Old Testament references to the personal lives of these men. In light of that silence, some have asserted that Christian theology has given too large a place to the Fall.[13] In my view, such a claim ignores the enormous redemptive-historical significance that the New Testament assigns to the Fall, especially in Romans 5.

Let us begin here with a summary of what the apostle Paul teaches about the subject. Two passages require no further comment beyond what I have already written about them. "I feel a divine jealousy for you, since I betrothed you to one husband, to present you as a pure virgin to Christ. But I am afraid that as the serpent deceived Eve by his cunning, your thoughts will be led astray from a sincere and pure devotion to Christ" (2 Cor. 11:2–3). This is a direct reference to Genesis 3, the encounter between the serpent (Satan) and Eve.[14] The other passage is: "Adam was formed first, then Eve; and Adam was not deceived, but the woman was deceived and became a transgressor" (1 Tim. 2:13–14). In both cases Paul was writing about the Fall as a historically real event, not as a figurative story from some literary source (see the next section). In this respect, Paul was simply following the Master, who also viewed Adam and Eve as historically real figures (see §10.1.1).

In 1 Corinthians 15:45–47, the apostle Paul refers to the creation of Adam, again as a historical event: "Thus it is written, 'The first man Adam became a living [i.e., here, life-*receiving*] being' [Gen. 2:7]; the last Adam became a life-*giving* spirit. But it is not the spiritual that is first but the natural, and then the spiritual. The first man was from the earth, a man of dust; the second man is from heaven." In verses 21–22 in the same chapter, Paul also indirectly referred to the Fall: "For as by a man came death, by a

[13] From orthodox side, e.g., König (2006, 412–13).

[14] Cf. chapter 3, note 69.

man has come also the resurrection of the dead. For as in Adam all die, so also in Christ shall all be made alive." Notice this unambiguous statement: "By a man [who turns out to be Adam] came death," that is, by the fall of Adam, death entered into the (human) world. This is what God himself had said, "Of the tree of the knowledge of good and evil you shall not eat, for in the day that [or, when] you eat of it you shall surely die" (Gen. 2:17), and "By the sweat of your face you shall eat bread, till you return to the ground, for out of it you were taken" (3:19). God said it, Paul said it, so who are we to doubt that human death entered the world through the Fall of Adam and Eve?

10.2.3 LITERARY AND HISTORICAL

Let me return to an important point mentioned in the previous section. Would the arguments of Paul in Romans 5 and 1 Corinthians 15 lose their power if we had to assume that Christ is a historical figure but that Adam is only a literary figure? Various authors to be mentioned later in §10.4.1 answer this question in the negative; they hold that Adam is only a literary figure. If they reply that it is quite possible to compare a literary figure and a historical person, they are perfectly correct. For instance, I could say that there are a number of correspondences between Christ and the figure of the returning king in J. R. R. Tolkien's *The Lord of the Rings*.[15] Such a comparison does not at all require that the latter king be a historical figure. I could also say that the Good Samaritan in Luke 15 and the bridegroom in Matthew 25 clearly resemble Christ, although they are only literary figures, not historical persons.

I take great personal interest in the figure of Hamlet, as presented in the play by William Shakespeare. I have even translated this work into Dutch, and have provided an introduction and numerous annotations,[16] all purely for my own literary pleasure. Without difficulty, I could even give some sermons on Hamlet because of the many lessons that we can learn from him (sometimes in spite of him). What has often been suggested about Adam is certainly true about Hamlet; in my book I have defended the thesis that Hamlet is Everyman. I quoted several authors (Samuel T. Coleridge, Anatole France, C. S. Lewis) who have said the same thing or something similar. Due to Shakespeare's genius, Hamlet is the expression of something universal in humans, something that exists at all times and among all people.[17] In my sermons I could imagine myself saying: "Look

[15] Tolkien (2012).
[16] Shakespeare (2004).
[17] Cf. line 2.2.303–04, "What a piece of work is man!"

at Hamlet! What an example he is, both positive and negative! Look what he did there, and take that as an example. Or look how he behaved here, and learn from it not to behave like that." I could even make a comparison with Christ, and point out similarities, and especially profound differences, between what Christ did and what Hamlet did.

However, a few things I could never do. I could never say: what Hamlet did in the past has consequences for all of us because he is our remote ancestor. "In him," while we existed only in his loins, so to speak, we did the same foolish things that Hamlet did, and we are still doing them. We bear the positive consequences of the right things he did, and the negative consequences of the wrong things he did. Why can I not say such things? Because Hamlet is not our "father" at all; he is just a literary figure. Whatever he did—in Shakespeare's representation—has not the slightest effect upon me. Hamlet may be an *example*, but he is not a *type*. I can learn from him, but I did not inherit anything from him. His actions may be lessons, but they have no consequences for my personal circumstances. This is because the Shakespearean Hamlet never existed as a real historical person.[18] This is the great difference with Adam. Everything Paul tells us about him makes sense if, and only if, this Adam really was a historical person. Let us now look a little more closely at the chapter that should make this most clear to us.

10.3 ROMANS 5

10.3.1 ADAM AND CHRIST

The same thought touched upon in 1 Corinthians 15:22 Paul develops extensively in Romans 5:12–19.

> Therefore, just as sin came into the world through one man, and death through sin, and so death spread to all men [or, humans] because all sinned—for sin indeed was in the world before the law was given, but sin is not counted where there is no law. Yet death reigned from Adam to Moses, even over those whose sinning was not like the transgression of Adam, who was a type of the one who was to come. But the free gift is not like the trespass. For if many died through one man's trespass, much more have the grace of God and the free gift by

[18] I am not speaking of Hamlet's model, the legendary Scandinavian Amleth, who possibly goes back to a historical person; see *Gesta Danorum* (*Deeds of the Danes*), by Saxo Grammaticus.

> the grace of that one man Jesus Christ abounded for many. And the free gift is not like the result of that one man's sin. For the judgment following one trespass brought condemnation, but the free gift following many trespasses brought justification. For if, because of one man's trespass, death reigned through that one man, much more will those who receive the abundance of grace and the free gift of righteousness reign in life through the one man Jesus Christ. Therefore, as one trespass [or, the trespass of one] led to condemnation for all men [or, humans], so one act of righteousness [or, the act of righteousness of one] leads to justification and life for all men [or, humans]. For as by the one man's disobedience the many were made sinners, so by the one man's obedience the many will be made righteous.

Without discussing all the exegetical details of this passage, we must investigate what it teaches us about the significance of the Fall.[19] First, let us notice again that, not only according to Genesis 3 but also according to Paul, human death is not something natural that belongs to life as God created it.[20] Of course, this is a major problem for AEH advocates, who maintain that many early humans had died, or were dying, the moment when their (the advocates') so-called "Fall" occurred (whatever is meant by this). One might even ask AEH advocates whether their "Adam and Eve" were orphans at this point, or whether the couple's parents were still alive. Human death entering through Adam's sin is a thought that would have made no sense to this couple. If God had told Adam that he would die if he disobeyed, Adam could have replied that he would die anyway, since death had supposedly been part of hominid life for hundreds of thousands of years. Thus, Peter Enns wrote in flagrant contradiction to Paul, "Death is not the enemy to be defeated...death is not the unnatural state introduced by a disobedient couple in a primordial garden. Actually, it is the means that promotes the continued evolution of life on this planet and even ensures workable population numbers. Death may hurt, but it is evolution's ally."[21]

I firmly believe that the Bible teaches the opposite, and that we are bound to this: for humans, death is *not* a self-evident and automatic phenomenon.

---

[19] In addition to (other) commentaries on Romans, see Chafer (1983, 2:297–312); Smith (1953, 169–76); Berkouwer (1971, 490–522); Schlatter (1962); Oosterhoff (1972, 69–92); Harrison (1976, ad loc.); Wilson (1977); Hendriksen (1980, ad loc.); Erickson (2007, 648–56); Hughes (1989, 128–35, 174); Berkhof (1986, 203–205); Van Genderen (2008, 407–410); Blocher (1997, chapter 3); Schreiner (1998, ad loc.); G. Haslam in Nevin (2009, 57–72).

[20] About this subject, see extensively, Berkouwer (1962, chapter 7); cf. Noordegraaf (1990, 89–92).

[21] Enns (2012, 147).

The fact that believers look forward to a world without sin and death shows that death is not a natural aspect of life; as German dogmatician Wolfhart Pannenberg put it, "Christian hope expects a life without death (1 Cor. 15:52ff.)." Jesus and Paul were connecting with the common Jewish view as we find it expressed in Wisdom 1:13 (RSV), "God did not make death, and he does not delight in the death of the living."[22] Human death is an "essential consequence of sin," and not "a punishment that God has arbitrarily set and imposed."[23] Since the Fall, human sin and human death belong together in an organic way—so organically that we have come to believe that death is a natural part of human life—but before the Fall these did not belong together.

The point is not whether biologists could show that, *today*, death is an inevitable part of life, for humans just as it is for animals. The point is that, for the first humans, through the tree of life, physical death could have been postponed forever. We do not know how this could have worked, but this does not alter the fact itself. What we do realize is that for fallen humans, endless living on earth would have been disastrous. This is why the first humans had to be expelled from the Garden of Eden, and thereby their access to the tree of life blocked (Gen. 3:22–24). They died, not simply because this is natural for humans, but because they no longer had access to the tree of life. And this was not only a punishment, but also a blessing.

One more comment: let us not say that it would have been awkward if Adam and Eve, *apart from* a fall, would have had to continue living forever in the Garden of Eden. First, Adam and Eve were not destined to remain in Eden forever. They had been called to subdue the *earth* and to have dominion over it (Gen. 1:26, 28); moreover, *every* plant and *every* tree had been given to them (v. 29), not just those in Eden. Their domain was to be the entire earth, and with their descendants they were called to take charge of the entire earth. Second, if living forever in the first Paradise would have been awkward, then how awkward would it be to live forever in the eschatological Paradise (Rev. 2:7; 22:1–2)?

10.3.2 SIN AND DEATH

Pannenberg's point is well taken: the fact that believers look forward to a world without death shows that death does *not* belong to human life as originally created; on the contrary, it is an invading enemy that must be overcome. As Jesus said, "Whoever hears my word and believes him who sent me has eternal life. He does not come into judgment, but has passed

---

[22] Pannenberg (2010, 271); cf. Vogel (1949/50, 124); *contra* Barth (1960, III/2:639).

[23] Pannenberg (2010, 274).

from death to life" (John 5:24). "I am the resurrection and the life. Whoever believes in me, though he die, yet shall he live, and everyone who lives and believes in me shall never die" (11:25–26).

Paul reminds us that Christ "must reign until he has put all his enemies under his feet. *The last enemy to be destroyed is death*" (1 Cor. 15:25–26).

> This perishable body must put on the imperishable, and this mortal body must put on immortality. When the perishable puts on the imperishable, and the mortal puts on immortality, then shall come to pass the saying that is written: '*Death is swallowed up in victory*.' [Isa. 25:8] 'O death, where is your victory? O death, where is your sting?' [Hos. 13:14] The sting of death is sin, and the power of sin is the law. But thanks be to God, who gives us the victory through our Lord Jesus Christ" (vv. 53–57).

Here we learn unambiguously that Christ's triumph over death refers primarily to his victory over *physical* death.

God "saved us and called us to a holy calling, not because of our works but because of his own purpose and grace, which he gave us in Christ Jesus before the ages began, and which now has been manifested through the appearing of our Savior Christ Jesus, *who abolished death* and brought life and immortality to light through the gospel" (2 Tim. 1:9–10). Christ partook of blood and flesh, "that through death he might destroy the one who has the power of death, that is, the devil, and deliver all those who through fear of death were subject to lifelong slavery" (Heb. 2:14–15).

"And the sea gave up the dead who were in it, Death and Hades gave up the dead who were in them, and they were judged, each one of them, according to what they had done. Then *Death and Hades were thrown into the lake of fire*. This is the second death, the lake of fire" (Rev. 20:13–14). God "will wipe away every tear from their eyes, and *death shall be no more*, neither shall there be mourning, nor crying, nor pain anymore, for the former things have passed away" (21:4). Physical death will be destroyed by being thrown into the lake of fire, which itself is called "second death," that is, everlasting perdition for the wicked.

In 1 Corinthians 15:22, Adam and Christ are being compared only with regard to the matters of (physical) death and (physical) resurrection. But in Romans 5 it is the problem of sin that is being dealt with in particular. Through Adam's trespass, sin entered the world, and through sin came the judgment upon sin, namely, death. Over against this, we find Christ's act of righteousness—his work of redemption—through which there is now

God's grace and righteousness for all who believe in Christ. The *one* act of disobedience by Adam—eating of the forbidden tree—led to dramatic consequences (death, condemnation) for "the many," that is, the physical progeny of Adam. The *one* act of gracious obedience by Christ—"obedient to the point of death, even death on a cross" (Phil. 2:8)—led to glorious consequences (life, justification) for "the many," that is, the spiritual family of Christ, believers.

Through Adam's one sin, death spread to all his descendants (Rom. 5:12).[24] As we read in verse 14, "Death reigned from Adam to Moses, even over those whose sinning was not like the transgression of Adam." Here is how I and others understand this. Adam had transgressed a specific and explicit commandment of God (Gen. 2:17). But after Adam, and until the arrival of the Mosaic Law, many people sinned as well, even without transgressing specific explicit divine commandments.[25] Sin, as acting against God's will, is always sin, even if God has not (yet) delineated his will by means of specific commandments. But sin acquires the character of "trespass" only if explicit commandments (here, the Mosaic Law) are violated. Incidentally, the opposite is true as well: "When Gentiles, who do not have the law, by nature do what the law requires, they are a law to themselves, even though they do not have the law. They show that the work of the law is written on their hearts" (Rom. 2:14–15).

10.3.3 TYPE AND ANTITYPE

The contrast between Adam and Christ is remarkable: people fall under the judgment of death because they are sinners, even if they have not sinned the way Adam did—and people are declared righteous, even if they have not acted righteously the way Christ did. In both cases, the act of one person has enormous consequences for a multitude that is described as "the many" (Rom. 5:19). In this respect, Adam is a "type" (Gk. *typos*) of Christ (5:14c).[26] But although Paul does not mention this term, Adam is also an

---

[24] Was Enoch an exception? "Enoch walked with God, and he was not, for God took him" (Gen. 5:24). Heb. 11:5 explains, "By faith Enoch was taken up so that he should not see death, and he was not found, because God had taken him." But some of the rabbis believed that God took Enoch away through death; see Cohen (1983, 24). Enoch did not necessarily experience an "ascension," unlike Elijah, who did (2 Kings 2:1, 11).

[25] *Contra* van den Brink (2017, 240), who suggests that in Rom. 5:13 the "law" is some "moral law." Even less acceptable is the view of J. H. Walton in Barrett and Caneday (2013, 240), who asks: "What about Paul's statement that 'without the law there is no sin'?" I could not find such a verse. Presumably, the mistake originated from misunderstanding Rom. 5:13.

[26] A "type" is a kind of "model," says G. Haslam in Nevin (2009, 63), but not just an ahistorical "teaching model," as Versteeg (2012) emphasizes.

"antitype" (Gk. *antitypos*) of Christ,[27] expressed by the phrase "much more" in verses 15 and 17. That is, through his act of atonement Christ has gained immeasurably more than Adam lost through *his* act of falling into sin.[28]

In verse 16, Paul is no longer speaking about physical death only, but also about condemnation, or eternal judgment ("damnation," Rom. 3:8; 13:2; 14:23 KJV). For this verdict of condemnation that comes upon all sinners, one evil act was enough; but the justification that Christ brought about through his one act of redemption involves *all* the sins of those who believe. In my view, this explains the contrast in verse 16a: after Paul had offered a parallel in verse 15, he now says, "the free gift is not like the result of that one man's sin"—God's gift in Christ accomplished immeasurably more than Adam's one evil act.

The thread of verse 12 resumes in verse 18, but at the same time a summary of verses 15–17 is given. Here, the meaning of the Greek term *dikaiōma* is important: in verse 16, it means "justification," but in verse 17 it stands in contrast to "trespass" and thus refers to the "righteousness," that is, righteous act, of Christ. This refers to the "righteous work" of Christ, not to the "righteous life" of Christ (his law-keeping), as some have asserted. Similarly, the term "obedience" in verse 19 does not refer to the law-keeping life of Christ but to his obedient self-surrender in death (cf. John 10:17–18; Phil. 2:7–8; Heb. 5:8–9).[29]

In summary, we put this question to AEH advocates: Why did Jesus Christ have to die on the cross if human physical death is not a consequence of sin, but instead a purely natural phenomenon? Why would Jesus have to die to take away something that is just as much part of my being human as my character and appearance? Paul is clear on this: Adam introduced death as an evil reality, Jesus vanquished death *because* it is an evil reality and not a natural reality. James 1:15 says, "Desire when it has conceived gives birth to sin, and sin when it is fully grown brings forth death." Paul writes, "The wages of sin is death" (Rom. 6:23), and "The sting of death is sin" (1 Cor. 15:56). As John Mahoney asked,

> If the first man is not historical and the fall into sin is not historical, then one begins to wonder why there is a need for our Lord to come and undo the work of the first man. In other words, apart from an affirmation of the historical Adam and his fall into sin on behalf of

---

[27] We find this term in 1 Pet. 3:21 (NKJV), "There is also an antitype which now saves us—baptism."

[28] Cf. Ouweneel (2008, §8.4.1).

[29] See Ouweneel (2018d, chapter 4).

the entire human race, the entire rationale for the plan of redemption and the sending of God the Son begins to slip through our fingers.[30]

Until now, I have seen no argument from any AEH advocate that could change this picture. The moment we assert that human death is a natural reality, a natural part of human life, we affect Christ's work of redemption.[31] The same Lamb of God who takes away the sin of the world (John 1:29), through the same work of redemption takes away death as well: "Truly, truly, I say to you, if anyone keeps my word, he will never see death" (8:51). In the AEH paradigm, death is life's natural partner; in the biblical paradigm, death is life's enemy that must be defeated: "The last enemy to be destroyed is death" (1 Cor. 15:26).

10.4 UNHISTORICAL ADAM

10.4.1 DENIS LAMOUREUX

Because they struggle with the notion of the historical Adam, AEH advocates have repeatedly tried to cast doubt on the historicity of the Adam as he is described in Romans 5. For instance, I vigorously disagree with the assertion of British theologian James Dunn concerning Paul's argument in Romans 5: "The effect of the comparison between the two epochal figures, Adam and Christ, is not so much to historicize the individual Adam as to bring out the more than individual significance of the historic Christ."[32] This statement contains little that is sensible. First, *of course*, Paul is not aiming at "historicizing the individual Adam" because Adam *was* already historical. He simply assumed the historical individual Adam, like all Jews and Christians of his time did. There was not the slightest need to bring out Adam's historicity in a special way. Second, everything that Paul says about the "individual significance of the historic Christ" (why the word "historic" here?) is parallel to similar aspects of the individual significance of the historic Adam.

Canadian theologian Denis Lamoureux made a claim similar to that of Dunn about Romans 5:12.

---

[30] Mahoney (2011, 76).

[31] Cf. C. J. Collins in Barrett and Caneday (2013, 164).

[32] Dunn (1988, 290); cf. Barrett (1968, 351): "Sin and death, traced back by Paul to Adam, are a description of humanity as it empirically is. For this reason the historicity of Adam is unimportant." A remarkable *non sequitur*!

> The context and intention of these passages are not debating the historicity of Adam.... Moreover, it must be remembered that Paul employs the science-of-the-day in his inspired letters. He used the 3-tier universe[33] in one of the most significant passages in the New Testament—the kenosis of Jesus (Phil. 2:5–11). Consistency demands that since this apostle holds an ancient understanding of the structure of the universe, then he undoubtedly accepted an ancient view of human origins—*de novo* creation.[34]

This is an astonishing statement for several reasons.

(a) Notice that two different views are actually intertwined here. One view is that for Paul the question of Adam's historicity is not very relevant (as Dunn also suggested). The other view is that Paul *may* find Adam's historicity very relevant but we need not take him very seriously here because Paul's idea is just part of his "ancient understanding of the structure of the universe."

(b) Lamoureux calls Paul's worldview "ancient science,"[35] but in fact it has nothing to do with science at all (cf. §3.5). In practice, we all still function with the 3-tier view of our world when we say, for instance, that God dwells "above" us (see the examples given in §10.6.1). It is part of our everyday observational worldview, which is why we still speak of sunrise and sunset.

(c) Jesus used the same 3-tier view of our world, too, as we will see. Is Lamoureux prepared to say the same about him? Namely, since Jesus "holds an ancient understanding of the structure of the universe, then he undoubtedly accepted" ancient, outdated views of origins. Lamoureux asks, "Are we to trust Calvin and Luther regarding the structure of the heavens and earth? No. Why then should we trust their sixteenth-century understanding of human origins?"[36] But then, why would we trust Luther and Calvin when they give us their sixteenth-century understanding of the justification by faith? Moreover, are we to trust *Jesus* regarding the size of seeds, the "death" of a seed in the ground, and of stars falling from the sky?[37]

---

33 This worldview sees the universe as consisting of three levels: what is above the earth, the earth itself, and what is under the earth; see Greenwood (2015, especially chapters 2 and 3).

34 Lamoureux (2008, 274; cf. also 2009, 84, 141); cf. Lamoureux in Barrett and Caneday (2013, 37–65).

35 In Barrett and Caneday (2013, 50); he also said that Paul "accepted an ancient biology of the origin of life" (62); no, Paul's view was not *biology* at all.

36 In Barrett and Caneday (2013, 88); he follows the same reasoning with respect to Paul (61–62).

37 The examples are those of Lamoureux himself; see in Barrett and Caneday (2013, 60); cf. Mark 4:31; John 12:24; Matt. 24:29.

No? But then, why should we trust Jesus' first-century understanding of eternal salvation? Why do AEH advocates frequently refer to Paul's ancient, outdated ideas—or Luther's and Calvin's, for that matter—and never to Jesus' own ancient, outdated ideas? The answer is not difficult to guess. Lamoureux tells us about what "consistency demands"; let him, then, be consistent in telling us whether his criticism of Paul, Luther, and Calvin applies to Jesus as well.

(d) No biblical doctrine is dependent on whether the 3-tier model is scientifically correct or not. The situation here is the same as with the Galileo conflict: no subject in systematic theology depends on whether the earth turns around the sun or the other way round, or whether the earth is flat or spherical.[38]

(e) However, in Romans 5, Paul's argument certainly depends on whether there was a historical Adam, directly created by God.[39] If there were no historical Adam (in the ordinary sense of Gen. 1–3), then why would there have been a historical Christ? If we cannot trust Paul when it comes to the historical veracity of Genesis 1–3, why should we trust him when it comes to the historical veracity of salvation in Christ? As Michael Reeves has argued, "With a mythical Adam, then, Christ might as well be...a symbol of divine forgiveness and new life [and nothing more]. Instead, the story Paul tells us of a *historical* problem of sin, guilt and death being introduced into the creation, a problem that required a *historical* solution."[40] He continued,

> The historical reality of Adam is an essential means of preserving a Christian account of sin and evil, a Christian understanding of God, and the rationale for the incarnation, cross and resurrection. His physical fatherhood of all humankind preserves God's justice in condemning us in Adam (and, by inference, God's justice in redeeming us in Christ) as well as safeguarding the logic of the incarnation. Neither belief can be reinterpreted without the most severe consequences.[41]

---

[38] J. H. Walton in Barrett and Caneday (2013, 69) used this argument with respect to Jesus' "unscientific" sayings.

[39] Cf. Wilson (1977, 86): "For those modern scholars who regard the Fall as a mythical statement of what is nevertheless a valid religious truth, Adam becomes a sort of religious Everyman. It would be frivolous to reproduce such a view as an exposition of the apostle's teaching in this place."

[40] In Nevin (2009, 45); italics added.

[41] In ibid., 56.

Both Dunn and Lamoureux treat the matter of historicity with shocking superficiality—for no other apparent reason than to accommodate the theory of evolution. This is the heartbreaking tragedy of the AEH position: its advocates pay far too high a price to purchase something without lasting value.

10.4.2 PETER ENNS

Similar ideas are defended by Peter Enns in his book with the pregnant title: *The Evolution of Adam*. Enns is convinced that the apostle Paul himself believed that Adam was a historical figure, but this was because Paul lived in antiquity; he did not know any better.[42] Enns asserts that Paul distorted the meaning of Genesis to adapt it to his own doctrine, "reworking the past to speak to the present."[43] He also tells us that God "adopted mythic categories" from the ancient world. Today, argues Enns, we may freely discard these myths, as long as we retain the kernel of truth they contain (whatever this may be).[44] But if this is so, why not, by the same argument, discard the "myth" of the resurrection of Jesus? Why is *this* not a piece of ancient science or ancient mythology? (Antiquity knows many resurrection myths, all of which are discarded by modern science.) Where does Enns draw the line? Joel Beeke comments, "These are clear and sobering examples of how denying the reality of Adam [in the biblical sense] puts one on a trajectory to deny the full trustworthiness of the Holy Scriptures."[45]

Again, I wonder whether Peter Enns, who believes in Jesus' resurrection, would dare to say the same about Jesus: Were Jesus' views also just a product of antiquity? Was he also trapped in ancient beliefs about the origin of the world and of humanity? Did he also distort the meaning of Genesis to adapt it to his own doctrine? Did he not know the historical and doctrinal truth about creation and the Fall? But it was in, through, and for *him* that God created this very world (Col. 1:16). With a variation upon Genesis 18:25 (MEV, "Should not the Judge of all the earth do right?"), I would like to ask, Should not the Creator of all the earth know the truth about his own creation? Or to put it another way, would Jesus be able to rise from the dead (Matt. 28:7; John 10:18; 1 Thess. 4:14), but not be able to tell the truth about Adam and the Fall?

The key questions I wish to ask Enns, as well as Lamoureux,[46] are these: If what Paul has to say about Adam and the Fall is due to his ancient, and

---

[42] Enns (2012, xvi–xvii, 139).
[43] Ibid., 113; cf. 82.
[44] Enns (2005, 53).
[45] J. Beeke in Phillips (2015, 38).
[46] Lamoureux (2008; 2009).

therefore outdated, views,

(a) then why does this same argument not apply to Jesus as well?

(b) then why could Paul's views on salvation not be due to his ancient, and therefore outdated, views as well?

What *a priori* hermeneutical rule do we employ for determining which parts in Paul's thinking are ancient and outdated, and which are not? In the light of modern science, the idea of a bodily resurrection is just as absurd as, if not more absurd than, the idea of a historical Adam.[47] Why should we believe Paul on the former but not on the latter point?[48] Is this not purely arbitrary? My fear is that eventually people like Enns and Lamoureux, or at least their followers, will acknowledge the validity of my question—and abandon the notion of a bodily resurrection as well.

Michael Reeves commented that

> it would be the height of rhetorical folly for Paul to draw a parallel between Adam and Christ here [i.e., in 1 Cor. 15] if he thought Adam was not a historical, but a mythical figure. For if the two could be paralleled, then Christ's resurrection could be construed mythically [as well], and then Paul's entire letter [1 Cor.] would lose its point, purpose and punch.[49]

To this I would add that the situation would be still worse if Paul himself erroneously believed that the mythical Adam was a historical figure; if he were wrong on Adam, he could also be very wrong on many other things, such as the resurrection, salvation, and the second coming of Christ.

10.4.3 CELIA DEANE-DRUMMOND

The powerful impact that words can have is shown by British theologian Celia Deane-Drummond, who states, "A prehistorical figure [i.e., Adam] cannot be compared easily with the historically real figure of Christ, and inasmuch as the effects of Christ are transhistorical, they are transhistorical in a different manner compared with Adam's prehistory."[50] This is an interesting playing with words. Why is Adam prehistorical, Christ historical, and the effects of his work transhistorical? Is this because Christ's existence can be verified by historical sources (which some liberals would deny anyway), and Adam's existence cannot? The term "prehistorical," invented by

---

47 This observation comes from Shapiro (2016).

48 Cf. R. B. Gaffin in Versteeg (2012, xvi–xxiv).

49 In Nevin (2009, 46).

50 C. Deane-Drummond in Cavanaugh and Smith (2017, 47).

historians to describe the time in human history before the invention of a writing system, has no place here. Human history begins with the first humans on earth; it begins in Genesis 1:27. It is an egregious prejudice to call Adam "prehistorical," thereby to assign him to a dark period when saga and myth still dominated.

As a term, the use of "protohistorical" would be less problematic because that term literally indicates only that we are dealing here with the first part of human history, which is substantively undeniable. Yet, we must be careful with this term because, just as with the term "prehistorical," it may create the suggestion of something less reliable, something less historical than what is ordinarily historical.

Deane-Drummond is an AEH advocate whose language clearly illustrates how her ideological colleagues struggle with the term "historical." She writes, "My own view is that the significance of Genesis is 'historical,' but without implying literal figures of Adam and Eve or a literal paradisiacal state before the Fall."[51] So Genesis 1–3 is "historical," but without the well-known historical persons and a historical garden? Is it history without history? This is like saying, I believe in American history, but without George Washington and Abraham Lincoln. This entire position seeks to retain the theory of evolution while suggesting a very special kind of "historicity" for Genesis 1–3, perhaps to reassure traditional Christians. Who is deceiving whom here? I prefer the outright honesty of the liberal theologians, who did not call these stories "historical" but plainly identified them as myth.[52]

10.5 "BECAUSE ALL SINNED"

10.5.1 *EPH' HŌI*

In what way did the sin of Adam have consequences for his descendants? In other words, what is the meaning of the expression "because all sinned" in Romans 5:12b? Throughout the centuries, this one clause has been explained in various ways, sometimes with profound consequences for systematic theology. Humanly speaking, we might regret that Paul interrupts his own argument here[53] (some have even viewed the entire passage in vv. 13–17 as a parenthesis), and that he does not express himself a little

---

[51] Ibid., 36.

[52] One famous example from the discipline of biology was Ayala (1995; cf. 2009 for his Roman Catholic anti-creationism).

[53] Grammarians describe this as an example of *anacoluthon*, which refers to a sentence or construction that lacks grammatical sequence.

more clearly. When it comes to the various interpretations of the phrase "because all sinned" (Gk. *eph' hōi pantes hēmarton*), we can distinguish among and arrange six options.[54]

1–3. In the phrase *eph' hōi* (lit., "on which/whom"), *hōi* is a masculine pronoun:
Option 1. *hōi* refers back to "death," more or less as follows: "death…in which all sinned";
2–3. *hōi* refers back to "one man":[55]
Option 2. *epi* = more or less *en*, "in": "one man…in whom all sinned";
Option 3. *epi* = "through, because of": "one man…through (or, because of) whom all sinned."
4–6. *Eph' hōi* = "because":
Option 4. *hēmarton* means: all people participate in Adam's sin ("because [in Adam] all sinned");
5–6. *hēmarton* means: all people have personally sinned in their lives:
Option 5. namely, just like (or, according to the example of) Adam ("as is evident from the fact that all sinned"; thus, people's own responsibility is involved);
Option 6. namely, as a consequence of original sin, inherited from Adam ("because [through original sin] all are sinners"; here, people's own responsibility is not involved).

All these varieties have had their respectable representatives;[56] I would mention that Augustine defended Option 1,[57] John of Damascus defended Option 3, Albrecht Bengel defended Option 4,[58] Pelagius defended Option 5, and Calvin defended Option 6. Some defended a combination of

---

54 The clearest surveys, which I am presenting here in my own way, are to be found in Cranfield (1975, 274–81) and Medema (1985, 78–79).

55 Cf. the Vulgate: *per unum hominem…in quo omnes peccaverunt.*

56 For complete references, see Ouweneel (2008, §11.1.2).

57 In *De peccatorum meritis et remissione* I.10.11, he translates in *quo* (thus the Vulgate) as "in whom/which," which he explains as "in sin" (but *hamartia* is feminine), or "in Adam"; see Cranfield (1975, 276).

58 Bengel (1862, 516); his is the well-known statement: "all have sinned in the sinning Adam" (Lat. *omnes peccarunt Adamo peccante*).

Options 5 and 6.[59] Harrison did not wish to distinguish sharply between Options 4, 5, and 6.[60]

The idea that we have sinned "in Adam," or more broadly, that a person may experience certain things "in" a remote ancestor, is itself not foreign to the Bible. Thus, we read, "Levi himself, who receives tithes, paid tithes through Abraham, for he was still in the loins of his ancestor when Melchizedek met him" (Heb. 7:9–10; see §§4.6.3 and 9.5.2).[61] When Abram gave tithes to Melchizedek, Levi gave them too, as it were, for he was in the "loins" of his forefather Abraham. "In Abraham" Levi gave tithes to, and received blessing from, Melchizedek. In a similar way, we might imagine that "in Adam" we ate of the tree of knowledge, or more broadly, when Adam ate, this had consequences for all his offspring. This notion is entirely supported in the rest of Romans 5 (vv. 15–19): what Adam did had consequences for "the many" who belong to him (i.e., his natural descendants), just as what Christ did had consequences for "the many" who belong to *him* (his spiritual family).

10.5.2 EVALUATION

Elsewhere I have given a more extensive evaluation of the various views.[62] Here I will provide my conclusion. In my view, Paul wishes to avoid the thought that God would be cruel or unrighteous if he were to let us bear the consequences of Adam's sin. People are not condemned only because they have sinned "in Adam" but also because they added their own sins to this sin. They are neither responsible for Adam's sin, nor for inheriting Adam's sinful nature. But they *are* responsible for what they did and do with this sinful nature (at least from the age when they know their right hand from their left; cf. Jonah 4:11), for the sins that were and are produced by this sinful nature. Sins are the evil things we think, say, and do. *We* do them—we cannot blame our sinful nature as if sins were the involuntary, automatic products of our sinful nature.[63] That would be a denial of our very humanity. Humans must take, and do take, responsibility for what they think, say, and do; taking responsibility belongs to being human.

Animals (including the supposed early hominids) did and do not have such responsibility—only humans do. This responsibility was not the product of

---

[59] Kelly (n.d., 7:134); Grant (1901, 217); Cranfield (1975, 278–79); Hendriksen (1980, ad loc.); Medema (1985, 79).

[60] Harrison (1976, 62).

[61] Cf. Berkouwer (1971, 441–43); Robinson (1982, 296–97).

[62] Ouweneel (2008, §11.1.3).

[63] A striking example of this attitude is Aaron: "So I said to them, 'Let any who have gold take it off.' So they gave it to me, and I threw it into the fire, and out came this calf" (Exod. 32:24). In other words, I could not help it.

human evolution—how should one imagine such a process?—but of Adam's creation in the image of God (*contra* AEH advocates). Similarly, neither was our sinful nature the product of human evolution; it was the product of Adam's Fall (again *contra* AEH advocates). At a certain moment in time and space, sin entered this world; no AEH approach can do justice to this dramatic event, no matter what Smith, van den Brink, and others have proposed. Here, "sin" is not a sinful act but a Satanic power. Beginning with Romans 5:12, Paul's subject is no longer so much our sinful acts but rather sin as a dark power that has entered the world, and that is manifested in sinful human nature. Sin entered through the doorway of Adam's Fall.

Therefore, Calvin may have expressed himself too strongly in saying, "this sinning [in Rom. 5:12c] means: being corrupt and criminal,"[64] but this aspect is certainly included in it. As the Heidelberg Catechism puts it (Q/A 7), "7. *From where, then, does this depraved nature of man come?* From the fall and disobedience of our first parents, Adam and Eve, in Paradise (Genesis 3; Rom. 5:12, 18–19), whereby our nature became so corrupt that we are all conceived and born in sin (Ps. 51:5)."[65] The message of Romans 5:12 is this: "in Adam" we have all sinned, not only because we participated in Adam's personal sin, being in his "loins," but also because we added to this our own actual sins produced by the sinful nature we inherited from Adam. This is similar to the expression "in Adam all die" (1 Cor. 15:22): people not only participate in Adam's personal death—whatever this may entail—but they also personally undergo (notice: Paul uses the present tense) physical death.

What justice can AEH advocates do to Paul's picture? The apostle Paul is convinced that, before the Fall, no humans ever sinned and ever died (actually, there *were* only two humans on earth), but after the Fall *all* humans sin, and *all* humans eventually die. Any description of the Fall must do justice to these two great facts—one prelapsarian, one postlapsarian—not just for historical reasons, but also for theological reasons.

10.5.3 AMBIGUITY

If I understand the apostle Paul correctly, he shows us, on the one hand, that we participated in Adam's sin, but on the other hand, that people are not condemned for this but rather for their actual sins. Through Adam, we received a sinful nature; through this sinful nature, we personally sin. On the one hand, we may say that, historically speaking, our misery comes

---

[64] Calvin, *Comm. Romans* (ad loc.).

[65] Dennison (2008, 2:272); italics original. See Ouweneel (2016, ad loc.).

from Adam; on the other hand, morally speaking, our misery comes from our actual sins.

Paul, who was "educated at the feet of Gamaliel" (Acts 22:3), must have been familiar with the pseudepigraphic Jewish literature,[66] in which we find the same ambivalence.[67] Thus we read in 2 Baruch 54:15 (cf. §10.5.1 Options 4 and 6), "Adam first sinned and brought untimely death upon all,"[68] but to this it adds (cf. Option 5): although this is so, "yet of those who were born from him each one of them has prepared for his own soul torment to come." And verse 19 (cf. Option 5), "Adam is therefore not the cause, save only of his own soul, but each of us has been the Adam of his own soul." 2 Baruch 17:3, 19:8, and 23:4 speak of Adam as the one who brought death over all humanity (cf. Option 4).

On the one hand, 2 Esdras [= 4 Ezra] 7:48 says, "O Adam, what have you done! For though it was you who sinned, you are not fallen alone, but we all that come of you" (cf. Options 4 and 6).[69] But on the other hand, verse 49 says, "For what profit is it to us, if there is promised us an immortal time, whereas we have done the works that bring death?" (cf. Option 5). Please notice again that all these Jewish sources agree on one point: human death has entered the world through, and because of, *Adam's sin*. In other words, it is the conviction of the Old and New Testaments, and of Jewish pseudepigraphic literature, that if the first humans had not sinned, *there would have been, and there would now be, no human death.*

The same ambiguity—we die because of Adam, *and* we die because of our actual sins—is found in the rabbinic literature. On the one hand, Rabbi Jehuda says, "You are children of the first man who has brought death upon you as a punishment, as well as upon all his descendants."[70] On the other hand, Numbers Rabbah 19:18 says that all people die because all people sin. Both are true for Paul as well. Both Paul and the later rabbis did not at all doubt that people die because they are sinners. This is in flagrant contradiction with the AEH position, which asserts that people die because they are creatures.

A Jewish writing called Testament of Abraham tells us that Adam sits at the gate of Hades, "watching with tears the multitude of souls passing

---

[66] See extensively, J. B. Green in Cavanaugh and Smith (2017, 98–105).

[67] Cranfield (1975, 280n2); see further Strack (1922, 3:227–29).

[68] http://www.pseudepigrapha.com/pseudepigrapha/2Baruch.html.

[69] http://www.sacred-texts.com/bib/apo/es2007.htm#048 (modernized spelling).

[70] Sifre on Deut. 323 (138b) on chapter 32:32. Versteeg (2012, 52) observes, "The expression 'death came upon him [Adam] and upon his descendants' is almost a stereotype in rabbinic literature," with reference to Brandenburger (1962, 60).

through the wide gate to meet their punishment, and with joy the few entering the narrow gate to receive their reward." The Jewish Encyclopedia quotes these words, and adds this rabbinic statement:

> No man dies without a sin of his own. Accordingly, all the pious, being permitted to behold the Shekinah [i.e., the glorious presence of God] before their death, reproach Adam [as they pass him by at the gate] for having brought death upon them; to which he replies: "I died with but one sin, but you have committed many: on account of these you have died; not on my account."[71]

Here we find the same ambiguity again: on the one hand, Adam is blamed for being the origin of sin and death; on the other hand, the wicked as well as the pious die because of their own sins. Here again, Adam is seen as the person through whom sin as well as death entered the human race.

Regardless of their portrait of a historical Fall, AEH advocates cannot circumvent this objection: according to AEH advocates, as well as liberal theologians, human death is natural, whereas according to the Bible, and all Jewish and Christian tradition,[72] it is unnatural. According to AEH advocates, we die because we are humans, whereas according to the Bible, and all Jewish and Christian tradition, we die because we are *fallen* humans. No matter how close to Scripture AEH advocates wish to appear, they can never obscure this major feature. They can never agree with the apostles and the rabbis, and with centuries of Jewish and Christian expositors, that *human death entered the world, not at creation, but at the Fall.*

10.6 WRAPPING UP

10.6.1 THE RELIABILITY OF JESUS AND PAUL

We have heard the testimony of Genesis and several other Old Testament passages pertaining to the creation of the first humans and their Fall. We have heard the testimony of Jesus himself, and if anyone should know about those early events, it is he who was "in the bosom of the Father" (John 1:18 NKJV), the One in, through, and for whom God created the world (Col. 1:16). He is the Son, whom God "appointed the heir of all things, through

---

[71] http://www.jewishencyclopedia.com/articles/758-adam.

[72] Cf. Wisdom 2:23 (GNT), "When God created us, he did not intend for us to die; he made us like himself" (RSV, "God created man for incorruption, and made him in the image of his own eternity").

whom also he created the world. He is the radiance of the glory of God and the exact imprint of his nature, and he upholds the universe by the word of his power" (Heb. 1:2–3). He was there when Adam and Eve were created—their creation was accomplished through him—and he was there when they fell into sin. The latter definitely did *not* happen through him, but he was the involuntary witness of it.

Isaiah told about the glory of YHWH (Isa. 6), and the apostle John tells us that "Isaiah said these things because he saw his [i.e., *Christ's*] glory and spoke of him" (John 12:41). Why not read the same into Genesis 3? It was *God* who spoke to Adam (vv. 9–19), but this is the *Triune* God; and the Triune God is manifested through the Logos (John 1:1–3). Thus, we could say with equal validity that the pre-incarnate Christ called to Adam and spoke to him. Let no one tell us that Jesus, the incarnate Son of God, did not know about Adam's creation and Fall, or knew less than AEH advocates claim can be known about it. As the Son of Man he might not know everything about the future (Mark 13:32),[73] but *the Son of God was personally present at Adam's creation and Adam's fall.* As Greg Haslam wrote, "Neither Jesus nor his apostles could be mistaken in affirming the historical accuracy of Genesis. Jesus was personally present at creation. Paul was inspired by the Holy Spirit."[74]

Given the claims of AEH advocates, what remains of the Bible's reliability due to the inspiration of the Spirit (see 2 Tim. 3:16; 1 Pet. 1:10–11; 2 Pet. 1:20–21)? Matthew, Luke, John, Paul, and so on, wrote God's words. Or are we to think that the apostle Paul was just a "man of his time," his understanding darkened by his mythical worldview, not really aware of the events of early history? This was the apostle who had never known "Christ according to the flesh" (cf. 2 Cor. 5:16), but who had seen the glorified Christ (Acts 22:18; 1 Cor. 9:1), and had been caught up to the third heaven and to Paradise (2 Cor. 12:2–3). Would this man have continued to believe in myths (cf. §10.4.1), for example, by holding to a 3-tier worldview (Phil 2:10, "in heaven and on earth and under the earth")?[75] Is God himself then also trapped in an outdated worldview when he speaks of things "in heaven above...in the earth beneath, or...under the earth" (Exod. 20:4)? If we cannot trust Paul when he speaks about Adam, how do we know whether we can trust him when he speaks about Christ?

Peter Enns made similar claims about Paul's supposed ignorance:

---

73 van den Brink (2017, 122n28) (ab)uses this verse to extend Jesus' ignorance to the ordinary things of life.

74 In Nevin (2009, 71).

75 Enns (2012, 94).

> Paul shared with his contemporaries certain assumptions about the nature of the physical reality, assumptions that we now know are no longer accurate. The real issue before us is not whether Paul shared those assumptions, but what the implications are for how we read Paul, especially his view of Adam.[76]

These implications turn out to be negative, of course: "Paul's understanding of Adam as the cause [of sin and death] reflects his time and place."[77] Thus, because of Paul's outdated 3-tier worldview, we cannot take him seriously when he speaks of Adam!

But what about Jesus himself, who also seemed to believe in that ancient, outdated 3-tier worldview? Why do BioLogos people repeatedly tell us about the scientific ignorance of Paul—and thus about his historical ignorance concerning Adam—but never about the supposed scientific and historical ignorance of Jesus? Are they afraid, because the wish to appear as orthodox gospel believing Christians, to attack the person of Jesus? But Jesus spoke of angels "descending" from heaven (John 1:51), and of himself as the One who had "descended" from heaven (3:13), and of the bread that "came down from heaven" (6:41–42, 50–51, 58).

Paul and Jesus are in good company because the apostle John also uses the expressions "in heaven," "on earth," and "under the earth" (Rev. 5:3, 13). Matthew speaks of the dove "descending" from the "heavens" (Matt. 3:16). Luke also held to an ancient, outdated science because he tells us that Jesus was "lifted up" to heaven, and the apostles followed him with their eyes (Luke 24:51; Acts 1:9–11). Jesus himself, when addressing the Father, "lifted up his eyes to heaven" (John 17:1; cf. 11:41)—and, according to AEH advocates, was totally mistaken in doing so.[78] Saying or thinking that heaven is "above" betrays a lamentably outdated worldview. And worst of all: Jesus intimated that one day he would come "on the clouds of heaven" (Matt. 24:30; 26:64). If this is not an ancient and outdated picture, then what is? So much for Jesus.

Apparently, *all* the New Testament writers, and even Jesus himself, are bound to a 3-tier picture of the universe. This is the picture that Enns, Lamoureux, van den Brink, and others use to dissuade us from taking these

---

[76] Ibid., 93–94.

[77] Ibid., 123.

[78] I cite no Old Testament references here because, from the AEH vantage point, Old Testament writers must have been bound even more narrowly to a 3-tier view of the universe (cf. in the Pentateuch alone: Gen. 6:17; 7:19; 49:25; Exod. 20:4; Deut. 4:18, 39; 5:8; 11:21; 28:23; 33:13).

Bible authorities seriously with regard to their view of the universe. And consequently, we cannot take other things seriously either, like the belief in a historical Adam. But where is the boundary here? By what hermeneutical rule is this boundary determined? Why do AEH advocates not draw any further inferences? They claim that if we accept a 3-tier universe, we must be wrong about Adam, too. Why not conclude that if the apostles were wrong about Adam, they were also wrong about the miracles of Jesus, including the resurrection? Are such miracles not part of the ancient worldview, too, and therefore outdated, since miracles have no place in modern science?

Were we to apply this same approach to AEH advocates themselves, then whenever we hear them speak about a sunrise or sunset, or use expressions like "taken up in glory" (cf. 1 Tim. 3:16), we would need to consider *them* to be proponents of an ancient, outdated worldview, and we would need to refuse to take *them* seriously in anything they write.

The truth is that Jesus and Paul knew about the historical Adam, and AEH does *not* know what to do with this same Adam. Let them be honest about it. Edward Young argued that it is far more honest to say, "Genesis purports to be a historical account, but I do not believe that account," than to say, "I believe that Genesis is profoundly true" in some theological sense, but not in any literal-historical sense.[79] Peter Enns wrote, "Admitting the historical and scientific problems with Paul's [and Jesus'!?] Adam does not mean in the least that the gospel message is therefore undermined."[80] Enns will have to forgive us for believing the opposite to be the case. If you start to meddle with the beginnings of human history, what guarantee do we have that, eventually, you will not meddle with the core and the conclusion of human history as well? For modern science, it is *all* entirely unacceptable. As Philip G. Ryken rightly observed, "To deny the historical Adam [as presented by Genesis, not as invented by AEH advocates!] is to stand against the teaching of Moses, Luke, Jesus, and Paul."[81]

10.6.2 THREE KINGS[82]

In a very general sense, the kingdom of God is simply God's general government over all created things, from the foundation of the world until eternity (cf. Exod. 15:18, "The LORD reigns [or, is King] for ever and ever"). Apart from this general providence of God, we may distinguish three beings

---

[79] Young (1976, 19).
[80] Enns (2012, 123).
[81] In Barrett and Caneday (2013, 268); also cf. J. MacArthur in Chou (2016, chapter 13).
[82] See Ouweneel (2017, chapter 5).

who may be called the only worldwide kings that the earth (or all creation) has ever seen.

(a) *The first Adam*. More specifically than God's general providence, the kingdom of God is the manifestation of his counsel to put the entire created world under the feet of the first man, the "first Adam" (cf. Ps. 8:6; 1 Cor. 15:25, 27, 45; Eph. 1:22; Heb. 2:8), and to entrust world dominion to his care (Gen. 1:28, "Fill the earth and subdue it. Rule over...every living creature"). In this book, we have devoted significant attention to Adam and Eve as God's viceroys or vice-gerents over all creation.

(b) *Satan*. Adam and Eve utterly failed in their regal task, for through their Fall they surrendered their rule over the world to the power of sin, death, and Satan (Gen. 3). This is an aspect of the Fall that is not often underscored but is very important.[83] Indeed, Satan could truly say to Jesus that all the authority and splendor of the kingdoms of this world had been "given" to him (Luke 4:5–6). These had been "given" to him by Adam or by God, and in a sense we could say by both: Adam "gave" it inadvertently, and God providentially allowed that surrender of rule. Notice that Jesus did not deny Satan's claim. On the contrary, on another occasion, he recognized that there is something in this world that can be called the "kingdom" of Satan (Matt. 12:26). Three times Jesus called Satan "the ruler of this world" (John 12:31; 14:30; 16:11). At the same time, Jesus made clear that, with his coming into the world, the power of Satan was soon to be vanquished.

(c) *The last Adam*. Thank God, Jesus could say that, through his coming into this world and his manifestation of the power of God, the kingdom of God had arrived (Matt. 12:28). Since Calvary Satan has been a sentenced rebel who will never be able to compete with this kingdom, no matter how much he is still agitating, "prowling around like a roaring lion" (1 Pet. 5:8). What the "first Adam" has ruined, the "last Adam" is restoring (cf. 1 Cor. 15:45–47; then vv. 24–28). In his hands is the "restoration of all things" (Acts 3:21 NKJV). If we look at Psalm 8 in light of Hebrews 2, this transition from the first to the last Adam is depicted beautifully: the Son of Man, under whose feet all created things are put, is no longer (the first) Adam, but: "We do see Jesus, who was made lower than the angels for a little while, now crowned with glory and honor because he suffered death, so that by the grace of God he might taste death for everyone" (Heb. 2:9).

---

[83] See Chesterton (1985, 160): Genesis shows us "that happiness is not only a hope, but also in some strange manner a memory; and that we are all kings in exile."

Please notice that Jesus' domain of power cannot possibly be smaller than that of Adam. If the first Adam had dominion over the sea, the heavens, and the earth (Gen. 1:26, 28), the last Adam cannot rule over less. Therefore, he himself could say, "All authority in heaven and on earth has been given to me" (Matt. 28:18; cf. Ps. 72:8 CEV: "Let his kingdom reach from sea to sea, from the Euphrates River across all the earth"). He is vaguely depicted in the mysterious language of Revelation 10:1–3.

> Then I saw another mighty angel coming down from heaven, wrapped in a cloud, with a rainbow over his head, and his face was like the sun, and his legs like pillars of fire.... And he set his right foot on the sea, and his left foot on the land, and called out with a loud voice, like a lion roaring.[84]

In addition to seeing Satan, who is not a human being, we again witness the enormous significance of the juxtaposition of the first and the last Adam. Alistair Donald observes,

> The parallelism of New Testament teaching is between Adam and Christ as federal representatives of humanity. In the theology being expounded by [Denis] Alexander[85] [and other AEH advocates] it is very difficult to understand how one among several millions and who had been predated by others could properly be said to represent the whole {i.e., the humanity of the time]. This is not secondary, it is central to our understanding of Christian faith. The New Testament argues that we can have confidence that we are included in the atoning work of Christ because it is evident that we are included in Adam.[86]

Joel Beeke wrote in a similar vein,

> Paul is not simply using the history of Adam as an instructive parable or a cautionary tale. He is describing the two great figures in history upon whom everything hangs. If there was no real Adam [in any truly biblical sense], then Paul's theology collapses. The apostle would then be profoundly mistaken, not just in his understanding of Adam, but in his doctrine of Christ's work. On the contrary, we believe that Paul

[84] This is the view of many commentators; cf. http://biblehub.com/revelation/10-1.htm.
[85] Alexander (2014).
[86] In Nevin (2009, 23).

was inspired by God,[87] an apostle whose message did not come from man, but was revealed to him by Christ himself (Gal. 1:12).[88]

10.6.3 THE NEW CREATION

The kingdom of God is one of several great truths that show us the profound importance of the first chapters of Genesis. Too many Christians limit their Bible reading to mainly the New Testament, and might even believe they are following the apostle Paul in wishing to "know nothing except Jesus Christ and him crucified" (1 Cor. 2:2). Such people are not likely to read books like this one. But too many people who are genuinely interested in the doctrines of Scripture often tend to neglect the Old Testament. This is why chapters like Romans 5 and 1 Corinthians 15 (and many more—think of the entire book of Hebrews) are so important, because they build a bridge between the very first chapters of the Bible and New Testament doctrine.

Greg Haslam observes that "The NT endorses the accuracy of Genesis directly and indirectly over *200 times*, and cites Genesis 1–11 *107 times*. Jesus refers to Genesis twenty-five times to reinforce important doctrines."[89] These numbers are certainly impressive. If, for many evangelical Christians, the Gospels are far more important than the Old Testament, they should at least realize how foundationally important Genesis was to Jesus and to the New Testament writers in general.

Consider the concept of the "new creation" in the New Testament: "If anyone is in Christ, he is a new creation. The old has passed away; behold, the new has come" (2 Cor. 5:17). "For neither circumcision counts for anything, nor uncircumcision, but a new creation" (Gal. 6:15). Consider as well some closely related verses on the "new man" (NKJV): "you put on the new man which was created according to God, in true righteousness and holiness" (Eph. 4:24); you "have put on the new [man] who is renewed in knowledge according to the image of Him who created him" (Col. 3:10). What sense would it make to speak of a new creation if we do not have a clear picture of the old one? If we think that the old creation was the product of an evolutionary process that took billions of years, what about the new creation? Or conversely, if Saul of Tarsus—and millions of people after him—could be changed in one moment from an old creation into a new creation, simply by the Word of God, through regeneration and the

---

[87] This is a mild inaccuracy: the Bible does not say that the Bible *authors* were inspired but that *Scripture* was inspired (God-breathed; 2 Tim. 3:16).

[88] J. Beeke in Phillips (2015, 36).

[89] In Nevin (2009, 58).

renewal of the Holy Spirit, why could this Word and Spirit not have done the very same thing in Genesis 1, namely, call forth each part of the creation at once?

To state the matter in one simple sentence: If conception in the womb occurs in a moment, if regeneration occurs in a moment, if the new creation occurs in a moment, if Jesus' creation of new wine occurred in a moment (John 2:3–10), then the old creation also occurred in a moment (cf. Ps. 33:6, 9).

Paul verbalizes a real gospel comparison here: "God, who said, 'Let light shine out of darkness,' [Gen. 1:3] has shone in our hearts to give the light of the knowledge of the glory of God in the face of Jesus Christ" (2 Cor. 4:6). How much time was needed for God to call forth the light in the midst of the darkness? Just as long as it takes him to bring a person from darkness into the light of Christ (cf. Acts 26:18; Col. 1:12–13; 1 Pet. 2:9). How much time was needed for God to prepare the first creation for the appearance of the first humans? The time he took was equivalent to the time it will take him to remove the old heaven and earth (Rev. 20:11), and to introduce the new (21:1)—a creation in which the Son will be subjected to the Father, and God will be all in all (1 Cor. 15:28).

If for God "a thousand years are but as yesterday" (Ps. 90:4; cf. 2 Pet. 3:8), these are the years that God's impatient, suffering people must wait before the time of their redemption arrives. The last two thousand years of church history are "a long time" in the language of Jesus' parable (Matt. 25:19). But the point is that God thinks in thousands of years. In his redemptive history there is no place for billions of years. Those who wish to introduce them are violating biblical history; biblical history cannot absorb such a time shock. The Bible knows of an eternity from before the foundation of the world (cf. John 17:24; Eph. 1:4; 1 Pet. 1:20; Rev. 13:8), and of an eternity after the "last day" (cf. John 11:24; cf. the "forevers" in Rev., e.g., 1:6; 5:13; 7:12; 11:15; 14:11; 19:3; 20:10; 22:5). However, Scripture does not know of an "eternity" in between, but at most of some 6,000–10,000 years.

So what shall we do? Hold to the Word of the Most High. "All Scripture is breathed out by God and profitable for teaching, for reproof, for correction, and for training in righteousness, that the man of God may be complete, equipped for every good work" (2 Tim. 3:16–17).

"O Timothy, guard the deposit entrusted to you. Avoid the irreverent babble and contradictions of what is falsely called 'knowledge,' for by professing it some have swerved from the faith. Grace be with you" (1 Tim. 6:20–21).

Do not overlook this one fact, beloved, that with the Lord one day is as a thousand years, and a thousand years as one day. The Lord is not slow to fulfill his promise as some count slowness, but is patient toward you, not wishing that any should perish, but that all should reach repentance. But the day of the Lord will come like a thief, and then the heavens will pass away with a roar, and the heavenly bodies will be burned up and dissolved, and the earth and the works that are done on it will be exposed.

Since all these things are thus to be dissolved, what sort of people ought you to be in lives of holiness and godliness, waiting for and hastening the coming of the day of God, because of which the heavens will be set on fire and dissolved, and the heavenly bodies will melt as they burn! But according to his promise we are waiting for new heavens and a new earth in which righteousness dwells (2 Pet. 3:8–13).

APPENDIX 1
TWELVE APPROACHES

Here is a summary of twelve evolutionist and creationist approaches to Genesis 1–3.[1]

1 – PURELY EVOLUTIONIST

This view could be naturalist or agnostic or atheist evolutionism: acceptance of the theory of general (particles-to-people) evolution, rejection of any supernatural influences. Genesis 1–3 is mythical.

PURELY CREATIONIST

This category entails the acceptance of the literal-historical meaning of Genesis 1–3, and rejection of the theory of general evolution. Here are several varieties of creationist views:

2 – YOUNG EARTH CREATIONISM

This entails the belief that the earth, and thus also humanity, are young (perhaps 6,000–10,000 years); the creation days were ordinary days of twenty-four hours.

OLD EARTH CREATIONISM

This sub-category entails the belief that the earth is old (billions of years); humanity may also be (much) older than 6,000–10,000 years but not

[1] See especially Rau (2012); I am reworking his material.

necessarily so. The earth's old age may be accounted for as follows:

3 – THE GAP THEORY

This view holds that there is a gap between the creation (Gen. 1:1) and the recreation of the world (Gen. 1:2–2:3). The geological ages belong to this gap.

4 – THE INTERMITTENT DAY VIEW

In this view, the creation days were ordinary days of twenty-four hours but were separated by long periods (the geological ages).

5 – THE DAY-AGE CREATION VIEW

Here, each creation "day" is viewed as a long period of time, like a geological age.

6 – THE FIGURATIVE DAY VIEW

The "days" in Genesis 1 are to be taken figuratively, as an aspect of the literary composition that this chapter constitutes.

HYBRID VIEWS

This category entails commitments drawn from both creationist views (involving belief in divine creation) and evolutionist views (involving accepting the theory of general evolution; terms used include *theistic evolutionism* as well as *evolutionary creationism*). Here are several varieties:

7 – DEISM

This view entails the belief that God created the world but let it run on its own without his interference.

PLANNED EVOLUTION

Within this category, people view Adam and Eve not as literal individuals; the supposed process of general evolution was initiated and guided by God, and led to a specific goal. There are two varieties.

8 – ADAMIC SYMBOLISM

This view holds that Adam and Eve were purely figurative.

ADAM AND EVE CAME FROM A GROUP OF HOMINIDS (AEH)

These are the varieties in this sub-category.

9 – AEH LONG

Adam and Eve came from a group of hominids some hundreds of thousands of years ago.

10 – AEH SHORT

Adam and Eve came from a group of hominids some tens of thousands of years ago (i.e., more within a biblical scope of time).

DIRECTED EVOLUTION

This category contains views of Adam and Eve as literal individuals: the supposed process of general evolution was initiated and guided by God, and led to a specific goal. These are the varieties.

11 – DIRECTED EVOLUTION LONG

Adam and Eve lived many hundreds of thousands of years ago.

12 – DIRECTED EVOLUTION SHORT

Adam and Eve lived many tens of thousands of years ago (i.e., more within a biblical scope of time).

Within the categories and sub-categories identified above, I distinguish a total of twelve different views, without at all claiming that this list is exhaustive. Please note the following.

1. Views 8–12 would all fit the broad and varied BioLogos spectrum of views.[2]
2. Only view 1 is purely evolutionist; advocates of views #8–12 usually dislike the label of "evolutionist."
3. Only view 2 is purely creationist in the narrow sense that this term has acquired in the twentieth century. However, old earth creationism has also claimed for itself the term creationism for decades. Original creationism always involved the rejection of the theory of general evolution; in this sense, the term "old earth creationism" is correct, whereas the term "evolutionary creationism" is confusing.

---

[2] See extensively Haarsma and Haarsma (2011).

4. Only views 2–6 accept a historical Adam according to the traditional reading of Genesis 1–3.
5. Views 11–12 also accept a kind of historical Adam (which renders view 11 quite suspect, of course) as part of a synthesis between Genesis 1–3 and modern science.
6. The position of AEH (an acronym used throughout this book) in the stricter sense entails views 11 and 12; in the wider sense it entails views 9–12.

Let me conclude my stating my own preference. I prefer view 2, although my arguments in this book leave room for views 3–6. Nothing in this book compels people to choose among these specific views.

APPENDIX 2
CONCORDISM AND PERSPECTIVISM

Basically there are two varieties of these confusing "-isms":

CONCORDISM

This view holds that there is "concord" between biblical statements and the external world.

PERSPECTIVISM

This view holds that, in biblical statements, it is the perspective that matters, not the concrete literal-historical form.

Applied to Genesis 1–3, this results in the following views.

#1 — LITERAL(ISTIC) CONCORDISM

Genesis 1–3 corresponds word for word with real persons and real events from the early history of the cosmos and humanity.

Positive aspect: understands Genesis 1–3 literally.
Negative aspect, from the standpoint of views #2 and #3: understands Genesis 1–3 in a literalistic way, not (sufficiently) allowing for metaphors, anthropomorphisms, and literary composition techniques, as well as minimizing the faith message.

#2 – MEDIATING FORM: HISTORICAL CONCORDISM OR MODERATE PERSPECTIVISM
Genesis 1–3 is basically historical.

Positive aspect: takes Genesis 1–3 literally, but allows for metaphors, anthropomorphisms, and literary composition techniques, and emphasizes the faith message.
Negative aspect, from the standpoint of view #1: does not understand Genesis 1–3 literally enough.
Negative aspect from the standpoint of view #3: understands Genesis 1–3 too literally.

#3 – EXTREME PERSPECTIVISM
Accepts only the faith message of Genesis 1–3, not (all) the literal historical aspects.

Positive aspect: concerned with the faith message of Genesis 1–3, not with the supposed literal-historical framework.
Negative aspect from the standpoint of views #1 and #2: (largely) rejects the literal-historical contents of Genesis 1–3, and in this way in fact also loses the faith message.

In addition to this, we encounter *scientism*, the view that overestimates the scientific (theoretical) way of observing and thinking at the expense of the everyday (non-theoretical, practical) way of observing and thinking.

One specific form of *scientism* arises when ordinary observational, practical statements about nature, the cosmos, and history are mistaken for scientific statements.

- These statements arise from ancient science if they involve an ancient, and thus outdated worldview.
- These statements arise from modern science if they involve contemporary, and thus supposedly trustworthy scientific models and theories.

By contrast, with respect to real *scientists*:

- those holding view #1 interpret biblical statements in Genesis 1–3 as modern scientific statements, and therefore accept them literally.
- those holding view #3 interpret biblical statements in Genesis 1–3 as ancient science, and therefore reject them (and they reject views #1 and #2 as well).

APPENDIX 3

PARALLELS BETWEEN ADAM, NOAH, ABRAHAM, ISRAEL, AND THE NEW TESTAMENT CHURCH

The objective of this Appendix is to underscore parallelisms between Adam, Noah, Abraham, Israel, and the church. If Abraham, Israel, and the New Testament church are historical, then so too are Adam and Noah.

	ADAM	NOAH	ABRAHAM	ISRAEL	NT CHURCH
1	Selected from all creatures (Gen. 1)	Selected from all humanity (Gen. 6)	Selected from all nations (Gen. 12:1–3)	Selected from all nations (Exod. 19:3–6)	Selected as a people from among the Gentiles (Acts 15:14; Titus 2:14)
2	His story begins with dry land drawn from the water (Gen. 1:6–7)	His new story begins with dry land drawn from the water (Gen. 8)	Described as the one coming from beyond the River (Josh. 24:3) (this is one reason why he may be described as "Hebrew": coming from "beyond")	Passed through the Red Sea (Exod. 14) and the Jordan River (Josh. 3)	Its origin is similar to that of Israel: "baptized in the cloud and in the sea" (1 Cor. 10:2; also cf. Heb. 3–4)

	ADAM	NOAH	ABRAHAM	ISRAEL	NT CHURCH
3	"Be fruitful" (Gen. 1:28)	"Be fruitful" (Gen. 8:17; 9:1, 7)	The promise of fruitfulness (Gen. 17:6; cf. 28:3; 47:27; 48:4)	The promise of fruitfulness (Exod. 1:7; Lev. 26:9)	The promise of spiritual fruits (Rom. 6:22; 7:4; Gal. 5:22; Eph. 5:9; Phil. 1:11; Col. 1:6, 10; Heb. 12:11; James 3:17)
4	Placed in the Garden of Eden (Gen. 2)	Planted a vineyard after the Flood (Gen. 9:20)	Came to Canaan (Gen. 12:4–9), which was like Eden (Gen. 13:10; cf. Isa. 51:3; Ezek. 36:25)	Came to Canaan	In Christ it seated in the "heavenly places" (Eph. 2:4–6) (one could also say that it began, so to speak, in a garden; John 19:41; 20:15)
5	Under God's commandment (Gen. 2:16–17)	Under God's commandments (Gen. 9:1–17)	Under God's laws (Gen. 26:5; cf. 17:1)	Under God's Law (Exod. 20–Deut. 27)	Under the Law of Christ (Gal. 6:2; cf. John 14:15, 21; 15:10; Rom. 8:4; 13:8–10)
6	He soon fell (Gen. 3)	He soon fell (Gen. 9:20–21)	He soon fell went to Egypt (Gen. 12:10–20), obtained Hagar there, fathered Ishmael through her (Gen. 16), laughed out of unbelief (Gen. 17:17)	They soon fell (Exod. 32–34)	It soon fell (Acts 5:1–11; 20:29–30).

	ADAM	NOAH	ABRAHAM	ISRAEL	NT CHURCH
7	His fall led to God's curse (Gen. 3:17)	His fall led to God's curses (Gen. 9:25	The reference to "curse" is only for Abram's enemies (Gen. 12:3; cf. 27:29)	There are curses for Israel if they do not serve the Lord (Lev. 26:14–39; Deut. 28:15–68)	There are "curses" (though never so called) upon nominal Christianity, the false church (e.g., Rev. 17 and 18)
8	…but also to blessing (Gen. 3:21)	…but also to blessing (Gen. 9:26)	The blessing is for those who bless Abraham (Gen. 12:2–3; 18:18; 22:18), and for his offspring (17:16, 20; 22:17)	There are blessings for Israel if they do serve the Lord (Lev. 26:1–13; Deut. 28:1–14)	There are blessings upon the true church (Eph. 1:3, 6), including the blessing of Abraham (Gal. 3:14)
9	The Adamic covenant (if we may read Hos. 6:7 this way)	The Noahic covenant (Gen. 9:8–17)	The Abrahamic covenant (Gen. 15 and 17)	The Sinaitic covenant (Exod. 19–24) and the Palestinian covenant (Deut. 29); eventually the New Covenant (Jer. 31:31–34)	The New Covenant (2 Cor. 3:6; Heb. 8)
10	Sacrifice: implied in making the garments of skins (Gen. 3:21)	Sacrifice: on Noah's altar (Gen. 8:20–22)	Sacrifice: on Abraham's altar (Gen. 12:7–8; 13:4, 18; 22:9)	The entire sacrificial system (Lev.)	The sacrifice of Christ (Heb. 9); our sacrifices (1 Pet. 2:5, 9; Heb. 13:15)

	ADAM	NOAH	ABRAHAM	ISRAEL	NT CHURCH
11	Covenant sign: the tree of life	Covenant sign: the rainbow (Gen. 9:13)	Covenant sign: circumcision (Gen. 17:11)	Covenant sign: the Sabbath (Exod. 31:13; Ezek. 20:12, 20)	Covenant sign: the Lord's Supper (Luke 22:20; 1 Cor. 11:25)
12	Eschato-logical meaning: the last Adam (1 Cor. 15:45–47)	Eschato-logical meaning: the new world (2 Pet. 3:5–13)	Eschato-logical meaning: the patriarchal promises fulfilled (e.g., Micah 7:19–20)	Eschato-logical meaning: "all Israel saved" (Rom. 11:25–27)	Eschato-logical meaning: "always with the Lord" (1 Thess. 4:17)

APPENDIX 4
PARALLELS BETWEEN GENESIS 1–2 AND JOHN 1–2

GENESIS 1–2	JOHN 1–2
"In the beginning God" (1:1)	"In the beginning God" (1:1)
DAY 1	
(1:2–5): light enters the world	(1:1–18): the Light enters the world
DAY 2	
(1:6–8): separation of the waters	(1:19–28): separation through water (baptism)
DAY 3	
(1:9–13): the first life	(1:29–34) "*Next day*": life through the Lamb's death
DAY 4	
Parallel with Day 1 (1:14–19): the light bearers	Parallel with Day 1 (1:35–41) "*Next day*": the first disciples "light bearers"
DAY 5	
Parallel with Day 2 (1:20–23): life in the skies	Parallel with Day 2 (1:42–52) "*Next day*": angels in the skies
DAY 6	
Parallel with Day 3 (1:24–31): the wedding of Adam and Eve	Parallel with Day 3 "*Third day*" (2:1–11): the wedding at Cana
DAY 7	
(2:1–3): rest	(2:12): rest

Notice the triple expression "next day" in John 1:29, 35, 43 (which implies *four* days), and the "third day" in 2:1. Only 1:1–18 does not mention or imply the word "day," although the first words ("In the beginning") and the reference to "light" (vv. 4–5, 7–9) suggest it. If the passage is taken to represent a "day" as well (though only in the figurative sense), there is a total of six days, so that 2:1 refers to the sixth day. In my view, it is called a "third day" because of the implied parallels between the days of creation (1 and 4; 2 and 5; 3 and 6).

APPENDIX 5

PARALLELS BETWEEN THE SEVEN "SACRAMENTS" IN THE OLD AND NEW TESTAMENTS

	"SACRAMENT"	COVENANT	VISIBLE	A SIGN OF...	ESCHATOLOGICAL MEANING
1	The tree of life	Adamic	a tree	life (an eternal future of bliss)	Revelation 2:7; 22:1–3
2	The rainbow	Noahic	a sky phenom-enon	life (a new future for the world)	Genesis 9:13–15; 2 Peter 3:5–7
3	Circumcision	Abrahamic	a physical operation	death (God's judgment on human flesh)	Deuteronomy 30:6
4	The Sabbath	Sinaitic	a day	life (rest, renewal, cf. Heb. 4:9)	Isaiah 56:1–8
5	Water baptism	New Covenant	water	death (cf. Rom. 6:3–4; Col. 2:12)	Ezekiel 36:24–29
6	The Lord's Supper	New Covenant	bread and wine	death (cf. 1 Cor. 11:26)	Matthew 26:29
7	Anointing the sick	New Covenant	oil	life (a new future for the sick)	Isaiah 33:5–6, 24

N.B. If six of the seven "sacraments" arise from real, historical phenomena, then so too does the very first one.

APPENDIX 6

CAVEAT EMPTOR: STUDY BIBLES AND THE HISTORICAL EVENTS OF GENESIS 1–11

BY NELSON D. KLOOSTERMAN

Among the multitude of Bible translations, versions, and editions that are being published throughout the world today, the genre of "study Bible" poses a particular challenge to Christian believers. This is not the place for either reviewing those Bibles or enumerating those challenges. But in connection with our defense in this book of the historicity of the creation, of Adam and Eve as the first humans created directly by God, and of humanity's Fall into sin, we do wish to alert our readers to one particular challenge this genre presents, and we do so by using three specific study Bibles as illustrations.

The first study Bible is called *The Jesus Bible, NIV Edition*.[1] And the single particular challenge involves this Bible's *deafening silence* about the historicity of the events recorded in Genesis 1–11.

Before the Bible reader begins to read the text of Genesis itself, the editors provide this graphic:

Creation	Abram Goes to Canaan	Jacob and His Family Go to Egypt
Unknown	**c. 2091 BC**	**c. 1876 BC**

[1] *The Jesus Bible, NIV Edition*. The subtitle is: "sixty-six books. one story. all about one name." This study Bible contains written contributions from L. Giglio, M. Lucado, J. Piper, R. Zacharias, and R. Alcorn. Emphasis added to this graphic.

The discerning reader knows that there are at least two ways to attack truth. One way is overt contradiction of the truth by advocating opposite claims. Throughout this volume, we have been battling that strategy being employed by AEH advocates. A second way to attack the truth is to silence it, to hide it, to omit mentioning it. Regrettably, this study Bible employs the second approach.

How does it do that? Though some may think this a small matter, the graphic reproduced above omits mention, under "Creation," of Adam and Eve, and of their Fall into sin. By omitting this mention, and by assigning no date to an empty category known as "Creation," readers are given the stark visual impression that "there was no there there"—that nobody real, nobody similar to, say, Abram and Jacob, lived "back then."

Fortunately, the study notes themselves are better than the graphic, since at various points the notes clearly acknowledge the historical character of the unique divine origin of creation, of the direct special creation of Adam and Eve, and of humanity's Fall into sin.

Things are quite different, however, with *The Chronological Study Bible, New International Version*.[2] The study notes of this Bible are far more explicit in sharing with readers the conclusions of the historiographic science of archaeology—and in placing these historiographic conclusions over the text of Scripture.

Under the heading entitled "Epoch One: Before the Patriarchs," we are given a table showing time periods, beginning with "Old Stone Age, before 10,000 B.C." The editors introduce this chart by saying, "Very early dates are based on theories of evolution and geology, and interpreters of the Bible differ on how such dates relate to the creation accounts in Genesis."[3]

Again, the discerning reader immediately spots the confusion of categories. "Theories of evolution and geology" do not belong to the same class, the same category. Although these two are often related, geology relies on scientific techniques to explain structures and processes of change in the solid features of the Earth. Theories of evolution, by contrast, employ imagination to reconstruct the data and conclusions of geology. Geology practices scientific techniques, whereas theories of evolution employ metaphysics.

Next, we are told that "scholars have placed the first human settlements as early as 7,000 to 8,000 years before Christ." If this is true, then did "human settlements" not exist before that time, during the periods these

---

[2] Thomas Nelson/HarperCollins Christian Publishing, Inc., 2014

[3] Ibid., 1.

editors have dubbed the "Old Stone Age, before 10,000 B.C.," or the "Middle Stone Age, 10,000 to 8000 B.C."? We are puzzled when we are told that during the Old Stone Age, "people lived in caves or temporary shelters," and that during the Middle Stone Age, "real settlements first appeared, and there was an evolution in the arts of civilization."[4] How, if no "real settlements" existed in these periods of time, were people able to invent, produce, and employ instruments for hunting and gathering?

Things become no more clear when the editors move to consider "The Biblical literature." Remember, the editors have decided to let archaeology tell Bible readers that there was some kind of human existence some time before 10,000 B.C. Both biblical and extra-biblical accounts speak of a creation, a flood, and a tower.

> The major narratives of this primeval history give an account of creation, a great Flood, and the tower at Babel. The creation account (Ge 1–3) describes the creation of all things, including humankind. The *newly created humans* rebel against God, resulting in their expulsion from the Garden of Eden.[5]

Take note of, but do not misunderstand, the italicized words. The editors are not at all suggesting that these events all *happened* within a short time span, but they are suggesting that *the narrative perspective gives the appearance* of "newly created humans" rebelling against God soon after their creation.

How do we know this is the editors' intention?

Immediately after the text of Genesis 2:25, they inform us, by way of a "Time Capsule," that experts have dated the existence, from Japan to Australia to Ukraine, of stone tools, cooking ovens, flutes, and ancient cave paintings as having appeared between 26,000 and 10,000 B.C.[6] The obvious question is: Why do these phenomena not point to some form of human settlement and cultural interaction? The next "Time Capsule" appears in the text immediately after Genesis 3:8, the account of humanity's Fall into sin. This table dates the beginning of human settlements in 8000 B.C. The juxtaposition of these "Time Capsules" leads the careful, but now confused, reader to ask: Did the rebellion of "newly created humans" actually occur soon after their creation?

---

4 Ibid.

5 Ibid., 2; italics added.

6 Ibid., 5.

At a minimum, this additional material, including dates based on *theories* of evolution and geology, and printed within the Bible itself, generates serious confusion and unclarity about the historical reality of the divine creation of the world, of the direct special creation of humanity, and of humanity's Fall into sin.

Our third and final illustration is provided by a study Bible entitled *NIV Cultural Backgrounds Study Bible: Bringing to Life the Ancient World of Scripture.*[7] Of all three study Bibles, this is by far the most dangerous for Christian believers. The title of "study Bible" lends an aura of credibility, authority, and sanctity to this book that it scarcely deserves.

Briefly, with regard to the argument being advanced in this "Bible," the publisher puffs the pedigree of the book's editors:

> Both scholars [Walton and Keener] have published heavily documented works that support the sort of background that is provided here on a more accessible level. Both have been studying, writing and lecturing around the world about the field of the Bible cultural backgrounds for the duration of their decades-long careers as academics.[8]

Read carefully: "Scholars…heavily documented…more accessible level… around the world…decades-long careers as academics."

With these accolades you the reader are being put on notice that a new priestly class is now functioning (a verb very dear to John Walton when it comes to Gen. 1–2) as mediators between Holy Scripture and the Christian believer. That class of mediators consists of scholars who explain the language, literature, and culture of the Bible. Content that used to belong to the literary genre of books known as "commentaries" is today enfolded and embedded *within Scripture itself*! From now on, with this edition of "Scripture" in hand, the reader can and need no longer read the Bible for himself or herself. In effect, the scholarly culture of the "authors" becomes the unavoidable lens through which the Bible text must be read.

Most unsatisfactory, then, in addition to omissions similar to those study Bibles we have previously mentioned, is the imposition on the reader of

---

7 Edited by C. S. Keener and J. H. Walton; Grand Rapids, MI: Zondervan, 2016. To the dismay of some, including this author, who believe that God through the human writers was the Author of the Bible, it may be considered a deeply troubling, even embarrassing, publishing moment when Walton and Keener are identified, in a separate introductory section, as the *authors* of this study Bible (ibid., viii).

8 Ibid.

debatable and doubtful theological claims in this *NIV Cultural Backgrounds Study Bible*. Consider:

> The creation of humans from dust is similar to what is found in ancient Near Eastern mythology.... Here in [Genesis] 2 there are archetypal elements that are identifiable. Man is made from the dust, and since he will also return to dust (3:19), all people can be seen as created from the dust (see Ps. 103:14). The creation of Eve from Adam's side (Ge 2:21–23) likewise expresses a relationship between man and woman that permeates the race. In these Adam and Eve are archetypes representing all of humanity in their creation, just as they do in their sin and their destiny (death) in ch. 3. Their function as archetypes does not suggest that they are not historical individuals; it only suggests that they function more importantly as representatives of the race.[9]

No clear, explicit distinction is made here between the Bible's creation account and "what is found in ancient Near Eastern mythology." Adam is an archetype, not a unique creation by direct divine action. The claim that "Adam and Eve are archetypes representing all of humanity...in their sin" does serious injustice to the Bible's doctrine of original sin, since nowhere does the Bible teach that *Eve* functioned as such a representative, but only Adam.

But note very carefully the authors' closing statement from the citation: "Their function as archetypes does not suggest that they [Adam and Eve] are not historical individuals; it only suggests that they function more importantly as representatives of the race." Here we are being treated—in a *study Bible*, no less!—to a profound ambiguity and an illegitimate choice.

The profound ambiguity is this: the authors assure us that the term *archetype* does not suggest Adam and Eve were not historical—*but the authors never insist that the term requires that Adam and Eve were historical!* In fact, the reader will scour in vain the multitudinous scholarly comments in this study "Bible" to find a full-throated unqualified affirmation of the historical divine special creation of Adam and Eve.

Here is the illegitimate choice: Adam and Eve function *more importantly*, we are told, *not* as historical humans, but as representatives of the race. *But who says that the representative function of Adam and Eve is more important that their historicity?* Why are this comparison—"more importantly"—and

[9] Ibid., 8–9.

its explicit choice even being offered to us readers? Answer: let the reader not forget that one of the "authors" of this "Bible" is one, John Walton, who has gained theological notoriety by making, rehearsing, and restating this very claim throughout the evangelical world. According to Walton, the most important, if not the only important, feature of the Genesis creation narrative is *not* the historicity of the events recorded, but the function of those events. *That* now becomes the lens through which the readers of this "Bible" must read the text of sacred Scripture!

It's one thing for people like me to discard his claim, as I am free to do on the basis of valid exegetical and theological objections; but now I have to read *his* objectionable claim in my "Bible"?

Our conclusion regarding study Bibles and what they do to Genesis 1–11 is this: as the ancient Latin adage puts it: *Caveat emptor*—let the buyer beware!

BIBLIOGRAPHY

Aalders, G. Ch. 1932. *De goddelijke openbaring in de eerste drie hoofdstukken van Genesis*. Kampen: Kok.

______. 1981. *Genesis*. Translated by W. Heynen. Vol. 1. Bible Student's Commentary. Grand Rapids, MI: Zondervan.

Ager, D. 1995. *The New Catastrophism: The Importance of the Rare Event in Geological History*. Cambridge: Cambridge University Press.

Alexander, D. 2014. *Creation or Evolution: Do We Have to Choose?* 2nd ed. Oxford: Monarch Books.

Alexander, T. D. and D. W. Baker, eds. 2003. *Dictionary of the Old Testament: Pentateuch*. IVP Bible Dictionary Series. Downers Grove, IL: InterVarsity Press.

Anderson, A. S. 1976. "The Seed of the Woman." *Third National Creation Science Conference*. 99–104. St. Paul, MN: Bible-Science Association/Twin-City Creation-Science Association.

Anderson, B. W. (1967) 1987. *Creation versus Chaos*. Philadelphia, PA: Fortress Press.

Andrews, E. H. 1981. *God, Science, and Evolution*. El Cajon, CA: Creation Life Publishers.

______, W. Gitt, and W. J. Ouweneel, eds. (1986) 2000 (repr.). *Concepts in Creationism*. Welwyn Garden City: Evangelical Press.

Arnold, B. T. and B. E. Beyer. 1999. *Encountering the Old Testament: A Christian Survey*. Grand Rapids, MI: Baker Academic.

Ash, D. and P. Hewitt. 1991. *Science of the Gods: Reconciling Mystery and Matter*. Bath: Gateway Books.

Ashton, J. F. 2012. *Evolution Impossible: 12 Reasons Why Evolution Cannot Explain the Origin of Life on Earth*. Green Forest, AR: Master Books.

Atkinson, D. 1990. *The Message of Genesis 1–11: The Dawn of Creation*. The Bible Speaks Today. Vol. 1. Downers Grove, IL: IVP Academic Press.

Augustine. 1982. *The Literal Meaning of Genesis*. Translated by J. H. Taylor. 2 vols. New York: Paulist Press.

Axe, D. 2017 (repr.). *Undeniable: How Biology Confirms Our Intuition That Life Is Designed*. New York: HarperOne.

Ayala, F. J. 1995. "The Myth of Eve: Molecular Biology and Human Origins." *Science*, New Series 270 no. 5244 (Dec. 22): 1930–36.

______. 2009. *Darwin's Gift to Science and Religion*. Washington DC: Joseph Henry Press.

Barcellos, R. C. 2013. *Better Than the Beginning: Creation in Biblical Perspective*. Palmdale, CA: Reformed Baptist Academic Press.

Barr, J. 1993. *The Garden of Eden and the Hope of Immortality*. Minneapolis, MN: Fortress Press.

Barrett, C. K. 1968. *1 Corinthians*. Harper's New Testament Commentary. Peabody, MA: Hendrickson.

Barrett, M. and A. B. Caneday, eds. 2013. *Four Views on the Historical Adam*. Grand Rapids, MI: Zondervan.

Barrow, J. D. and F. J. Tipler. (1986) 1988 (rev. ed.). *The Anthropic Cosmological Principle*. Oxford: Oxford University Press.

Barth, K. 1936–1988. *Church Dogmatics*. Translated by G. W. Bromiley. Vols. I/1–IV/4. Edinburgh: T. and T. Clark.

Bates, G. H. n.d. (ca. 1890). *Gems from Genesis to Revelation*. Glasgow: Pickering and Inglis.

Batto, B. F. 1992. *Slaying the Dragon: Mythmaking in the Biblical Tradition*. Louisville, KY: Westminster John Knox.

Bavinck, H. 2002–2008. *Reformed Dogmatics*. Edited by John Bolt. Translated by John Vriend. 4 vols. Grand Rapids, MI: Baker Academic.

Behm, J. 1964. "*Kardia*." In *Theological Dictionary of the New Testament*. Edited by G. Kittel et al. Translated by G. W. Bromiley. 10 vols. Grand Rapids, MI: Eerdmans. Logos edition. 3:605–13.

Bell, E. A., P. Boehnke, T. M. Harrison, and W. L. Mao. 2015. "Potentially Biogenic Carbon Preserved in a 4.1 Billion-Year-Old Zircon." *Proceedings of the National Academy of Sciences of the United States of America* 112.47: 14518–21.

Bengel, J. A. (1742) 1862 (repr.). *Gnomon Novi Testamenti*. London: Williams and Norgate.

Berg, L. S. 1969. *Nomogenesis or Evolution Determined by Law*. Cambridge, MA: MIT Press.

Bergman, J. 2010. "Why Orthodox Darwinism Demands Atheism." *Answers Research Journal* 3:147–152.

Berkhof, H. 1986. *Christian Faith: An Introduction to the Study of the Faith.* Translated by S. Woudstra. Grand Rapids, MI: Eerdmans.
Berkhof, L. 1949. *Systematic Theology*. Grand Rapids, MI: Eerdmans. https://www.biblicaltraining.org/library/systematic-theology-louis-berkhof.
Berkouwer, G. C. 1952. *The Providence of God*. Translated by L. B. Smedes. Studies in Dogmatics. Grand Rapids, MI: Eerdmans.
______. 1962. *Man: The Image of God*. Translated by D. W. Jellema. Studies in Dogmatics. Grand Rapids, MI: Eerdmans.
______. 1971. *Sin*. Translated by P. C. Holtrop. Studies in Dogmatics. Grand Rapids, MI: Eerdmans.
Berlinski, D. 2009 (repr.). *The Devil's Delusion: Atheism and Its Scientific Pretensions*. New York: Basic Books.
Bingham, G. C. 1981. *Man the Steward of Creation: Studies in Human Stewardship under the Mandate and Gifts of God*. Adelaide: New Creation Publications.
Blackwell, R. J. 1991. *Galileo, Bellarmine, and the Bible*. Notre Dame, IN: University of Notre Dame Press.
Blocher, H. 1984. *In the Beginning: The Opening Chapters of Genesis*. Downers Grove, IL: IVP Books.
______. 1997. *Original Sin: Illuminating the Riddle*. Leicester: Apollos.
Bloom, J. A. 1997. "On Human Origins: A Survey." *Christian Scholars Review* 27.2: 181–203 (updated version 2012: http://www.asa3.org/ASA/education/origins/humans-jb.htm).
Bloore, J. 1938. *Genesis*. Bible Handbooks. New York: Loizeaux Brothers.
Böhl, F. M. Th. 1923. *Genesis. Tekst en uitleg: practische bijbelverklaring*. Vol. 1. Groningen: J. B. Wolters.
Bonhoeffer, D. 1995. *Ethics*. New York: Touchstone.
______. (1937) 1997 (repr.). *Creation and Fall*. Minneapolis, MN: Fortress Press.
Borger, P. 2009. *Terug naar de oorsprong: Of hoe de nieuwe biologie het tijdperk van Darwin beëindigt*. Houten: CBC 't Gulden Boek.
Bowler, P. J. 2009. *Evolution: The History of an Idea*. 4th ed. Berkeley, CA: University of California Press.
______. 2013. *Darwin Deleted: Imagining a World Without Darwin*. Chicago: University of California Press.
Bozarth, G.R. 1979. "The Meaning of Evolution." *American Atheist* (Sept. 20, 1979): 30.
Brand, L. and A. Chadwick. 2017. *Faith, Reason, and Earth History: A Paradigm of Earth and Biological Origins by Intelligent Design*. 3rd ed. Berrien Springs, MI: Andrews University Press.
Brandenburger, E. 1962. *Adam und Christus: Exegetisch-religionsgeschichtliche*

Untersuchung zu Römer 5:12–21 (1. Kor. 15). Neukirchen Kreis Moers: Neukirchener Verlag.

Bratsiotis, P. N. 1951/52. "Genesis 1 26 in der orthodoxen Theologie." *Evangelische Theologie* 11:289–97.

Brentano, F. (1874) 1995. *Psychology from an Empirical Standpoint*. 2nd ed. London: Routledge.

Brockman, J. 1995. *The Third Culture: Beyond the Scientific Revolution*. New York: Touchstone.

Brown, C. 2006. *Miracles and the Critical Mind*. Pasadena, CA: Fuller Seminary Press.

Brown, C. G. 2013. *The Healing Gods: Complementary and Alternative Medicine in Christian America*. Oxford: Oxford University Press.

Brown, P. D. and R. Stackpole, eds. 2014. *More Than Myth? Seeking the Full Truth about Genesis, Creation, and Evolution*. Leicester (UK): Chartwell Press.

Brown, W. P. 2010. *The Seven Pillars of Creation: The Bible, Science, and the Ecology of Wonder*. New York: Oxford University Press.

Bruce, F. F. 1990. *A Mind For What Matters: Collected Essays of F. F. Bruce*. Grand Rapids, MI: Eerdmans.

Brunner, E. 1952. *The Christian Doctrine of Creation and Redemption*. Translated by O. Wyon. Philadelphia, PA: The Westminster Press.

Buijs, G. J., P. Blokhuis, S. Griffioen, and R. Kuiper, eds. 2005. *Homo Respondens: Verkenningen rond het mens zijn*. Amsterdam: Buijten and Schipperheijn.

Bultmann, R. K. 1971. *The Gospel of John: A Commentary*. Louisville, KY: Westminster John Knox.

______. 2012 (repr.). *Jesus Christ and Mythology*. Norwich (UK): Hymns Ancient and Modern Ltd.

Buswell, J. O. 1962. *A Systematic Theology of the Christian Religion*. Vol. 1. Grand Rapids, MI: Eerdmans.

Calvin, J. 1960. *Institutes of the Christian Religion*. Edited by John T. McNeill. Translated by Ford Lewis Battles. 2 vols. Library of Christian Classics 20–21. Philadelphia: Westminster Press.

______. 2005. *Commentary on Jeremiah and Lamentations*. Translated by J. Owen. Vol. 1. Grand Rapids, MI: Christian Classics Ethereal Library. http://www.ccel.org.

Capra, F. (1975) 2010. *The Tao of Physics: An Exploration of the Parallels Between Modern Physics and Eastern Mysticism*. 5th ed. Boston, MA: Shambhala.

______. (1982) 1984. *The Turning Point: Science, Society, and the Rising Culture*. 2nd ed. New York: Bantam Books.

Carlson, R. F. and T. Longman III. 2010. *Science, Creation, and the Bible: Reconciling Rival Theories of Origins*. Downers Grove, IL: IVP Academic.

Carroll, L. (1872) 1999. *Through the Looking-Glass*. Mineola, NY: Dover Publications.

Carroll, S. B. 2016. *Endless Forms Most Beautiful: The New Science of Evo Devo*. New York: W. W. Norton.

Carson, D. A. 1990 (repr.). *The Gospel According to John*. Pillar New Testament Commentary. Grand Rapids, MI: Eerdmans.

Carter, R., ed. 2014. *Evolution's Achilles Heels*. Powder Springs, GA: Creation Book Publishers.

______. 2017. "Reading Evolution into the Scriptures." *Journal of Creation* 31.2: 41–46.

Cassuto, U. 1961. *From Adam to Noah: A Commentary on the Book of Genesis I–VI (Part I)*. Jerusalem: Magnes Press (Hebrew University).

______. 1983. *The Documentary Hypothesis and the Composition of the Pentateuch*. Jerusalem: Magnes Press (Hebrew University).

Cavanaugh, W. T. and J. K. A. Smith, eds. 2017. *Evolution and the Fall*. Grand Rapids, MI: Eerdmans.

Chafer, L. S. 1983. *Systematic Theology*. 15th ed. 8 vols. Dallas, TX: Dallas Seminary Press.

Chesterton, G. K. 1985. *As I Was Saying*. Edited by Robert Knille. Grand Rapids, MI: Eerdmans.

Cheyne, T. K. (1904) 1970 (repr.). *Critica Biblica, or, Critical Linguistic, Literary and Historical Notes on the Text of the Old Testament Writings: Ezekiel and Minor Prophets*. Amsterdam: Philo Press.

Chou, A., ed. 2016. *What Happened in the Garden? The Reality and Ramifications of the Creation and Fall of Man*. Grand Rapids, MI: Kregel Academic.

Clark, R. E. D. 1966. *Darwin, Before and After: The Story of Evolution*. Exeter: Paternoster.

Clifford, R. J. 1994. *Creation Accounts in the Ancient Near East and in the Bible*. Catholic Biblical Quarterly Monograph 24. Washington, DC: Catholic Biblical Association of America.

Coates, C. A. 1920. *An Outline of Genesis*. Kingston-on-Thames: Stow Hill Bible and Tract Depot.

Cohen, A. 1983 (repr.). *The Soncino Chumash*. The Soncino Books of the Bible. London/New York: Soncino Press.

______. 1985 (repr.). *The Psalms*. The Soncino Books of the Bible. London/New York: Soncino Press.

Collins, C. J. 2003. *Science and Faith: Friends of Foes?* Westchester, IL: Crossway Books.

______. 2006. *Genesis 1–4: Linguistic, Literary, and Theological Commentary.* Phillipsburg, NJ: P&R Publishing.

______. 2010. "Adam and Eve as Historical People, and Why It Matters." *Perspectives on Science and Christian Faith* 62.3 (September): 147–65.

______. 2011. *Did Adam and Eve Really Exist? Who They Were and Why You Should Care.* Wheaton, IL: Crossway Books.

Collins, F. S. 2006. *The Language of God: A Scientist Presents Evidence for Belief.* New York: Free Press.

Collins, J. J. 1990. "Inspiration or Illusion: Biblical Theology and the Book of Daniel." *Ex Auditu* 6:29–38.

Couliano, I. P. 1991. *The Tree of Gnosis: Gnostic Mythology from Early Christianity to Modern Nihilism.* San Francisco, CA: HarperCollins.

Coyne, J. A. 2009. "Creationism for Liberals." *The New Republic.* August 11: 34–43.

Cranfield, C. E. B. 1975. *The Epistle to the Romans.* International Critical Commentary. Edinburgh: Clark.

Dalley, S. 1989. *Myths from Mesopotamia: Creation, the Flood, Gilgamesh, and Others.* Oxford World's Classics. Oxford: Oxford University Press.

Darby, J. N. n.d.-1. *The Collected Writings of J. N. Darby.* Kingston-on-Thames: Stow Hill Bible and Tract Depot.

______. n.d.-2 *Synopsis of the Books of the Bible.* Kingston-on-Thames: Stow Hill Bible and Tract Depot.

Darlington, C. D. 1960. *Darwin's Place in History.* Oxford: Blackwell.

Darwin, C. R. (1859) 2003. *On the Origin of Species.* London: John Murray. 150th Anniversary Edition. New York: Signet.

______. (1871) 1997 (repr.). *The Descent of Man.* Westminster, MD: Prometheus Books.

______. 1989. *The Voyage of the Beagle: Charles Darwin's Journal of Researches.* Abridged edition. Edited by J. Browne and M. Neve. London: Penguin Classics.

Davidheiser, B. 1969. *Evolution and Christian Faith.* Grand Rapids, MI: Baker Book House.

Davies, P. 1999. "Life Force." *New Scientist* 163 (2204): 27–30.

Davison, J. A. 2005. "A Prescribed Evolutionary Hypothesis." *Rivista di Biologia/Biology Forum* 98: 155–66.

Dawkins, R. 1986. *The Blind Watchmaker: Why the Evidence of Evolution Reveals a Universe Without Design.* New York: W.W. Norton.

______. 2009. *The Greatest Show on Earth: The Evidence for Evolution.* New York: Free Press.

De Graaff, F. 1987. *Jezus de Verborgene: Een voorbereiding tot inwijding in de mysteriën van het evangelie*. Kampen: Kok.

De Haan, M. R. 1995. *Portraits of Christ in Genesis*. 3rd ed. Grand Rapids, MI: Kregel Classics.

Dekker, C., R. Van Woudenberg, and G. van den Brink. 2007. *Omhoog kijken in platland: Over geloven in de wetenschap*. Kampen: Ten Have.

Demarest, B. 1997. *The Cross and Salvation: The Doctrine of Salvation*. Wheaton, IL: Crossway Books.

Dembski, W. and J. Witt. 2010. *Intelligent Design Uncensored: An Easy-to-Understand Guide to the Controversy*. Downers Grove, IL: IVP Books.

Dennison, J. T., Jr., ed. 2008–2014. *Reformed Confessions of the 16th and 17th Centuries in English Translation*. 4 vols. Grand Rapids, MI: Reformation Heritage Books.

Denton, M. 1986. *Evolution: A Theory in Crisis*. Chevy Chase, MD: Adler and Adler.

______. 1998. *Nature's Destiny: How the Laws of Biology Reveal Purpose in the Universe*. New York: Free Press.

______. 2016. *Evolution: Still A Theory in Crisis*. Seattle, WA: Discovery Institute Press.

Dobzhansky, Th. 1959. "Blyth, Darwin, and Natural Selection." *The American Naturalist* 93 (870): 204–206.

Dooyeweerd, H. 1960/1975. *In the Twilight of Western Thought: Studies in the Pretended Autonomy of Philosophical Thought*. Philadelphia: P&R Publishing.

______. (1953) 1984 (repr.). *A New Critique of Theoretical Thought*. 4 vols. Jordan Station, ON: Paideia Press.

______. 2003. *Roots of Western Culture: Pagan, Secular, and Christian Options*. Lewiston, NY: E. Mellen.

Driesch, H. (1914) 2010. *The History and Theory of Vitalism*. Charleston, SC: Nabu Press.

Duffield, G. P. and N. M. Van Cleave. 1996. *Woord en Geest: Hoofdlijnen van de theologie van de Pinksterbeweging*. Kampen: Kok/Rafaël Nederland.

Earle, R. 1978. *1 Timothy*. Expositor's Bible Commentary 11. Grand Rapids, MI: Zondervan.

Ebersberger, I., Galgoczy, P., Taudien, S., Taenzer, S., Platzer, M., Von Haeseler, A. 2007. "Mapping Human Genetic Ancestry." *Mol. Biol. Evol.* 10 (Oct. 24, 2007): 2266–76.

Eiseley, L. 1979. *Darwin and the Mysterious Mr. X: New Light on the Evolutionists*. London: Dent.

Ellul, J. 1990. *The Technological Bluff*. Grand Rapids, MI: Eerdmans, 1990.

Enns, P. 2005. *Inspiration and Incarnation: Evangelicals and the Problem of the Old Testament*. Grand Rapids, MI: Baker Academic.

______. 2012. *The Evolution of Adam: What the Bible Does and Doesn't Say about Human Origins*. Grand Rapids, MI: Brazos Press.

Enoch, H. (1966) 1976 (rev. ed.). *Evolution or Creation*. Welwyn Garden City: Evangelical Press.

Erickson, M. J. (1998) 2007. *Christian Theology*. 10th ed. Grand Rapids, MI: Baker Book House.

Estelle, B. D. 2012. "Should We Still Believe in a Historical Adam?" *New Horizons* 33.3: 9, 20.

Falk, D. R. 2004. *Coming to Peace with Science: Bridging the Worlds Between Faith and Biology*. Downers Grove, IL: IVP Academic.

Falkenberg, S. 2002. *Biblical Literalism*. New Reformation. archive.org/web/ 20080615062211/http://www.newreformation.org/literalism.htm.

Fee, G. D. 1987. *The First Epistle to the Corinthians*. New International Commentary on the New Testament. Grand Rapids, MI: Eerdmans.

Fesko, J. V. 2007. *Last Things First: Unlocking Genesis 1–3 with the Christ of Eschatology*. Fearn (UK): Mentor.

Feuerbach, L. (1841) 2008. *The Essence of Christianity*. Translated by G. Eliot. New York: Cosimo Classics.

Fishman, J. "Part Ape, Part Human: A New Ancestor Emerges from the Richest Collection of Fossil Skeletons Ever Found." *National Geographic* 22.2: 120–33.

Fodor, J. and M. Piattelli-Palmarini. 2010. *What Darwin Got Wrong*. London: Profile Books.

Forrest, B. and P. R. Gross. 2007. *Creationism's Trojan Horse: The Wedge of Intelligent Design*. 2nd ed. Oxford: Oxford University Press.

Frair, W. 2000. "Baraminology: Classification of Created Organisms." *Creation Research Society Quarterly* 37.2: 82–91. http://www.creationresearch.org/crsq/articles/ 37/37_2/baraminology.htm.

Francke, J. 1974. *De morgen der mensheid: Hoe professor dr. B. J. Oosterhoff Genesis 2 en 3 leest*. Enschede: Boersma.

Frey, H. 1962. *In den beginne: Verklaring van de hoofdstukken 1–11 van het boek Genesis*. Franeker: Wever.

Friedman, R. M. 2001. *The Politics of Excellence: Behind the Nobel Prize in Science*. New York: Henry Holt and Co.

Gaffin, R. B., Jr. 2012. "All Mankind, Descending from Him...?" *New Horizons* 33.3: 3–5.

______. 2015. *No Adam, No Gospel: Adam and the History of Redemption*. Phillipsburg, NJ: P&R Publishing.

Garrett, D. 2000. *Rethinking Genesis: The Sources and Authorship of the First Book of the Pentateuch*. Fearn (UK): Mentor.

Gasman, D. 2004. *The Scientific Origins of National Socialism*. London: Routledge.

Gauger, A., D. Axe, and C. Luskin. 2012. *Science and Human Origins*. Seattle, WA: Discovery Institute Press.

Gavrilets, S. 2003. "Perspective: Models of Speciation: What Have We Learned in 40 Years?" *Evolution* 57.10: 2197–215.

Geisler, N. L. 1983. *Explaining Hermeneutics: A Commentary*. Oakland, CA: International Council on Biblical Inerrancy

______. 2011. *Systematic Theology*. Minneapolis, MN: Bethany House.

Giberson, K. 2009. *Saving Darwin: How to Be a Christian and Believe in Evolution*. 2nd ed. New York: HarperOne.

Ginzberg, L. 1969. *Bible Times and Characters from the Creation to Jacob. Vol. 1 of Legends of the Jews*. Philadelphia, PA: The Jewish Publication Society of America.

Gispen, W. H. 1974. *Genesis*. Commentaar op het Oude Testament. Kampen: Kok.

Gitt, W. 1993. *Did God Use Evolution?* Bielefeld: Christliche Literatur-Verbreitung.

______. 1999. *The Wonder of Man*. Bielefeld: Christliche Literatur-Verbreitung.

______. 2006. *In the Beginning Was Information*. Green Forest, AR: Master Books.

Glas, G. 1995. "Ego, Self, and the Body: An Assessment of Dooyeweerd's philosophical Anthropology." In *Christian Philosophy at the Close of the Twentieth Century: Assessment and Perspective*. Edited by S. Griffioen and B. M. Balk. 67–78. Kampen, Kok.

______. 1996. "De mens: Schets van een antropologie vanuit reformatorisch wijsgerig perspectief." In *Kennis en werkelijkheid*. Edited by R. Van Woudenberg. 86–142. Kampen: Kok.

______. 2006. "Persons and Their Lives: Reformational Philosophy on Man, Ethics, and Beyond." *Philosophia Reformata* 71: 31–57.

Godet, F. L. 1998. *Commentary on Romans*. Grand Rapids, MI: Kregel.

Goldingay, J. 1994. *Models for Scripture*. Carlisle: Paternoster Press.

Gooding, D. W. and J. C. Lennox. (1992) 2014a. *The Definition of Christianity*. Coleraine (N. Ireland): Myrtlefield House.

______ and J. C. Lennox. (1997) 2014b. *Christianity: Opium or Truth?* Coleraine (N. Ireland): Myrtlefield House.

Gould, S. J. 1977. *Ever Since Darwin: Reflections in Natural History*. New York: W.W. Norton.

______. 2001. "Introduction." In *Evolution: The Triumph of an Idea*, by C. Zimmer. ix–xiv. New York: Harper Collins.

______. 2002. *The Structure of Evolutionary Theory.* Cambridge, MA: Belknap Press.

______ and N. Eldredge. 1977. "Punctuated Equilibria: The Tempo and Mode of Evolution Reconsidered." *Paleobiology* 3.2: 115–51.

Grant, F. W. 1890. *The Numerical Bible: The Pentateuch.* New York: Loizeaux Brothers.

______. 1901. *The Numerical Bible: Acts to II Corinthians.* New York: Loizeaux Brothers.

______. 1956 (repr.). *Genesis in the Light of the New Testament.* New York: Loizeaux Brothers.

Grudem, W. 1994. *Systematic Theology: An Introduction to Biblical Doctrine.* Grand Rapids, MI: Zondervan.

Guldberg, H. 2010. *Just Another Ape?* London: Societas.

Gunkel, H. 1997. *Genesis.* Translated by M. E. Biddle. Macon, GA: Mercer University Press.

Guthrie, D. (1970) 1990 (rev. ed.). *New Testament Introduction.* Downers Grove, IL:: IVP Academic.

Haarsma, D. B. and L. D. Haarsma. 2011. *Origins: Christian Perspectives on Creation, Evolution and Intelligent Design.* Grand Rapids, MI: Faith Alive Christian resources.

Haeckel, E. (1877; Eng.: 1879) 2015. *Freedom in Science and Teaching.* CreateSpace Independent Publishing Platform.

Hamilton, V. P. 1990. *The Book of Genesis Chapters 1–17.* New International Commentary on the Old Testament. Grand Rapids, MI: Eerdmans.

Harinck, G., ed. 2001. *De kwestie-Geelkerken: Een terugblik na 75 jaar.* Barneveld: De Vuurbaak.

Harlow, D. C. 2010. "After Adam: Reading Genesis in an Age of Evolutionary Science." *Perspectives on Science and Christian Faith* 62.3 (September): 179–95.

Harrison, E. F. 1976. *Romans.* Expositor's Bible Commentary 10. Grand Rapids, MI: Zondervan.

Harrison, G. A., J. S. Weiner, J. M. Tanner, and N. A. Barricot. 1964. *Human Biology.* Oxford: Oxford University Press.

Hartley, J. E. 1988. *The Book of Job. New International Commentary on the Old Testament.* Grand Rapids, MI: Eerdmans.

Hegel, G. W. F. (1822–1831) 2004. *The Philosophy of History.* Translated by J. Sibree. Mineola, NY: Dover Publications.

Henderson, J. L. and M. Oakes. (1963) 1990. *The Wisdom of the Serpent: The Myths of Death, Rebirth, and Resurrection.* Princeton, NJ: Princeton University Press.

Hendriksen, W. 1980. *Exposition of Paul's Epistle to the Romans*. Vol. 1. New Testament Commentary. Grand Rapids, MI: Baker.
Hess, R. S. 1990. "Splitting the Adam: the usage of adam in Genesis i-v." In *Studies in the Pentateuch*. Edited by J. A. Emerton. Vetus Testamentum Supplements 41. 1–15. Leiden: Brill.
Heyns, J. A. 1988. *Dogmatiek*. Pretoria: NG Kerkboekhandel.
Ho, M.-W. and S. W. Fox. 1988. *Evolutionary Processes and Metaphors*. New York: John Wiley and Sons.
______ and P. T. Saunders, eds. 1984. *Beyond Neo-Darwinism: An Introduction to the New Evolutionary Paradigm*. London: Academic Press.
Hoek, J. 1988. *Zonde: Opstand tegen de genade*. Kampen: Kok.
Hoekema, A. A. 1986. *Created in God's Image*. Grand Rapids, MI: Eerdmans.
Hoyle, F. and C. Wickramasinghe. 1984. *Evolution from Space: A Theory of Cosmic Creationism*. New York: Touchstone.
Hsü, K. J. 1989. "Evolution, Ideology, Darwinism and Science." *Klinische Wochenschrift* 67: 923–28.
Hughes, P. E. 1989. *The True Image: The Origin and Destiny of Man in Christ*. Grand Rapids, MI: Eerdmans.
Hugo, V. 1972. *Choses vues 1847–1848*. Paris: Gallimard.
Huijgen, A. 2011. *Divine Accommodation in John Calvin's Theology: Analysis and Assessment*. Göttingen: Vandenhoeck and Ruprecht.
Huizinga, J. (1924) 1999 (repr.). *The Waning of the Middle Ages*. Mineola, NY: Courier Corporation.
Huntemann, G. 1977. *Am Anfang die Wahrheit: Die fünf Bücher Mose über das Woher und Wohin von Welt und Gemeinde*. Neuhausen-Stuttgart: Hänssler.
Huxley, J. S. 1955. "Evolution and Genetics." In *What Is Science?* Edited by J. R. Newman. 256–93. New York: Simon and Schuster.
______. 1960. "The Evolutionary Vision." In *Issues in Evolution*. Edited by S. Tax. 249–61. Chicago: University of Chicago Press.
Huxley, T. H. (1889) 1992. *Agnosticism and Christianity and Other Essays*. Westminster, MD: Prometheus Books.
______. 1894. *Collected Essays*, IV: *Science and Hebrew Tradition*. London: Macmillan.
Hyers, C. 1984. *The Meaning of Creation: Genesis and Modern Science*. Atlanta, GA: John Knox Press.
Jacob, B. (1934) 1974 (repr.). *Das erste Buch der Tora Genesis: Übersetzt und erklärt*. Berlin: Schocken Verlag.
Jacob, F. 1979. "L'évolution sans projet." In *Le Darwinisme aujourd'hui*, by F. Chapeville et al. Paris: Éditions du Seuil.

Jastrow, R. 1978. *God and the Astronomers*. New York: W.W. Norton.

Johnson, P. E. 1991. *Darwin on Trial*. Washington, DC: Regnery Gateway.

Johnston, R. K. 1978. “The Role of Women in the Church and Home: An Evangelical Testcase in Hermeneutics.” In *Scripture, Tradition, and Interpretation*. Edited by W. W. Gasque and W. S. La Sor. 234–59. Grand Rapids, MI: Eerdmans.

Junker, R. and S. Scherer. 2013. *Evolution: Ein kritisches Lehrbuch*. 7th ed. Gießen: Weyel.

Kamphuis, J. 1985. *Uit verlies winst: Het beeld van God en het komende Koninkrijk*. Barneveld: Vuurbaak.

Kant, I. (1755) 2000. *Universal Natural History and Theory of the Heavens*. Translated by St. Palmquist. Aldershot: Ashgate.

Keller, T. n.d. *Creation, Evolution, and Christian Laypeople*. The BioLogos Foundation. https://biologos.org/uploads/projects/Keller_white_paper.pdf.

Kelly, D. F. 2015. *Creation and Change: Genesis 1.1–2.4 in the Light of Changing Scientific Paradigms*. Fearn (UK): Mentor.

Kelly, W. 1870. *Lectures Introductory to the Study of the Acts, the Catholic Epistles, and the Revelation*. London: W. H. Broom.

______, ed. n.d. (repr.). *The Bible Treasury*. Winschoten: H. L. Heijkoop.

Kerkut, G.A. 1960. *The Implications of Evolution*. Oxford: Pergamon Press.

Kidner, D. 1967. *Genesis: An Introduction and Commentary*. Downers Grove, IL: IVP Books.

Kiel, Y. 1997. *Sefer Bereshit 1–17. Da'at Miqra'*. Jerusalem: Mossad Harav Kook.

Kimura, M. 1983. *The Neutral Theory of Molecular Evolution*. Cambridge: Cambridge University Press.

Kinney, L. W. 2013 (repr.). *Types and Mysteries in the Gospel of John*. Neptune, NJ: Loizeaux Brothers.

Kirschner, M. W. and J. C. Gerhart. 2006. *The Plausibility of Life: Resolving Darwin's Dilemma*. New Haven, CT: Yale University Press.

Kitchen, K. A. 1995. “The Patriarchal Age: Myth or History?” *Biblical Archeology Review* 21 (March-April): 2 (http://cojs.org/the_patriarchal_age-_myth_or_history/).

______. 2006. *On the Reliability of the Old Testament*. Grand Rapids, MI: Eerdmans.

Kittel, G. et al., eds. 1964–1976. *Theological Dictionary of the New Testament*. Translated by G. W. Bromiley. 10 vols. Grand Rapids, MI: Eerdmans.

Kleinert, P. 1868. *Obadjah, Jonah, Micha, Nahum, Habakuk, Zephanjah*. Bielefeld/Leipzig: Velhagen and Klasing.

Klinghoffer, D., ed. 2015. *Debating Darwin's Doubt: A Scientific Controversy that Can No Longer Be Denied*. Seattle, WA: Discovery Institute Press.

König, A. 2006. *Die Groot Geloofswoordeboek*. Vereeniging: Christelike Uitgewersmaatskappy.

Koonin, E. V. 2011. *The Logic of Chance: The Nature and Origin of Biological Evolution*. Upper Saddle River, NJ: FT Press.

Kroeze, J. H. 1962. *Strijd bij de schlepping*. Den Haag: Van Keulen.

Kruyswijk, A. 1962. *Geen gesneden beeld…* Franeker: T. Wever.

Kuyper, A. 1894. *Encyclopaedie der Heilige Godgeleerdheid*. 3 vols. Amsterdam: J. A. Wormser.

______. 1923. *De engelen Gods*. 2nd ed. Kampen: Kok.

______. 2016. *Common Grace: God's Gifts for a Fallen World*. Vol. 1: *The Historical Section*. Edited by J. J. Ballor and S. J. Grabill. Translated by N. D. Kloosterman and E. M. van der Maas. Bellingham, WA/Grand Rapids, MI: Lexham Press/Acton Institute.

Kuyper, A., Jr. 1929. *Het beeld Gods*. Amsterdam: De Standaard.

Lakatos, I. and A. Musgrave. 1974. *Criticism and the Growth of Knowledge*. Cambridge: Cambridge University Press.

Lamoureux, D. O. 2008. *Evolutionary Creation: A Christian Approach to Evolution*. Eugene, OR: Wipf and Stock.

______. 2009. *I Love Jesus and I Accept Evolution*. Eugene, OR: Wipf and Stock.

______. 2015. "Beyond Original Sin: Is a Theological Paradigm Shift Inevitable?" *Perspectives on Science and Christian Faith* 67: 35–49.

Lapide, P. 2002. *The Resurrection of Jesus: A Jewish Perspective*. Eugene, OR: Wipf and Stock.

Laughlin, R. B. 2005. *A Different Universe*. New York: Basic Books.

Le Fanu, J. 2009. *Why Us? How Science Rediscovered the Mystery of Ourselves*. New York: Vintage.

Lennox, J. C. 2009 (rev.). *God's Undertaker: Has Science Buried God?* Oxford: Lion Hudson.

______. 2011. *Seven Days that Divide the World: The Beginning According to Genesis and Science*. Grand Rapids, MI: Zondervan.

Leupold, H. C. 1942. *Exposition of Genesis*. London: Evangelical Press.

Lever, J. 1956. *Creatie en evolutie*. Wageningen: Zomer and Keunings.

Lewin, R. 1980. "Evolutionary Theory Under Fire." *Science* 210 (Nov.): 883–87.

Lewis, C. S. 1952. *Mere Christianity*. London: Collins.

______. 1972. *God in the Dock: Essays on Theology and Ethics*. Grand Rapids, MI: Eerdmans.

______. (1940) 1996. *The Problem of Pain*. New York: Simon and Schuster.

Lewontin, R. C. 1972. "Testing the Theory of Natural Selection." *Nature* 236:181.

______. 1997. "Billions and Billions of Demons." *The New York Review of Books*. January 9 Issue (www.nybooks.com/articles/1997/01/09/billions-and-billions-of-demons/).

Lockyer, H. 1988. *All the Miracles of the Bible*. Grand Rapids, MI: Zondervan.

Long, V. P. 1994. *The Art of Biblical History*. Grand Rapids, MI: Zondervan.

Loonstra, B. 1999. *De Bijbel recht doen: Bezinning op gereformeerde hermeneutiek*. Zoetermeer: Boekencentrum.

Løvtrup, S. 1987. *Darwinism: The Refutation of a Myth*. New York: Springer/Croom Helm Publishers.

Lubenow, M. L. 2004 (rev. ed.). *Bones of Contention: A Creationist Assessment of Human Fossils*. Grand Rapids, MI: Baker Books.

Luther, M. 1897. "Assertio Omnium Articulorum...." In *Luthers Werke*. 91–151. Weimarer Ausgabe. Weimar: Böhlau Verlag.

Lyell, C. (1830–33) 1998 (abridged ed.). *Principles of Geology*. London: John Murray. London: Penguin Classics.

Lyell, K. 1881. *Life, Letters and Journals of Sir Charles Lyell, Bart*. London: John Murray.

Mackintosh, C. H. 1972 (repr.). *Genesis to Deuteronomy: Notes on the Pentateuch*. Neptune, NJ: Loizeaux Brothers.

Madueme, H. and M. Reeves, eds. 2014. *Adam, the Fall, and Original Sin*. Grand Rapids, MI: Baker Academic.

Macbeth, N. 1972. *Darwin Retried*. Boston: Gambit.

McDowell. J. 1999. *The New Evidence That Demands a Verdict*. Nashville, TN: Thomas Nelson Publishers.

McGrath, G. B. 1997. "Soteriology: Adam and the Fall." *Perspectives on Science and Christian Faith* 49.4 (June): 252–63.

McIntosh, A. 2008. "The Downgrade Controversy of the 21st Century." *Evangelical Times*, October. https://www.evangelical-times.org/25029/the-downgrade-controversy-of-the-21st-century/.

Macquarrie, J. 1981. "Truth in Christology." In *God Incarnate: Story and Belief*. Edited by A. E. Harvey. 24–33. London: SPCK.

Mahoney, J. W. 2011. "Why an Historical Adam Matters for a Biblical Doctrine of Sin." *Southern Baptist Journal of Theology* 15 (Spring): 60–78.

Malthus, T. R. (1798) 1992. *An Essay On the Principle of Population*. Edited by D. Winch. Cambridge Texts in the History of Political Thought. Cambridge: Cambridge University Press.

Mann, C. 1991. "Lynn Margulis: Science's Unruly Earth Mother." *Science* 252.5004: 378–81.

Marxsen, W. 1970. *The Resurrection of Jesus of Nazareth*. Translated by M. Hohl. London: SCM Press.

Mathews, K. A. 1996. *Genesis 1–11:26*. New American Commentary. Nashville, TN: Broadman and Holman.

Mazur, S. 2015. *The Paradigm Shifters: Overthrowing "the Hegemony of the Culture of Darwin."* New York: Caswell Books.

Medema, H. P. 1985. *Door het geloof rechtvaardig: Bijbelstudies over de brief van Paulus aan de Romeinen*. Vaassen: Medema.

Mejsnar, J. A. *The Evolution Myth: Or the Genes Cry Out Their Urgent Song, Mister Darwin Got It Wrong*. Prague: Karolinum Press, Charles University.

Meyer, S. C. 2010 (repr.). *Signature in the Cell: DNA and the Evidence for Intelligent Design*. New York: HarperOne.

______. 2014 (rev. ed.). *Darwin's Doubt: The Explosive Origin of Animal Life and the Case for Intelligent Design*. New York: HarperOne.

Meyer-Abich, A. 1940. "Hauptgedanken des Holismus." *Acta Biotheoretica* 5.2: 85–116.

Miller, D. W., ed. 1985. *Popper Selections*. Princeton: Princeton University Press.

Miller, K. R. 2007. *Finding Darwin's God: A Scientist's Search for Common Ground Between God and Evolution*. New York: Cliff Street Books.

______. 2009 (repr.). *Only a Theory: Evolution and the Battle for America's Soul*. New York: Penguin Books.

Milton, R. 2000. *Shattering the Myths of Darwinism*. Rochester, NY: Park Street Press.

Möller, H. 1977. *Der Anfang der Bibel (1. Mose 1–11)*. Berlin: Concordia-Verlag.

Morant, P. 1960. *Die Anfänge der Menschheit: Eine Auslegung der ersten elf Genesis-Kapitel*. Luzern: Räber and Cie.

More, L. T. 1925. *The Dogma of Evolution*. Princeton, NJ: Princeton University Press.

Moreland, J. P. 2007. *Kingdom Triangle: Recover the Christian Mind, Renovate the Soul, Restore the Spirit's Power*. Grand Rapids, MI: Zondervan.

______, S. C. Meyer, C. Shaw, A. K. Gauger, and W. Grudem, eds. 2017. *Theistic Evolution: A Scientific, Philosophical, and Theological Critique*. Wheaton, IL: Crossway Books.

Morris, H. M. 1976. *The Genesis Record: A Scientific and Devotional Commentary on the Book of Beginnings*. San Diego, CA: Creation-Life Publishers.

______. 2002. *The Biblical Basis for Modern Science*. Green Forest, AR: Master Books.

______ and G. E. Parker. (1982) 1987. *What Is Creation Science?* 2nd ed. Green Forest, AR: Master Books.

______ and J. C. Whitcomb. (1961) 2011. *The Genesis Flood: The Biblical Record and Its Scientific Implications*. 50th anniversary edition. Phillipsburg, NJ: P&R Publishing.

Morris, L. 1971. *The Gospel According to John*. New International Commentary on the New Testament. Grand Rapids, MI: Eerdmans.

Mortenson, T. 2016. *Searching for Adam: Genesis and the Truth About Man's Origin*. Green Forest, AR: Master Books.

______ and T. H. Ury, eds. 2008. *Coming to Grips with Genesis: Biblical Authority and the Age of the Earth*. Green Forest, AR: New Leaf Publishing.

Moule, H. C. G. 1891. *Romans*. Cambridge Bible for Schools and Colleges. Cambridge: Cambridge University Press.

Muether, J. R. and D. E. Olinger, eds. 2011. *Confident of Better Things: Essays Commemorating Seventy-Five Years of the Orthodox Presbyterian Church*. Willow Grove, PA: Committee for the Historian of the Orthodox Presbyterian Church.

Murphy, G. L. 2006. "Roads to Paradise and Perdition: Christ, Evolution, and Original Sin." *Perspectives on Science and Christian Faith* 58.2 (June): 109–18.

Murray, J. 1957. *Principles of Conduct: Aspects of Biblical Ethics*. Grand Rapids, MI: Eerdmans.

Nagel, T. 2012. *Mind and Cosmos: Why the Materialist Neo-Darwinian Conception of Nature is Almost Certainly False*. Oxford: Oxford University Press.

National Academy of Sciences/Institute of Medicine 2008. *Science, Evolution, and Creationism*. Washington, DC: National Academies Press.

Nei, M., Y. Suzuki, and M. Nozawa. 2010. "The neutral theory of molecular evolution in the genomic era." *Annual Review of Genomics and Human Genetics* 11:265–89.

Nevin, N. C., ed. 2009. *Should Christians Embrace Evolution? Biblical and Scientific Responses*. Phillipsburg, NJ: P&R Publishing.

Niditch, S. 1985. *Chaos to Cosmos: Studies in Biblical Patterns of Creation*. Scholars Press Studies in the Humanities. Durham, NC: Duke University Press.

Noble, D. 2008. *The Music of Life: Biology beyond the Genome*. Oxford: Oxford University Press.

Noordegraaf, A. 1990. *Leven voor Gods aangezicht: Gedachten over het mens-zijn*. Kampen: Kok.

Northcott, M. and R. J. Berry, eds. 2009. *Theology After Darwin*. Milton Keynes (UK): Paternoster.

Nunberg, G. 2009. *Going Nucular: Language, Politics, and Culture in Confrontational Times*. New York: Public Affairs.

Nussbaum, A. 2002. "Creationism and Geocentrism among Orthodox Jewish Scientists." *Reports of the National Center for Science Education* (January–April): 38–43.

Oosterhoff, B. J. 1972. *Hoe lezen wij Genesis 2 en 3? Een hermeneutische studie*. Kampen: Kok.

Osborn, H. F. (1905) 2007. *From the Greeks to Darwin: An Outline of the Development of the Evolution Idea*. Whitefish, MT: Kessinger.

Ouweneel, W. J. 1974. *Kanttekeningen bij Genesis één*. Winschoten: Uit het Woord der Waarheid.

______. 1975a. *Operatie Supermens: Een bijbels-biologische blik op de toekomst*. Amsterdam: Buijten and Schipperheijn/Groningen: De Vuurbaak.

______. 1975b. "Homeotic Mutants and Evolution." *Creation Research Society Quarterly* 12(3): 141–54.

______. 1982. *Wij zien Jezus: Bijbelstudies over de brief aan de Hebreeën*. 2 vols. Vaassen: Medema.

______. 1984a. *Psychologie: Een christelijke kijk op het mentale leven*. Amsterdam: Buijten and Schipperheijn.

______. 1984b. *Evolution in der Zeitenwende: Biologie und Evolutionslehre—Die Folgen des Evolutionismus*. Neuhausen-Stuttgart: Hänssler Verlag.

______. 1986. *De leer van de mens*. Amsterdam: Buijten and Schipperheijn.

______. 1987. *Woord and Wetenschap: Wetenschapsbeoefening aan de Evangelische Hogeschool*. Amsterdam: Buijten and Schipperheijn.

______. 1998. *De zevende koningin: Het eeuwig vrouwelijke en de raad van God*. Metahistorische triologie. Vol. 2. Heerenveen: Barnabas.

______. (2003) 2004. *Geneest de zieken! Over de bijbelse leer van ziekte, genezing en bevrijding*. 4th ed. Vaassen: Medema.

______. 2007a. *De Geest van God: Ontwerp van een pneumatologie*. EDR 1. Vaassen: Medema.

______. 2007b. *De Christus van God: Ontwerp van een christologie*. EDR 2. Vaassen: Medema.

______. 2008. *De schepping van God: Ontwerp van een scheppings-, mens- en zondeleer*. EDR 3. Vaassen: Medema.

______. 2010. *De kerk van God II: Ontwerp van een historische en praktische ecclesiologie*. EDR 8. Heerenveen: Medema.

______. 2012a. *De toekomst van God: Ontwerp van een eschatologie*. EDR 10. Heerenveen: Medema.

______. 2012b. *Het Woord van God: Ontwerp van een openbarings- en schriftleer*. EDR 11. Heerenveen: Medema.

______. 2014a. *Wisdom for Thinkers: An Introduction to Christian Philosophy*. St. Catharines, ON: Paideia Press.

______. 2014b. *Searching the Soul*. St. Catharines, ON: Paideia Press.

______. 2015. *What Then Is Theology?: An Introduction to Christian Theology*. St. Catharines, ON: Paideia Press.

______. 2016. *The Heidelberg Diary: Daily Devotions on the Heidelberg Catechism*. St. Catharines, ON: Paideia Press.

______. 2017. *The World Is Christ's: A Defense of Christian One-Kingdom Thinking*. Toronto: Ezra Press.

______. 2018a. *The Ninth King: The Last of the Celestial Empires: The Triumph of Christ over the Powers*. St. Catharines, ON: Paideia Press.

______. 2018b. *The Eternal Torah: An Evangelical Theology of Living Under God*. Vol. 1. *An Evangelical Introduction to Reformational Theology*. St. Catharines, ON: Paideia Press.

______. 2018c. *The Eternal Covenant: An Evangelical Theology of Living With God*. Vol. 2. *An Evangelical Introduction to Reformational Theology*. St. Catharines, ON: Paideia Press.

______. 2018d. *Eternal Righteousness: An Evangelical Theology of Living Before God*. Vol. 3. *An Evangelical Introduction to Reformational Theology*. St. Catharines, ON: Paideia Press.

______. 2018e. *Eternal Salvation: An Evangelical Theology of Christ Dying for Us*. Vol. 7. *An Evangelical Introduction to Reformational Theology*. St. Catharines, ON: Paideia Press.

Overhage, P. 1968. *Experiment Menschheit: Die Steuerung der menschlichen Evolution*. Frankfurt: Josef Knecht Verlag.

Paley, W. (1802) 2012. *Natural Theology, or, Evidences of the Existence and Attributes of the Deity, Collected from the Appearances of Nature*. Greenwood, WI: Suzeteo Enterprises.

Pannenberg, W. 2010. *Systematic Theology*. Translated by G. W. Bromiley. Vol. 2. Grand Rapids, MI: Eerdmans.

Pascal, B. (1670) 1995. *Pensées*. Translated by A. J. Krailsheimer. London: Penguin. http://www.gutenberg.org/ebooks/46921.

Paul, M. J. 2017. *Oorspronkelijk: Overwegingen bij schepping en evolutie*. Apeldoorn: De Banier.

Payne, A. 2015. *The First Chapters of Everything: How Genesis 1–4 Explains Our World*. Fearn (UK): Christian Focus Publications.

Peacocke, A. R. 1993. *Theology for a Scientific Age: Being and Becoming–Natural, Divine, and Human*. 2nd ed. Minneapolis, MN: Fortress Press.

Pentecost, J. D. 1981. *The Words and Works of Jesus Christ: A Study of the Life of Christ*. Grand Rapids, MI: Zondervan.

Peterson, D. 2012 (repr.). *The Moral Lives of Animals*. New York: Bloomsbury Press.

Petterson, M. S., F. Purnell Jr., and M. C. Carnes. 2014. *The Trial of Galileo: Aristotelianism, the "New Cosmology," and the Catholic Church, 1616–33*. New York: W. W. Norton.

Phillips, R. D. 2015. *God, Adam, and You: Biblical Creation Defended and Applied*. Phillipsburg, NJ: P&R Publishing.

Pietsch, T. W., ed. 2014. *Cuvier's History of the Natural Sciences: 24 Lessons from Antiquity to the Renaissance*. Paris: Publications scientifiques du Muséum national d'Histoire naturelle.

Plessner, H. (1928) 1975. *Die Stufen des Organischen und der Mensch: Einleitung in die philosophische Anthropologie*. 3rd ed. Berlin: De Gruyter.

Polanyi, M. 1973. *Personal knowledge: Towards a Postcritical Philosophy*. 2nd ed. London: Routledge and Kegan Paul.

Polkinghorne, J. 2011. *Reason and Reality: The Relationship Between Science and Theology*. London: SPCK.

Pookottil, R. 2013. BEEM: *Biological Emergence-based Evolutionary Mechanism: How Species Direct Their Own Evolution*. London: Fossil Fish Publishing.

Popma, K. 1972. *Harde feiten: Kanttekeningen bij het Genesisverhaal*. Franeker: Wever.

Proksch, D. O. 1913. *Die Genesis übersetzt and erklärt*. Kommentar Alten Testament. Leipzig: Deichertsche Verlagsbuchhandlung.

Popper, K. R. 1959. *The Logic of Scientific Discovery*. London: Routledge.

______. 1963. *Conjectures and Refutations: The Growth of Scientific Knowledge*. London: Routledge.

______. 1972. *Objective Knowledge: An Evolutionary Approach*. London: Routledge.

______. 1976. *Unended Quest: An Intellectual Autobiography*. London: Routledge.

Poythress, V. S. 2006. *Redeeming Science: A God-Centered Approach*. Westchester, IL: Crossway Books.

______. 2012. "Evaluating the Claims of Scientists." *New Horizons* 33.3: 6–8.

______. 2013. "Adam Versus Claims from Genetics." *Westminster Theological Journal* 75: 65–82.

______. 2014. *Did Adam Exist?* Phillipsburg, NJ: P&R Publishing.

Pruitt, T. 2009. *Was Adam an Historical Person?* http://www.alliancenet.org/mos/1517/ was-adam-an-historical-person#.WXcUv62iG_s.

______. 2010a. *On the Historicity of Adam and Eve*. http://www.alliancenet.org/mos/1517/ on-the-historicity-of-adam-and-eve#.WXcTqq2iG_s.

______. 2010b. *Does It Matter if Adam Was a Historical Person?* (2). http://www.alliancenet.org/mos/1517/does-it-matter-if-adam-was-an-historical-person-2#.WXcVja2iG_s.

Radmacher, E. D. and R. D. Preus, eds. 1984. *Hermeneutics, Inerrancy, and the Bible*. Grand Rapids, MI: Zondervan.

Rasimus, T. 2007. "The Serpent in Gnostic and Related Texts." In *L'Évangile selon Thomas et les textes de Nag Hammadi*. Edited by L. Painchaud and P.-H. Poirier. 417–71. Québec: Presses de l'Université Laval.

Rau, G. 2012. *Mapping the Origins Debate: Six Models of the Beginning of Everything*. Downers Grove, IL: IVP Academic.

Rice, J. R. 1975. *"In the Beginning": A Verse-by-Verse Commentary on the Book of Genesis*. Murfreesboro, TN: Sword of the Lord Publishers.

Richards, J. W., ed. 2010. *God and Evolution*. Seattle, WA: Discovery Institute Press.

Ridderbos, J. 1925. *Het verloren paradijs*. Kampen: Kok.

______. 1956. "Boom des levens." *Christelijke Encyclopedie*. Vol. 1: 705–707. Kampen: Kok.

Ridderbos, N. H. 1963. *Beschouwingen over Genesis 1*. 2nd ed. Kampen: Kok.

Robinson, T. 1982 (repr.). *Studies in Romans*. Expository and Homiletical Commentary. Grand Rapids, MI: Kregel.

Ross, A. P. 1997. *Creation and Blessing: A Guide to the Study and Exposition of Genesis*. Grand Rapids, MI: Baker Academic.

Ross, Hugh N. 2004. *A Matter of Days: Resolving A Creation Controversy*. Colorado Springs, CO: NavPress Publishing Group.

Rossiter, W. D. 2015. *Shadow of Oz: Theistic Evolution and the Absent God*. Eugene, OR: Pickwick Publications.

Safina, C. 2015. *Beyond Words: What Animals Think and Feel*. New York: Henry Holt and Company.

Sailhamer, J. H. 1990. *Genesis*. Expositor's Bible Commentary 2. Grand Rapids, MI: Zondervan.

______. (1996) 2011. *Genesis Unbound: A Provocative New Look at the Creation Account*. 2nd ed. Sisters, OR: Multnomah Books.

Sanford, J. C. 2014. *Genetic Entropy*. 4th ed. Waterloo, NY: FMS Publications.

Sarfati, J. 2004. *Refuting Compromise: A Biblical and Scientific Refutation of Progressive Creationism*. Green Forest, AR: Master Books.

______. 2008. *By Design*. Powder Springs, GA: Creation Book Publishers.

______. (1999) 2010a. *Refuting Evolution*. Powder Springs, GA: Creation Book Publishers.

______. 2010b. *The Greatest Hoax on Earth? Refuting Dawkins on Evolution*. Powder Springs, GA: Creation Book Publishers.

______. 2015. *The Genesis Account: A Theological, Historical, and Scientific Commentary on Genesis 1–11*. Powder Springs, GA: Creation Book Publishers.

Saxo Grammaticus. 2012. *The Nine Books of the Danish History: Gesta Danorum*. Edited by M. L. Stinson. CreateSpace Independent Publishing Platform.

Schaeffer, F. A. 1982. *The Complete Works: A Christian Worldview*. 5 vols.

Westchester, IL: Crossway Books.
Schelhaas, J. 1932. *De messiaansche profetie in den tijd vóór Israëls volksbestaan.* Hoogeveen: Slingenberg.
Schilder, K. 1939. *Heidelbergsche Catechismus.* Vol. 1. Goes: Oosterbaan and Le Cointre.
Schillebeeckx, E. 1989. *Mensen als verhaal van God.* Baarn: Nelissen.
Schindewolf, O.H. (1950) 1994. *Basic Questions in Paleontology: Geologic Time, Organic Evolution, and Biological Systematics.* Chicago: University of Chicago Press.
Schlatter, A. 1962. *Der Brief an die Römer.* Stuttgart: Calwer Verlag.
Schlink, E. 1942. "Gottes Ebenbild als Gesetz und Evangelium." In *Der alte und der neue Mensch: Aufsätze zur theologischen Anthropologie.* Beiträge zur evangelischen Theologie 8:68–87.
Schmidt, K. L. 1947. "Homo Imago Dei im Alten und Neuen Testament." *Eranos* 15:149–195.
Schreiner, T. R. 1998. *Romans.* Baker Exegetical Commentary on the New Testament. Grand Rapids, MI: Baker Academic.
Schubert-Soldern, R. 1962. *Mechanism and Vitalism: Philosophical Aspects of Biology.* Notre Dame, IN: University of Notre Dame Press.
Schuurman, E. 2003. *Faith and Hope in Technology.* Carlisle (UK): Piquant Editions.
Schwabe, C. 2001. *The Genomic Potential Hypothesis: A Chemist's View of the Origins, Evolution and Unfolding of Life.* Abingdon: CRC Press.
Schwegler, T. 1960. *Die biblische Urgeschichte im Lichte der Forschung.* München: Pustet.
Schweitzer, A. (1906) 2005. *The Quest of the Historical Jesus: A Critical Study of its Progress from Reimarus to Wrede.* Translated by W. Montgomery. Mineola, NY: Dover Publications.
Schwertley, B. 2000. *The Historicity of Adam.* http://www.mountainretreatorg.net/ apologetics/the_historicity_of_adam.shtml.
Shakespeare, W. 2004. *Hamlet.* Translated by W. J. Ouweneel. Soesterberg: Aspekt.
Shapiro, J.A. 2011. *Evolution: A View from the 21st Century.* Upper Saddle River, NJ: FT Press.
Shapiro, L. 2016. *The Miracle Myth: Why Belief in the Resurrection and the Supernatural Is Unjustified.* New York: Columbia University Press.
Sheldrake, R. 1991. *The Rebirth of Nature: The Greening of Science and God.* New York: Bantam Books.
Sherlock, C. 1997. *The Doctrine of Humanity.* Contours of Christian Theology. Downers Grove, IL: IVP Academic.

Sikkel, J. C. 1923. *Het boek der geboorten: Verklaring van het boek Genesis*. 2nd ed. Amsterdam: H. A. van Bottenburg.
Simpson, G. G. 1964. *This View of Life: The World of an Evolutionist*. New York: Harcourt, Brace and World.
______. 1967. *The Meaning of Evolution*. New Haven, CT: Yale University Press.
Singer, C. A. 1950. *A History of Biology*. New York: Henry Schumann.
Sinnott, E. W. 1966. *The Bridge of Life: From Matter to Spirit*. New York: Simon & Schuster.
Sirks, M. J. and C. Zirkle. 1964. The Evolution of Biology. New York: Ronald Press.
Slotki, I. W. 1983. *Isaiah*. Revised by A. J. Rosenberg. The Soncino Books of the Bible. London/New York: Soncino Press.
Smith, C. R. 1953. *The Bible Doctrine of Sin and of the Ways of God with Sinners*. London: Epworth Press.
Smith, D. L. 1994. *With Willful Intent: A Theology of Sin*. Wheaton, IL: Bridge Point.
Smith, Q. 1992. "A Big Bang Cosmological Argument for God's Nonexistence." *Faith and Philosophy* 9.2: 217–37.
Sneed, J. D. 1971. *The Logical Structure of Mathematical Physics*. Dordrecht: Reidel.
Snelling, A. 2014. *Earth's Catastrophic Past*. Dallas, TX: Master Books.
Spencer, H. 1864. *The Principles of Biology*. London: Williams & Norgate.
Spykman, G. J. 1992. *Reformational Theology: A New Paradigm for Doing Dogmatics*. Grand Rapids, MI: Eerdmans.
Stack, D. A. 2003. *The First Darwinian Left: Socialism and Darwinism, 1859–1914*. Cheltenham: New Clarion Press.
Steele, E. J. 1981. *Somatic Selection and Adaptive Evolution: On the Inheritance of Acquired Characters*. 2nd ed. Chicago: University of Chicago Press.
Stegmüller, W. 1996. "Der sogenannte Zirkel des Verstehens." In *Das Problem der Induktion: Humes Herausforderung und moderne Antworten*. Edited by W. Stegmüller. Darmstadt: Wissenschaftliche Buchgesellschaft.
Stewart, I. 2011. "Commandeering Time: The Ideological Status of Time in the Social Darwinism of Herbert Spencer." *Australian Journal of Politics* and History 57.3: 389–402.
Stott, J. R. W. (1972) 1999. *Understanding the Bible*. Expanded ed. Grand Rapids, MI: Zondervan.
Strack, H. L. and P. Billerbeck. (1922–1928) 1986–1997. *Kommentar zum Neuen Testament aus Talmud und Midrasch*. 6 vols. München: Beck.
Strauss, D. F. M. 1991. *Man and His World*. Bloemfontein: Tekskor.

______. 2009. *Discipline of the Disciplines.* St. Catharines, ON: Paideia Press.

______. 2010. *A Perspective on (Neo-)Darwinism.* https://www.researchgate.net/ publication/228370439_A_perspective_on_neo-_Darwinism_2010.

______. 2014. "Soul and Body: Transcending the Dialectical Intellectual Legacy of the West with an Integral Biblical View?" *Luce Verbi* 48.1:1–12. http://daniestrauss.com/selection/DS%202014%20on%20Soul%20and%20Body%20transcending%20the%20dialectical%20legacy%20with%20an%20integral%20biblical%20view.pdf

Sungenis, R. A. and R. J. Bennett. 2009. *Galileo Was Wrong: The Church Was Right.* 2 vols. Port Orange, FL: CAI Publishing.

Swaab, D. F. 2014. *We Are Our Brains: A Neurobiography of the Brain, from the Womb to Alzheimer's.* Translated by J. Hedley-Prole. New York: Spiegel and Grau.

Teilhard de Chardin, P. 1961. *The Phenomenon of Man.* New York: Harper and Row.

Teller, W. 1945. *Essays of an Atheist.* New York: Truth Seeker Company.

Thiselton, A. C. 2000. *The First Epistle to the Corinthians.* New International Greek Testament Commentary. Grand Rapids, MI: Eerdmans.

______. 2015. *Systematic Theology.* Grand Rapids, MI: Eerdmans.

Tinker, M. 2010. *Reclaiming Genesis: The Theatre of God's Glory–Or a Scientific Story?* Oxford: Monarch Books.

Tolkien, J. R. R. 2012. *The Lord of the Rings.* 50th anniversary ed. Wilmington, MA: Mariner Books.

Torrance, T. F. (1947) 1957. *Calvin's Doctrine of Man.* Grand Rapids, MI: Eerdmans.

Towner, P. H. 2006. *The Letters to Timothy and Titus.* New International Commentary on the New Testament. Grand Rapids, MI: Eerdmans.

Troost, A. 1989. "Kritiek van dr. J. D. Dengerink op de antropologie van H. Dooyeweerd." *Philosophia Reformata* 54.1: 65–82.

______. 2005. *Antropocentrische totaliteitswetenschap: Inleiding in de "reformatorische wijsbegeerte" van H. Dooyeweerd.* Budel: Damon.

Trueman, C. 2013. *Adam and Eve and Pinch Me.* http://www.alliancenet.org/mos/ postcards-from-palookaville/adam-and-eve-and-pinch-me#.WXcVEK2iG_s.

Van de Beek, A. 1984. *Waarom? Over lijden, schuld en God.* Nijkerk: Callenbach.

______. 1998. *Jezus Kurios: De Christologie als hart van de theologie.* Kampen: Kok.

______. 2005. *Toeval of schlepping? Scheppingstheologie in de context van het modern denken.* Kampen: Kok.

______. 2011. "Evolution, Original Sin, and Death" *Journal of Reformed Theology* 5: 206–20.

______. 2014. *Een lichtkring om het kruis: Scheppingsleer in christologisch perspectief.* Zoetermeer: Meinema.

van den Brink, G. 2000. *Oriëntatie in de filosofie.* Zoetermeer: Boekencentrum.

______. 2011. "Should We Drop the Fall? On Taking Evil Seriously." In *Strangers and Pilgrims on Earth: Essays in Honour of Abraham van de Beek.* Edited by E. Van der Borght and P. Van Geest. 761–78. Leiden/Boston: Brill.

______. 2015. "Erfzonde of oorsprongszonde? Over de waarde van een innovatief dogmatisch voorstel." In *Jan Hoek: Theoloog tussen preekstoel en leerstoel.* Edited by M. Van Campen and P. J. Vergunst. 101–113. Zoetermeer: Boekencentrum.

______. 2017. *En de aarde bracht voort: Christelijk geloof en evolutie.* Utrecht: Boekencentrum.

Van der Kamp, W. 1985. *Houvast aan het hemelruim.* Kampen: Kok.

Van der Zanden, L. n.d. *De mensch als beeld Gods.* Kampen: Kok.

VanDoodewaard, W. 2016. *The Quest for the Historical Adam.* Grand Rapids, MI: Reformation Heritage Books.

Van Genderen, J. and W. H. Velema. 2008. *Concise Reformed Dogmatics.* Translated by G. Bilkes and E. M. van der Maas. Phillipsburg, NJ: P&R Publishing.

Vanhoozer, K. J. 2009. *Is There a Meaning in This Text? The Bible, the Reader, and the Morality of Literary Knowledge.* Anniversary ed. Grand Rapids, MI: Zondervan.

______. 2012. *Remythologizing Theology: Divine Action, Passion, and Authorship.* Cambridge Studies in Christian Doctrine. Cambridge: Cambridge University Press.

Van Peursen, C. A. 1995. "Dooyeweerd en de wetenschappelijke discussie." In *Dooyeweerd herdacht.* Edited by J. de Bruijn. 79–83. Amsterdam: VU Uitgeverij.

Van Woudenberg, R. 2007. "Evolutionaire verklaringen van de moral." In *Omhoog kijken in platland. Over geloven in de wetenschap.* Edited by C. Dekker, R. Van Woudenberg, and G. van den Brink. 341–60. Kampen: Ten Have.

Venema, D. R. and S. McKnight. 2017. *Adam and the Genome: Reading Scripture after Genetic Science.* Grand Rapids, MI: Brazos Press.

Verbrugge, M. 1984. *Alive: An Enquiry into the Origin and Meaning of Life.* Vallecito, CA: Ross House Books.

Verkuyl, J. 1992. *De kern van het christelijk geloof.* Kampen: Kok.

Versteeg, J. P. 2012. *Adam in the New Testament: Mere Teaching Model or First*

Historical Man? 2nd ed. Phillipsburg, NJ: P and R Publishing.

Vogel, H. 1949/50. "Ecce Homo. Die Anthropologie Karl Barths." *Verkündigung und Forschung* 1949/1950: 102–128.

Von Bertalanffy, L. 1955. "Die Evolution der Organismen." In *Schöpfungsglaube und Evolutionstheorie*. Edited by D. Schlemmer. 53–68. Stuttgart: Alfred Kröner Verlag.

Von Meyenfeldt, F. H. 1950. *Het hart* (leb, lebab) *in het Oude Testament*. Leiden: E. J. Brill.

Von Rad, G. 1972. *Genesis: A Commentary*. Translated by J. H. Marks. Philadelphia, PA: Westminster Press.

Vriezen, Th. C. 1937. *Onderzoek naar de paradijsvoorstelling bij de oude Semietische volken*. Wageningen: Veenman.

Waltke, B. K. 2001. *Genesis: A Commentary*. Grand Rapids, MI: Zondervan.

______. 2007. *An Old Testament Theology: An Exegetical, Canonical, and Thematic Approach*. Grand Rapids, MI: Zondervan.

Walton, J. H. 2001. *Genesis*. NIV Application Commentary. Grand Rapids, MI: Zondervan.

______. 2009. *The Lost World of Genesis One: Ancient Cosmology and the Origins Debate*. Downers Grove, IL: IVP Academic.

______. 2015. *The Lost World of Adam and Eve: Genesis 2–3 and the Human Origins Debate*. Downers Grove, IL: IVP Academic.

Walvoord, J. F. and R. B. Zuck, eds. 1985. *The Bible Knowledge Commentary: Old Testament*. Wheaton, IL: Victor Books.

Ward, K. 1998. *Religion and Human Nature*. Oxford: Oxford University Press.

Watchman Nee. 1968. *The Spiritual Man*. New York: Christian Fellowship Publishers.

Watson, D. M. S. 1929. "Adaptation." *Nature* 124: 231–34.

Wax, T. 2011. *Counterfeit Gospels: Rediscovering the Good News in a World of False Hope*. Chicago, IL: Moody Publishers.

Wells, J. 2002. *Icons of Evolution: Science or Myth? Why Much of What We Teach about Evolution Is Wrong*. Washington, DC: Regnery Publications.

______. 2011. *The Myth of Junk DNA*. Seattle, WA: Discovery Institute Press.

Wenham, G. J. 1987. *Genesis 1–15*. Word Biblical Commentary. Vol. 1. Waco, TX: Word Books.

Wentsel, B. 1987. *Dogmatiek*. Vol. 3a: *God en mens verzoend: Godsleer, mensleer en zondeleer*. Kampen: Kok.

Westermann, C. 1984. *Genesis 1–11: A Commentary*. Translated by J. J. Scullion. Minneapolis, MN: Augsburg Publishing House.

Wilder-Smith, A. E. (1970) 1981. *The Creation of Life: Cybernetic Approach to Evolution*. Dallas, TX: Master Books.

Williams, A. and J. Hartnett. 2014. *Dismantling the Big Bang*. 4th ed. Green Forest, AR: Master Books.

Wilson, A. N. 2017. *Charles Darwin: Victorian Mythmaker*. London: John Murray.

Wilson, G. B. 1977. *Romans: A Digest of Reformed Comment*. Carlisle, PA: Banner of Truth Trust.

Wittgenstein, L. (1953) 2009. *Philosophical Investigations*. New York: Wiley-Blackwell.

Wolff, H. W. 1973. *Anthropologie des Alten Testaments*. München: Kaiser.

Wolters, A. M. 1985. *Creation Regained: Biblical Basics for a Reformational Worldview*. Grand Rapids, MI: Eerdmans.

Wommack, A. 2010. *Spirit, Soul and Body: What You Didn't Learn in Church*. Tulsa, OK: Harrison House.

Wood, T. C. and P. A. Garner, eds. 2009. *Genesis Kinds: Creationism and the Origin of Species*. Eugene, OR: Wipf and Stock.

Wright, C. 2001. "Hermeneutiek, Postmodernisme en de Waarheid." *Soteria* 18.1: 5–20.

Wright, N. T. 2014. *Surprised by Scripture: Engaging Contemporary Issues*. New York: HarperOne.

Young, D. A. 2012. *Good News for Science: Why Scientific Minds Need God*. Oxford, MS: Malius Press.

Young, E. J. 1966. *Genesis 3: A Devotional and Expository Study*. Carlisle, PA: The Banner of Truth Trust.

______. 1976. *In the Beginning: Genesis 1–3 and the Authority of Scripture*. Edinburgh: The Banner of Truth Trust.

Yount, Lisa 2008. *Antoine Lavoisier: Founder of Modern Chemistry*. Rev. ed. Berkeley Heights, NJ: Enslow Publishers.

Zenkowsky, B. 1951. *Das Bild vom Menschen in der Ostkirche: Grundlagen der orthodoxen Anthropologie*. Stuttgart: Evangelisches Verlagswerk.

______ and H. Petzold. 1969. *Das Bild des Menschen im Lichte der orthodoxen Anthropologie*. Marburg: Oekumenischer Verlag Dr. R.F. Edel.

Zimmerli, W. (1943) 1991. *1. Mose 1–11: Die Urgeschichte*. 5th ed. Zürcher Bibelkommentare. Zürich: Zwingli Verlag.

Zukav, G. 2001. *Thoughts From the Seat of the Soul*. New York: Touchstone.

SCRIPTURE INDEX

NAME INDEX

SUBJECT INDEX